# SOLAR DOMESTIC HOT WATER: A Practical Guide to Installation and Understanding

# SOLAR DOMESTIC HOT WATER: A Practical Guide to Installation and Understanding

Russell H. Plante



John Wiley & Sons
*New York • Chichester • Brisbane • Toronto • Singapore*

***Library of Congress Cataloging in Publication Data:***

Plante, Russell H. (Russell Howard), 1947–
Solar domestic hot water.

Includes index. 1. Solar heating. 2. Solar Water heaters.
I. Title

TH7413.P54 1983 696′.6 82–19966
ISBN 0-471-09592-3

Printed in the United States of America

10 9 8 7 6 5 4 3 2 1

*For Kathy, Nicholas, and Matthew*
*with love*

# PREFACE

The direct use of solar energy has never before been accepted on a worldwide basis. The costs of constructing devices or systems to use the sun's energy have always been greater than the costs of using the alternate energy sources available. In the immediate future, however, the price of energy will rise even more rapidly than in the recent past. Alternate sources of energy are dwindling, and their costs are increasing at rates of 10 percent or more every year. We must explore putting the sun's energy to practical use in supplying a percentage of our ever-increasing energy needs. The "first" most appropriate large-scale application of solar energy use concerns the heating of water for domestic use.

This book is written for the homeowner and for the beginning solar technology student. The purpose of the book is not to elaborate on conservation measures or to assess the variety of solar energy applications, but to develop an understanding of the practical applications of the already available components, types of systems, and methods pertinent to installation and operation of solar domestic hot water (DHW) systems. To date, most textbooks on this subject have either overtheorized or oversimplified the explanations required for understanding the practical application of solar technology. This book is designed to fill the gap between the highly academic presentations and the superficial literature currently available.

This book will guide the homeowner in the fundamentals of solar domestic hot water installation and understanding. In the past, homeowners have not been provided with adequate assurance or knowledge about solar hot water systems, their operation, and proper installation. Moreover, many homeowners may be quite capable of installing their own systems, thereby saving the installation costs, which may range from $800 to $1500 depending on the type of system, its size, and location. Such installations, of course, require the homeowner to be multidisciplined in carpentry, plumbing, and electricity, and to have a practical understanding of solar energy application, and to comply with all applicable local and state codes. If the homeowner plans to have the system installed, then he or she should be familiar with the various techniques that are pertinent to installation.

For the technology student, this book provides a stepping-stone from the basic theories and practical aspects of applying an alternate energy source to the more advanced courses encompassing aspects of engineering and mathematics that will subsequently increase the scope of solar energy utilization.

Time after time the same arguments against the use of solar hot water systems have appeared, and they have slowed the progress of the application of solar energy, resulting in the delay of our independence from oil. These arguments include:

1. Solar energy is impractical in many of the northern sections of the United States.
2. Solar domestic hot water systems are not a good investment.
3. Equipment will be cheaper later.
4. Solar technology is untried and not perfected.
5. Solar hot water systems are difficult to install.

All of the above statements are false. Consider the following:

1. Solar DHW production is feasible in every part of the country. The insolation received in the northeastern sections is only slightly less than the national average, still providing solar energy that is sufficient to provide as much as 100 percent of domestic water heating requirements.

2. Solar DHW heating is the most cost-effective use of solar energy and can be economically justified. The percentage savings can be more easily defined than other solar applications because there are fewer parameters to consider, and the demand for BTU consumption is more strictly defined. With current energy sources becoming increasingly scarce and expensive, a solar DHW system will increase the value of a home as it reduces the owner's utility bills. In most instances the payback is within 4 to 6 years, depending on the type of system and whether or not the system is self-installed.

3. Solar equipment is becoming *more* expensive because of increasing material prices and the cost of labor. Furthermore, federal and state tax credits make the present an opportune time to purchase equipment.

4. Solar domestic hot water technology is a low energy technology that has been used since the early 1900s in this country and even earlier in others. The technology for solar heating domestic water is well developed. Standard flat plate collectors, storage tanks, and control systems are commercially available. Faulty and careless workmanship, as well as improper use of components and materials, have been frequently encountered in the past, producing skepticism on the part of the consumer. Engineering detail and design integrity were also often ignored, resulting in either system inoperability or poor system performance. Today this unfortunate situation has changed as a result of the increased field expertise of the solar installer/dealer. The technology is now proven.

5. Solar hot water systems are not difficult to install. The information presented in this book will answer many questions about installation and will provide a systematic approach to solar domestic hot water. It is necessary that these systems be properly sized, sited, mounted, equipped, charged, and maintained to ensure their economic viability.

The use of solar energy to heat water for domestic use has been questioned because it is unfamiliar and not considered to be conventional.

Everyone is familiar with the older methods and feels "comfortable" with them. Now, however, it is time to understand the newer methods of employing alternate energies. Solar domestic hot water (DHW) is a viable first step in energy independence. While illustrating the types of systems, the components involved, and the procedures necessary for the installation, this book also elaborates on the functional aspects of component application. It is not only important to provide for proper installation and awareness of the materials necessary to make a system functional, but it is also important to understand why the system works the way it does. Solar domestic hot water provides not only a first step in the use of solar energy, it provides a means to learn what solar energy is about and what it is capable of accomplishing. The most vast, continuing energy resource available to humanity is the sun.

*Russell H. Plante*
*President*
*Applied Technologies, Inc.*

# ACKNOWLEDGMENTS

Several individuals helped to make this book possible and contributed in significant ways to its content. I especially thank the following individuals: My wife, Kathy, whose loving support and perseverance during the two years of manuscript preparation made this project possible; Page Mead, who was instrumental in directing me to John Wiley & Sons, and who provided me with invaluable support and encouragement; Donna Lebrun for being able to decipher my handwritten notes and her continued typing efforts during the preparation of the manuscript; Stephen Webber for his friendship and for providing the many excellent component and installation photographs that are important to system installation understanding; Bruce Gretter and Donald Jessen for their friendship and their support in installing the solar hot water system that made possible the pictorial sequences in Chapters 7 to 9; Chris Mello for his friendship and help in installing the solar hot water system necessary for the pictorial sequences in Chapters 7 to 9, his reviews and constructive comments on several chapters, and his roof ladder design concept, which is illustrated in Appendix A; Andy Galt for providing information about solar hot air systems from his practical field expertise as well as the photographs pertinent to Chapters 5 and 7; Harry Hopcroft for providing the information concerning the phase-change subambient systems discussed in Chapter 3.

Many corporations and government agencies supplied photographs and factual information for this book. I gratefully acknowledge the contributions of the following:

Applied Technologies, Inc.
Solar Engineering Division
Lyndon Way
Kittery, Maine 03904

ASHRAE
1791 Tullie Circle, N.E.
Atlanta, Georgia 30325

Atlantic Scientific Corporation
2718 So. Harbor City Blvd.
Melbourne, Florida 32901

Bray Oil Company
1925 North Marianna Avenue
Los Angeles, California 90032

B. H. Hoist Co.
Candy Industrial Park
4425 D-C Drive
Tyler, Texas 75703

Brookstone Company
Peterborough, New Hampshire 03458

Columbia Chase Corporation
Holbrook, Massachusetts 02343

Conservation and Renewable Energy Inquiry and Referral Service
(formerly the Solar Heating and Cooling Information Center)
Rockville, Maryland 20850

Copper Development Association, Inc.
Stamford, Connecticut 06905

Department of the Interior
U.S. Geological Survey
Arlington, Virginia 22202

Department of Energy
Washington, D.C. 20546

Dow Corning Corporation
Midland, Michigan 48640

Ford Products Corporation
Valley Cottage
New York, New York 10989

General Electric Co.
Solar Heating and Cooling Division
Philadelphia, Pennsylvania 19101

Granite State Solar Industries, Inc.
Dover, New Hampshire 03820

Grundfos Pump Corporation
2555 Clovis Avenue
Clovis, California 93612

Heliotrope General, Inc.
Spring Valley, California 92077

Independent Energy, Inc.
East Greenwich, Rhode Island 02818

Kalwall Corporation
Manchester, New Hampshire 03105

Kee Klamps Division
Gascoigne Industrial Products Ltd.
Buffalo, New York 14225

Lennox Industries, Inc.
1711 Olentangy River Road
Columbus, Ohio 43216

Mark Enterprises, Inc.
Woodbridge, Connecticut 06525

National Aeronautics and Space Administration
Washington, D.C. 20546

New England Energy Alternatives, Inc.
Stoughton, Massachusetts 02072

Northeast Regional Agricultural Engineering Service
Cornell University
Ithaca, New York 14853

Noranda Metal Industries, Inc.
Forge-Fin Division
Newton, Connecticut 06470

Olin Brass Corporation
East Alton, Illinois 62024

Oxford University Press
Walton Street
Oxford, England

Phelps Dodge Industries, Inc.
New York, New York 10022

Popular Science Magazine
380 Madison Avenue
New York, New York 10017

Radio Shack
Division of Tandy Corporation
Fort Worth, Texas 76102

Riemann-Gorger Hoist Co.
Buffalo, New York 14240

Resource Technology Corporation
New Britain, Connecticut 06051

Rho Sigma, Inc.
Van Nuys, California 91405

Solar Design Associates, Inc.
205 W. John Street
Champaign, Illinois 61820

Solar Engineering Magazine
Dallas, Texas 75229

Solar Specialties, Inc.
5891 Marion Drive
Denver, Colorado 80016

Sunworks
Somerville, New Jersey 08876

Terra-Light, Inc.
30 Manning Road
Billerica, Massachusetts 01821

The Lightning Protection Institute
Harvard, Illinois 60033

U.S. Department of Commerce
National Climatic Center
Asheville, North Carolina 28801

U.S. Department of Housing and
Urban Development
Washington, D.C. 20546

U.S. Solar Corporation
Hampton, Florida 32044

Vanderbeck Lightning Rod Co.
10 Sycamore Avenue
Hohokus, New Jersey 07423

Vaughn Corporation
386 Elm Street
Salisbury, Massachusetts 01950

Vistron Corporation
Filon Division
Hawthorne, California 90250

Yankee Resources
21 Potter Street
Brunswick, Maine 04011

I am appreciative of the help and thoughtful attention of the individuals who reviewed this manuscript. Thanks are owed to:

Mr. John Klima
Community College of Denver
Denver, Colorado

Mr. Russ Roy
Santa Fe Community College
Gainesville, Florida

Mr. Paul Hering
Universal Technical Institute
Phoenix, Arizona

Mr. Bill Abernathy
Orange Coast College
Costa Mesa, California

Professor Carl Childers
Texas Technical University
Lubbock, Texas

Professor Claude C. Hartman
Cerritos College
Norwalk, California

Richard Montgomery
Midland, Michigan

Professor Michael Garrison
The University of Texas at Austin
Austin, Texas

Professor Raymond P. Martin
Orange Coast College
Costa Mesa, California

Mr. Earl Stedman
Mississippi County Community College
Blytheville, Arkansas

Mr. George A. Cleland
Miami, Florida

# CONTENTS

# SOLAR DOMESTIC HOT WATER: A Practical Guide to Installation and Understanding

# Consider the Source

The sun is everyone's life. Without its energy, the past, present, and future of human beings would not be. In brief retrospect, our primordial star is theorized to have been formed from a cosmic cloud of individual particles in which matter was concentrated under the action of gravitational energy. This large mass of gas was then condensed and as the gravitational and kinetic energies of the particles were increased, the interior of the cloud was heated to extremely high temperatures. Finally, increased temperatures and pressures in the interior of this dense mass reached proportions capable of sustaining nuclear reactions. It is this energy that has sustained human existence.

The sun has provided us with stored chemical energy in the form of fossil fuels which is now rapidly being depleted, and this depletion is thus responsible for escalated social and economic costs. To curtail these adversities, the direct application of the sun's radiant energy to alternative conversion processes such as photovoltaic, photochemical, thermionic, thermoelectric, and heat must be continuously developed and utilized. An economic "first" application for existing solar energy alternatives involves using solar collectors to convert the sun's radiant energy into heat energy for domestic water heating. A basic understanding of the sun, its characteristics, its relationship with the earth, and the distribution of its energy to the earth should first be acquired to appreciate the further discussion of the types of solar DHW systems available, how they function, and how they can be installed and maintained.

## PHYSICAL CONSIDERATIONS

Our sun belongs to a class dwarf yellow stars of spectral type G. It has a diameter of approximately 864,000 miles, which is greater than three times the distance from the earth to the moon, and it rotates on its axis from west to east with a period of rotation at its equator of approximately 27 days. The sun's mass is approximately $2 \times 10^{30}$ kilograms, its volume is approximately $1.4 \times 10^{27}$cubic meters, and its density is approximately 1.4 grams per cubic centimeter. The sun is structured in several layers as depicted in Figure 1-1. Each layer has no distinct boundary but rather represents areas where certain dynamic processes occur.

From the interior outward the layers are as follows:

*Core*. This area is where the continuous thermonuclear reactions occur. Temperatures are in the neighborhood of 25,000,000° to 29,000,000° F

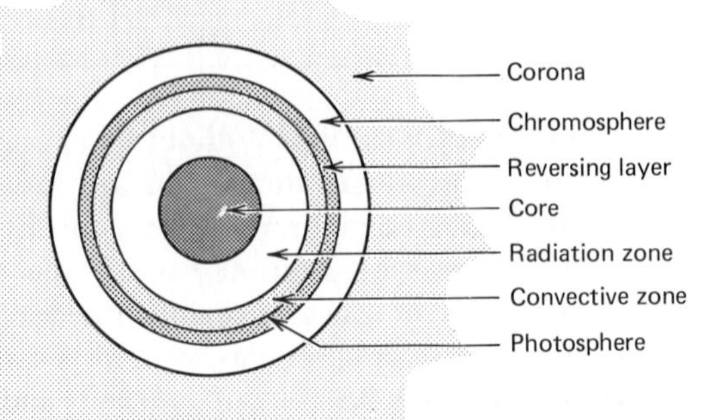


**Figure 1-1.** Structure of the sun.

and pressures are 100 billion times greater than atmospheric pressure on the earth. This hot interior region contains about 40 percent of the mass of the sun and provides over 90 percent of the total heat generation.

*Radiation zone.* Heat energy is transferred by waves of radiation traveling at the speed of light (186,000 miles per second).

*Convective zone.* This layer is approximately 50,000 miles thick and is in turmoil with violent convective currents as energy is transferred from the radiation zone.

*Photosphere.* The radiation that finally reaches the earth comes from this thin layer (approximately 200 to 300 miles thick). The temperature is about 11,000° F, and the region is composed of low density ionized gases that are opaque to visible light.

*Reversing layer.* This is a transparent region that is several hundred miles in depth and contains cooler gases at a temperature about 9,000° F. This layer absorbs and reradiates energy passing through it in all directions. Such absorption results in the characteristic dark-line spectra (Fraunhofer lines) which are seen against the continuous spectrum produced by the photosphere. Absorption spectra show that approximately 70 elements exist on the sun. The five most abundant elements are hydrogen (90 percent), helium (10 percent), oxygen, nitrogen, and carbon.

*Chromosphere.* This region extends about 6000 miles or more above the photosphere with a temperature of about 180,000° F in the upper regions. The most abundant gas in this region is hydrogen.

*Corona.* This is the final layer or region of the sun and can be seen during a total solar eclipse as a pale white halo extending one or two sun diameters beyond the sun. The temperature is approximately 4,000,000° F.

These seven layers compose the basic structure of our sun. The dynamic phenomena occurring at these various levels in the solar atmosphere are still under scrutiny in hopes that they can provide better understanding of the basic nuclear processes and hydrodynamics of the sun.

## SUN AND EARTH CONSIDERATIONS

For nearly 2,000 years, virtually everyone thought the earth was the center of the universe and that the sun and all of the other heavenly bodies revolved about our planet. We now know that the apparent motion of the sun across the sky is actually the result of the earth's own rotation. The earth spins on its axis at a rate of 360.99 deg in 24 hours, and, therefore, the sun appears to move across the sky at a rate of 15.04 deg per hour. The earth moves about the sun in an approximately circular path with the sun positioned slightly off center. This offset is such that the earth is closest to the sun around January 1 and farthest from the sun around July 1. Our changing seasons occur because the earth's rotational axis is tilted at approximately 23.5 deg with respect to the plane of the ecliptic containing our orbit. Figure 1-2 represents the earth-sun relationship as it would be viewed by an observer far out in space.

From Figure 1-2, it can be seen in the northern hemisphere that the north end of the axis is tilted away from the sun during the winter and that the north end of the axis is tilted toward the sun during the summer. In the southern hemisphere the seasons are reversed, and there is also a tendency

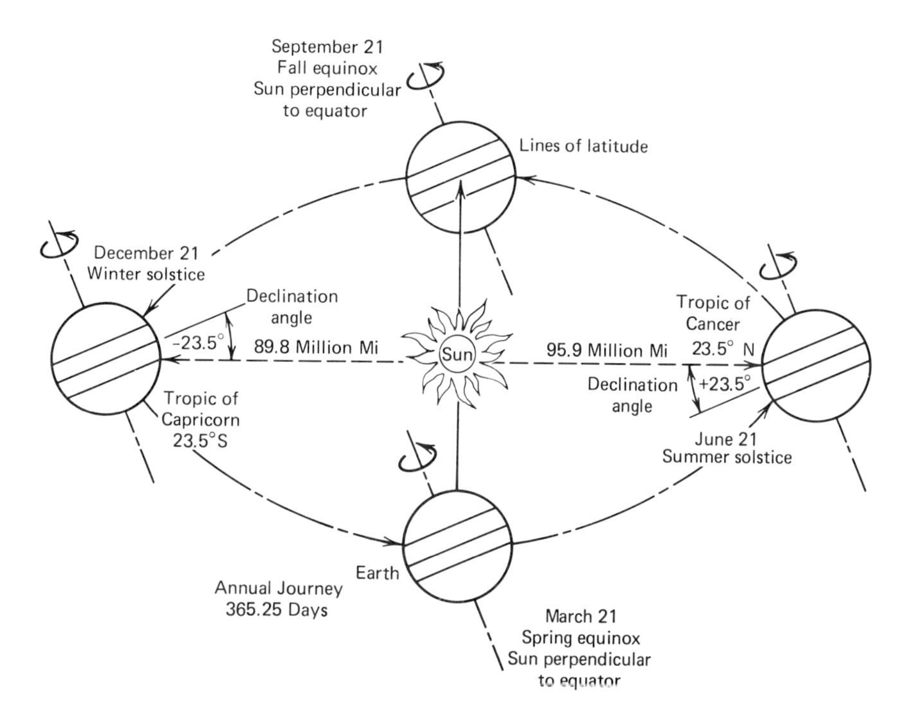


**Figure 1-2.** Earth-sun geometry.

for greater seasonal differences in temperature than in the northern hemisphere.

The angle of the earth's tilt with respect to the sun and the equatorial plane is called the ***declination*** angle. This angle varies day by day throughout the year from +23.5 deg on June 21 to −23.5 deg on December 21. When the earth's axis is perpendicular to the line joining the earth and the sun, day and night are of equal length (March 21 and September 21) and are called the ***spring and fall equinoxes***, respectively. When the angle of declination is at its greatest, +23.5 deg, a point in the northern hemisphere will have its longest period of daylight, called the ***summer solstice*** (June 21). When the angle of declination is at its lowest, −23.5 deg, a point in the northern hemisphere will have its longest period of darkness, called ***winter solstice*** (December 21).

Lines of ***latitude***, which are normally included on most maps, are actually designations of the angle between the equator and a line from the center of the earth to its surface, as shown in Figure 1-3. Latitudes vary from 0 deg at the equator to 90 deg at the earth's poles and run parallel with the equator. The latitude denoting the most northerly position of the sun, in which the declination angle is +23.5 deg, is known as the ***Tropic of Cancer***. The latitude denoting the most southerly position of the sun, in which the declination angle is −23.5 deg, is known as the ***Tropic of Capricorn***. For the purpose of determining time, the imaginary circle around the earth (zero deg latitude) has been divided into 24 segments of 15 deg each. These circular lines of ***longitude*** extend from the North Pole to the South Pole. Lines of longitude are defined to start (zero deg) at Greenwich, England.

The apparent position of the sun from any point on earth is defined by two angles. The angle of the sun's position in the sky with respect to the earth's horizontal is known as the solar ***altitude***. The position of the sun with respect to true south is referred to as the solar ***azimuth***. Figure 1-4 illustrates typical altitude and azimuth positions of the sun at the equinox and solstice days. When the sun's position is true south, the azimuth is zero and the altitude is a maximum. This instant of time is referred to as ***solar noon***.

The position of the sun in relation to specific geographic locations, seasons, and times of day can be determined by several different methods, each with varying degrees of accuracy. Model measurements by means of sun machines or shade dials, have the advantage of direct visual observations. Tabulative and calculative methods have the mathematical advantage

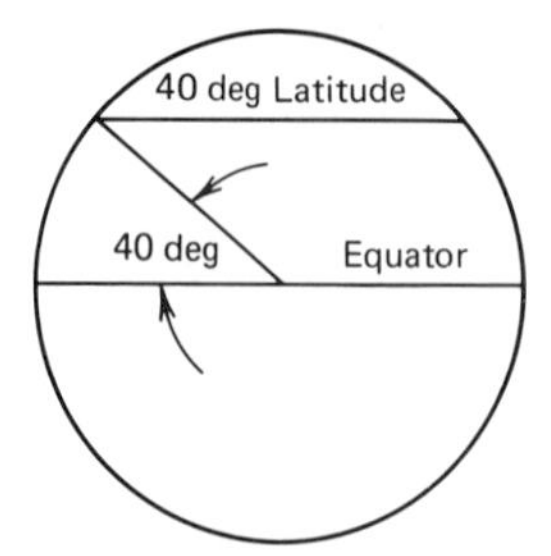


**Figure 1-3.** Lines of latitude.

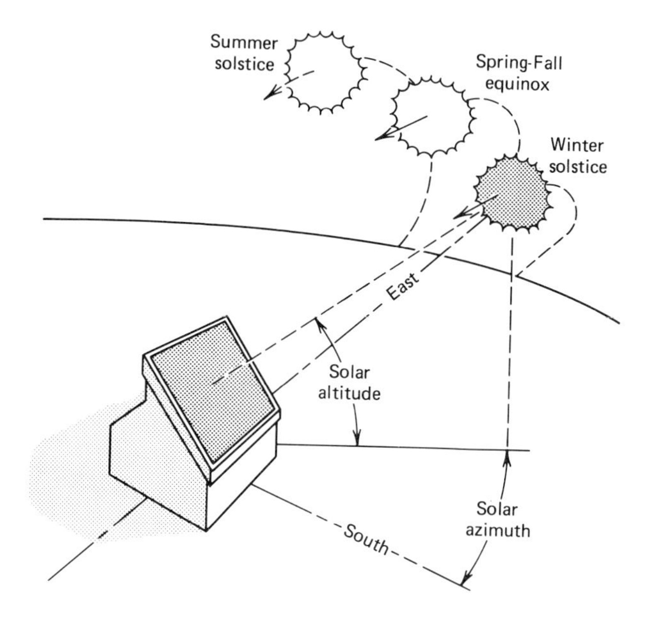


**Figure 1-4.** Altitude and azimuth. (Courtesy of Copper Development Assn., Inc., Stamford, Conn.)

of exactness. Graphic projection, however, can be more easily understood and can be correlated to both radiant energy and shading calculations. The ***sun path diagram*** method is a graphic projection depicting the path of the sun within the sky vault as projected onto a horizontal plane, as illustrated in Figure 1-5.

The horizon is represented as a circle with the observation point in the center. The sun's position at any date and hour can be determined from the diagram in terms of its altitude ($\alpha$) and its azimuth ($\beta$). Sun path diagrams are shown at 4 deg intervals from 24 deg north latitude to 52 deg north latitude in Figure 1-6. Equally spaced concentric circles represent the altitude angles ($\alpha$), and equally spaced radial lines represent the azimuth angles ($\beta$), at 10 deg intervals, respectively. The elliptical curves in the diagrams represent the horizontal projections of the sun's path on the twenty-first day of each

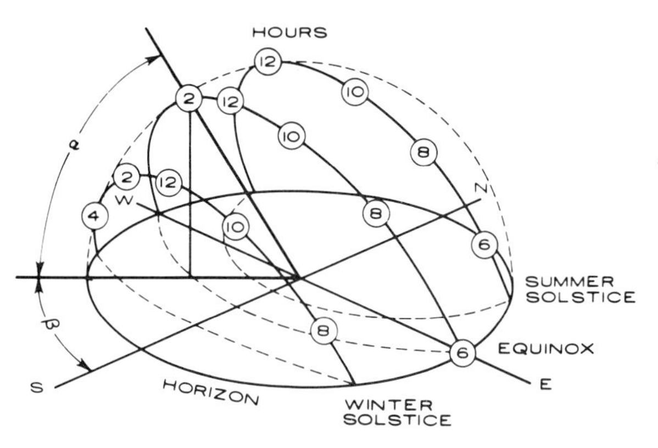


**Figure 1-5.** Sky-vault projection. (Reprinted with permission from Ramsey/Sleeper, *Architectural Graphic Standards*, 6th ed., John Wiley & Sons, Inc. © 1970.)

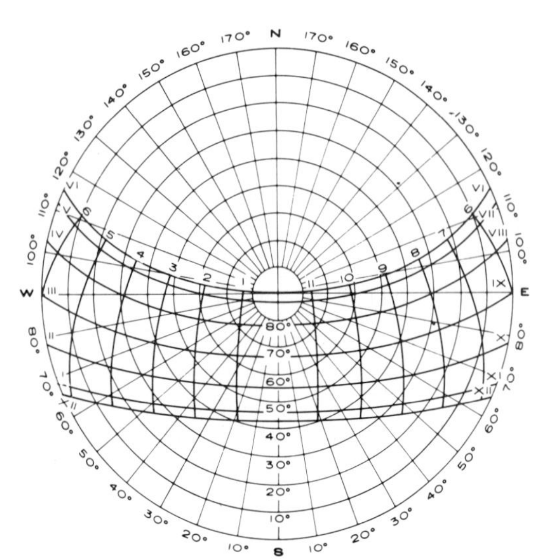

24°N LATITUDE

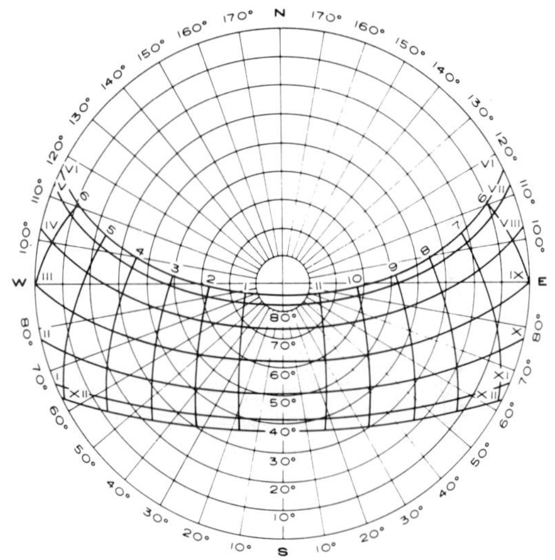

28°N LATITUDE

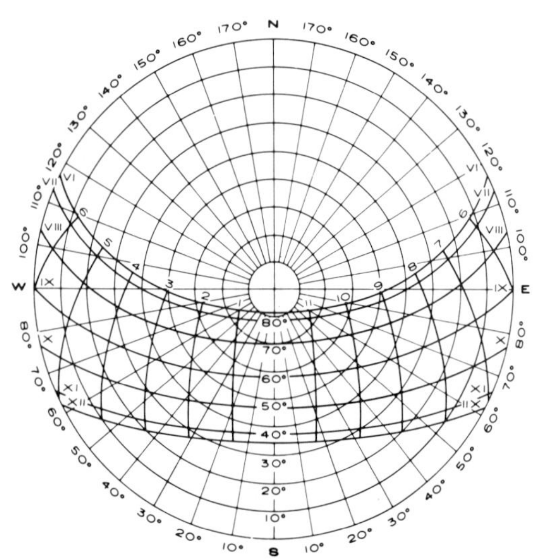

32°N LATITUDE

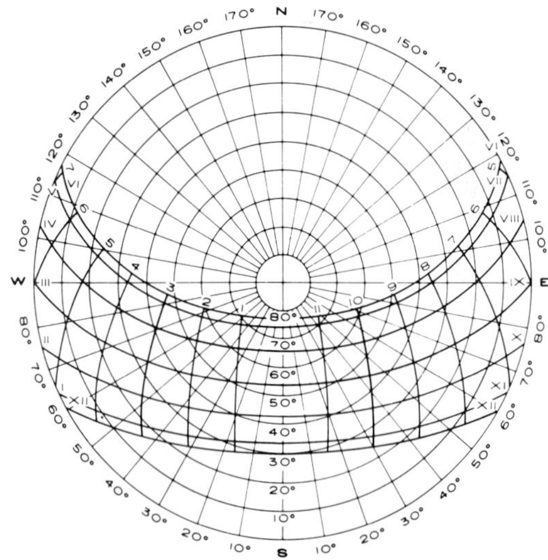

36°N LATITUDE

**Figure 1-6.** Sun path diagrams. (Reprinted with permission from Ramsey/Sleeper, *Architectural Graphic Standards*, 6th ed., John Wiley & Sons, Inc., © 1970)

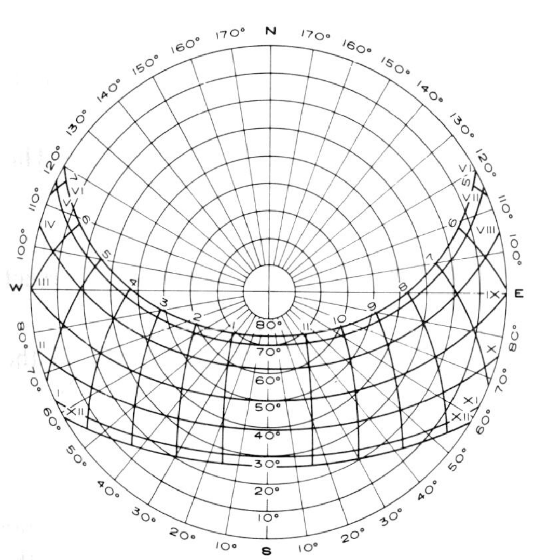


40°N LATITUDE

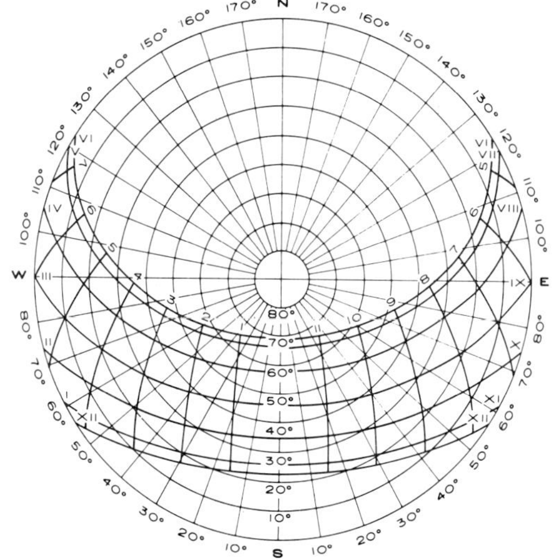


44°N LATITUDE

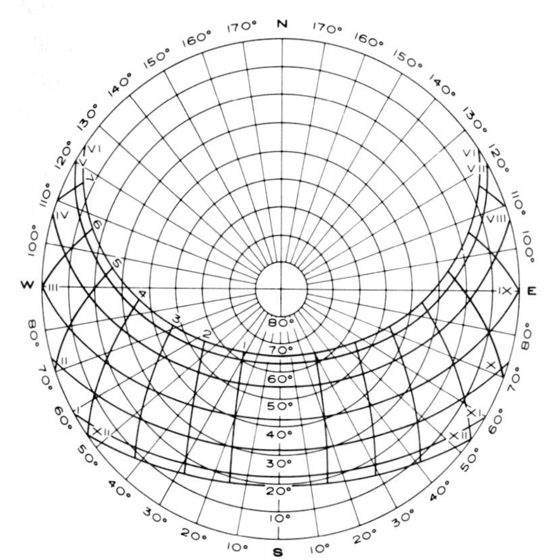


48°N LATITUDE

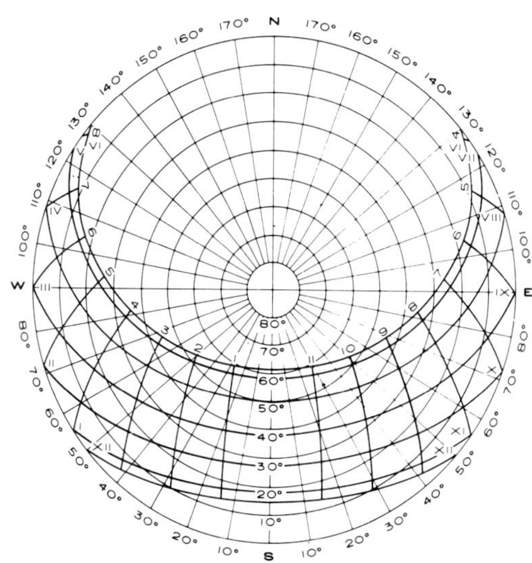


52°N LATITUDE

month. Roman numerals designate the month. A cross grid of curves graduates the hours indicated in arabic numerals. Let us illustrate the use of a sun path diagram in an example.

Find the sun's position in Billings, Montana, on February 21, at 2 P.M.

*Step 1.* Locate Billings, Montana, on a map to determine its latitude. The latitude is approximately 46 deg.

*Step 2.* In the 48 deg sun path diagram in Figure 1-6 select the February path (marked with II), and locate the 2-hour line. The two lines intersect at the approximate position of the sun.

*Step 3.* Read the altitude on the concentric circles (27 deg) and the azimuth (bearing angle) along the outer circle (32 deg west).

## ENERGY CONSIDERATIONS

### Type of Energy

The principal source of the sun's radiant energy is the fusion of hydrogen nuclei, which leads to the formation of helium. Nuclear fusion involves the combining of several small nuclei into one large nucleus with the subsequent release of huge amounts of energy. The nuclei of hydrogen are single particles called protons, each of which carries a positive electric charge. Similarly charged particles normally repel each other, but if the temperature is high enough, their motion can be sufficiently vigorous to allow them to approach very closely, and this short-range attractive force can result in fusion. For our sun, this fusion reaction is known as the proton-proton reaction. The conversion of hydrogen to helium actually involves three separate reactions. In summarizing this process we can say that four hydrogen nuclei combine to form helium and a large amount of energetic radiation called gamma rays. This outgoing radiation is a very high-energy electromagnetic radiation, which has the shortest wavelength known—approximately a hundred-millionth of a millimeter. As the initial gamma radiation strikes nuclei and electrons or is scattered in near-collisions, varying forms of energy result, which have less energy than the original gamma radiation but with energy in longer wavelengths such as X rays, ultraviolet, visible light, infrared (heat), and radio waves. These radiant energy waves can be arranged in increasing order of wavelength as illustrated in Figure 1-7.

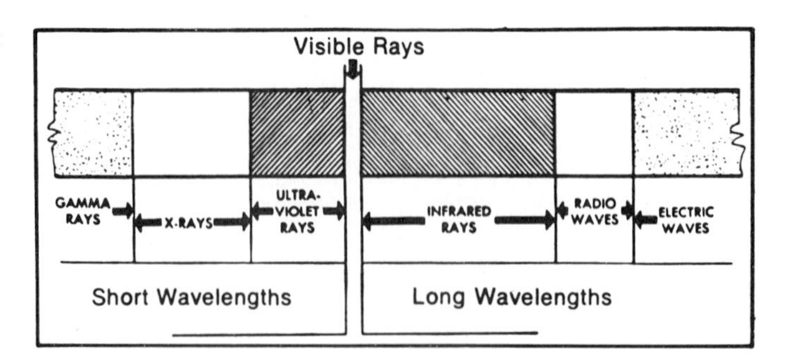


**Figure 1-7.** Electromagnetic spectrum. (*Source. Fundamentals of Solar Heating*, D.O.E., 1978)

The most general law of radiation is called Plank's law, in which the energy content of electromagnetic radiation is represented by $E = hf$ where $E$ is energy, $h$ is a constant (Plank's), and $f$ is the frequency of the wavelength. The relationship between wavelength $\lambda$, velocity $v$, and frequency $f$, is represented in the expression, $v = f\lambda$. Because the velocity of the electromagnetic radiation travels at the speed of light, $c$, and this speed is equivalent to the product of frequency and wavelength, Plank's law can be rewritten as $E = hc/\lambda$. The curve illustrated in Figure 1-8 illustrates the relationship between the amount of energy emitted and the wavelength for objects at a specific temperature. At any one temperature, a wide spectrum of wavelengths is produced. As the temperature increases so does the frequency, and this increased frequency results in a shorter wavelength. Therefore, the higher the temperature the more the maximum energy (peak of curve) shifts toward the shorter wavelength.

From this discussion, one can understand that we are concerned with three major energy regions of radiation from the sun. These regions include ultraviolet, visible, and infrared radiation. A solar collector absorbs radiation from all three of these regions, converting it into heat on the surface of the collector.

*Ultraviolet radiation.* Wavelength shorter than 0.4 micrometers not visible to the human eye, comprising approximately 9 percent of the solar energy available. A considerable amount of this energy is absorbed in the outer atmosphere and does not reach the earth's surface.

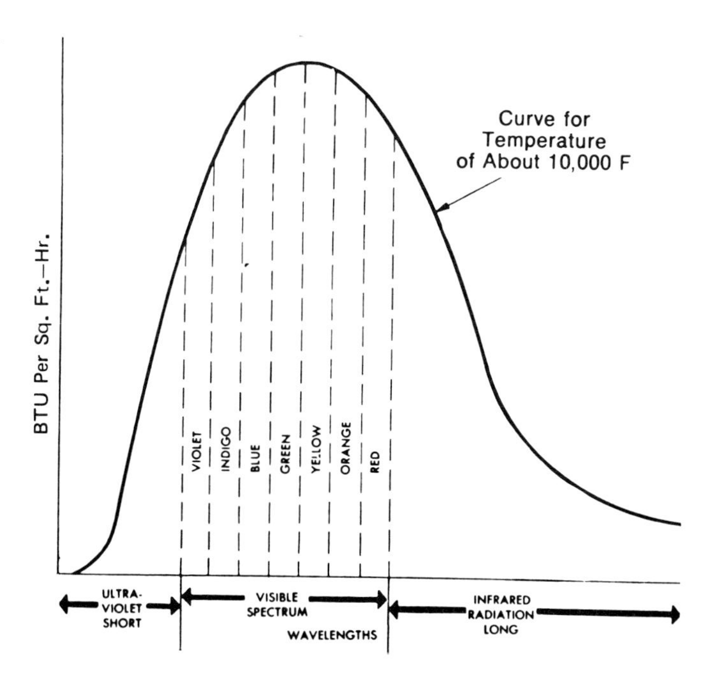


**Figure 1-8.** Radiant energy from the sun. (*Source. Fundamentals of Solar Heating*, D.O.E., 1978)

*Visible radiation.* Middle wavelengths visible to the human eye, comprising approximately 49 percent of the solar energy available. If this white light is passed through a prism thereby changing the speed of the wavelength travel, this region can be broken down into colors from violet to red.

*Infrared radiation.* Long wavelengths greater than 0.7 micrometers, emitted by materials with a temperature below 800° F, and comprising approximately 42 percent of the solar energy available. About 0.4 percent of this energy has a wavelength long enough such that it cannot pass through a thin layer of glass.

## Energy Distribution Through the Atmosphere

Only a very small proportion of the sun's total radiant energy reaches the earth. The intensity of this radiation just above the earth's atmosphere is quite constant. This ***solar constant*** is defined as the amount of energy incident on a unit area exposed normally to the sun's rays at an average sun-earth distance of approximately 92,956,000 miles. As the earth moves in its annual elliptical orbit around the sun, the total solar energy received varies by approximately ±3.5 percent. This solar constant is normally expressed in three different units of measure equivalent to the following:

436.5 BTU/ft$^2$-hr
1377 W/m$^2$
1.97 langley/min

These units, which provide a measure of solar intensity, can be expressed relative to one another as

$$1 \text{ langley/hr} = 3.693 \text{ BTU/ft}^2\text{-hr} = 11.65 \text{ W/m}^2$$

Before reaching the earth's atmosphere, there is very little loss in the amount of radiation emitted from the sun. Once the sun's energy enters the atmosphere the various wavelengths of energy are selectively depleted as illustrated in Figure 1-9. Some of the energy is reflected back into space as it

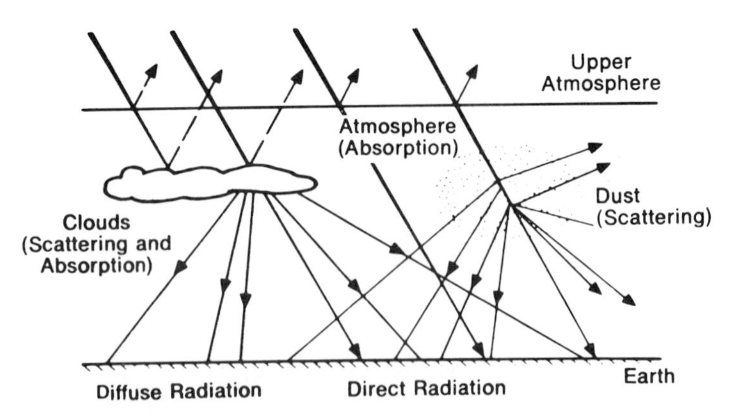


**Figure 1-9.** Atmospheric effects on solar radiation. (*Source. Fundamentals of Solar Heating,* D.O.E., 1978)

enters the upper atmosphere. Energy depletion continues with ultraviolet radiation being absorbed by the upper layer of ozone ($O_3$), resulting in only 1 to 3 percent of the total energy received on the earth's surface being ultraviolet. Only a small amount of absorption occurs in the visible region. Absorption of the longer wavelength radiation (infrared) is due principally to molecules of water and carbon dioxide. Scattering of the shorter wavelengths of visible light causes the sky to appear blue and the sun to appear yellow or orange. This scattering of light is called ***diffuse radiation***, and on cloudy days may represent all of the solar energy available for use. Most of the visible light, however, does manage to penetrate the atmosphere and is called ***direct beam radiation***.

The amount of reduction of direct beam radiation depends upon the length of the atmospheric path that it must traverse. Meteorologists describe this in terms of the air mass, $M$, as illustrated in Figure 1-10.

Air mass is the ratio of the actual path length $BC$ to the path length $BA$ that would exist if the sun were directly overhead. Mathematically, this can be approximated as

$$M = \frac{BA}{BC} = \frac{1}{\text{Cos } \theta}$$

Air mass 1 identifies the shortest path through the atmosphere and occurs when there is a 90-deg angle between the sun's radiation and the earth's surface. Air mass 2 exists when there is a 60-deg angle between the perpendicular to the earth's surface and the incoming angle of solar radiation. In this situation, the sun's ray's pass through twice as much atmosphere as in air mass 1. Similarly, for air mass 3, the sun's ray's pass through three times as much atmosphere as air mass 1. Air mass can also be simply expressed in terms of the sun's altitude above the horizon. In Figure 1-10, the altitude is

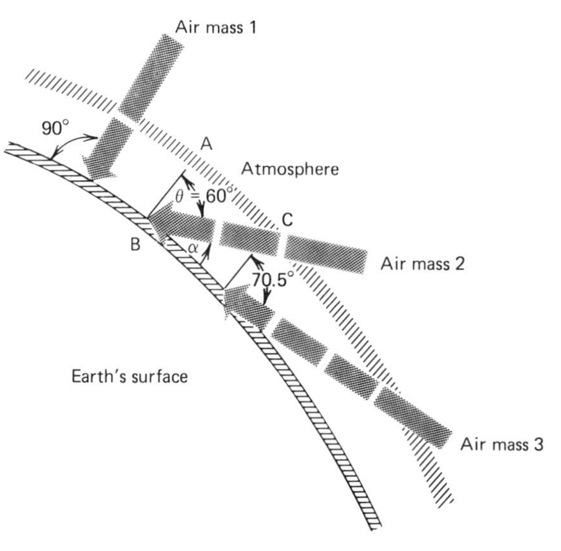


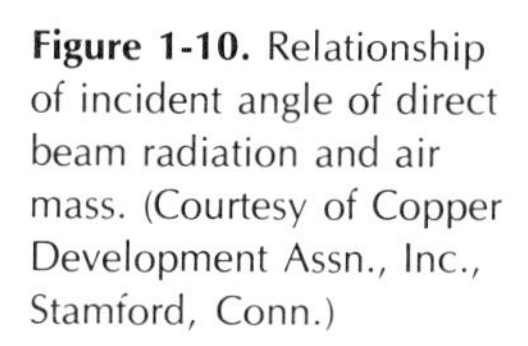
**Figure 1-10.** Relationship of incident angle of direct beam radiation and air mass. (Courtesy of Copper Development Assn., Inc., Stamford, Conn.)

represented by the Greek letter $\alpha$. The mathematical expression for air mass would be represented as

$$M = \frac{1}{\text{sine of the solar altitude}} = \frac{1}{\sin \alpha}$$

For example, on March 21 in Billings, Montana, if the solar altitude at noon is 42 deg, the air mass would be 1.5. These interactions of absorption, scattering processes, and direct radiation beam length of travel, contribute to the reduction in the total amount of energy received at air mass 0 (outside the earth's atmosphere). The energy received at the earth's surface can therefore be estimated to vary anywhere from 0 to 310 BTU/ft$^2$-hr.

## Energy Distribution Onto a Surface

The total solar energy received at the earth's surface is called ***insolation*** (not to be confused with the term ***insulation***). Insolation is comprised of direct radiation as well as diffuse radiation. The factors that affect the amount of insolation available at a particular location include latitude, time of day, time of year, cloud cover, shading obstructions, atmospheric turbidity, elevation, and orientation of the land surface. The amount of insolation available per unit area of ground surface is determined by the actual depletion of the solar constant energy, as previously discussed, and by the angle that the sun's rays make with a surface. When a beam of energy strikes a surface with a cross-sectional area of 1 ft$^2$ with an angle of incidence ($\measuredangle i$) of 0°, its energy is distributed over an area of 1 ft$^2$. The angle of incidence ($\measuredangle i$) is depicted in Figure 1-11 as the angle measured between the incoming beam of energy and a line drawn perpendicular to the surface that it strikes.

As this angle of incidence increases to $\measuredangle i_1$ and $\measuredangle i_2$, the beam's energy is decreased per unit area as a progressively larger area is covered. This illustrates the fact that the sun's apparent position in the sky is very important. The more perpendicular or normal the rays of energy are to a surface, the more energy there is per unit area. From our previous discussion of

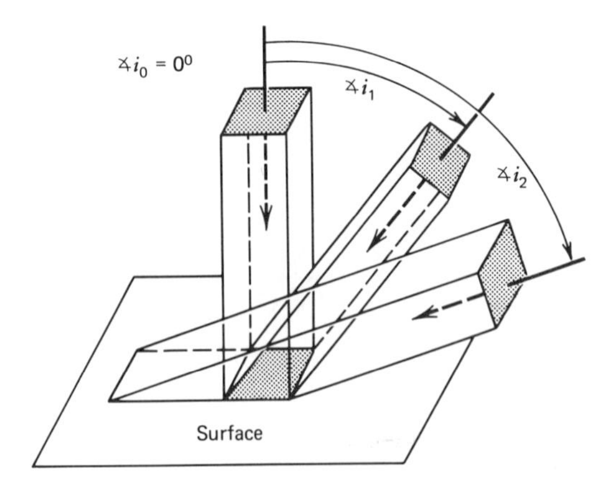


**Figure 1-11.** Energy distribution and angle of incidence. (*Source. Solar Energy Project*, D.O.E., 1979)

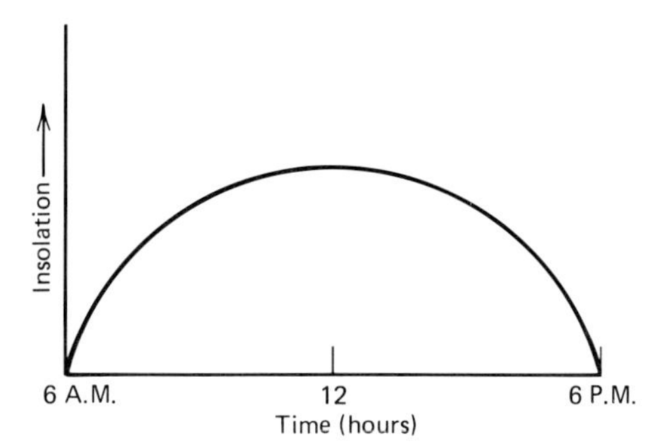


**Figure 1-12.** Insolation versus time.

latitude, we can understand that the higher the latitude, the more slanting are the sun's rays to the surface of the ground, resulting in less energy received per unit area. From our discussion of air mass, we can also understand that as the angle of incidence increases, there is also less energy received due to the greater depth of atmosphere encountered.

Insolation measured on a horizontal surface plotted against time of day is shown in Figure 1-12. As can be expected, the total insolation received is maximum at solar noon because the angle of incidence of the sun's rays is minimal.

Because the sun is low in the southern sky during winter in the northern hemisphere, more solar energy will strike a flat plate collector if it is tilted up from the horizontal at a steep angle toward the sun. During the summer, the same flat plate collector will intercept more solar radiation in a horizontal position. The angle of tilt, therefore, is very important in the overall collection of energy. Obviously, the optimum tilt occurs when the angle of the collector is the same as the incoming radiation, as illustrated in Figure 1-13.

In general, the complexity involved in making a collector adjustable in its tilt angle is not compensated by improved performance. It is preferable to select a fixed suitable angle based on the function for which the collector is intended to serve. A tilt angle equivalent to the latitude best serves the needs for solar domestic hot water.

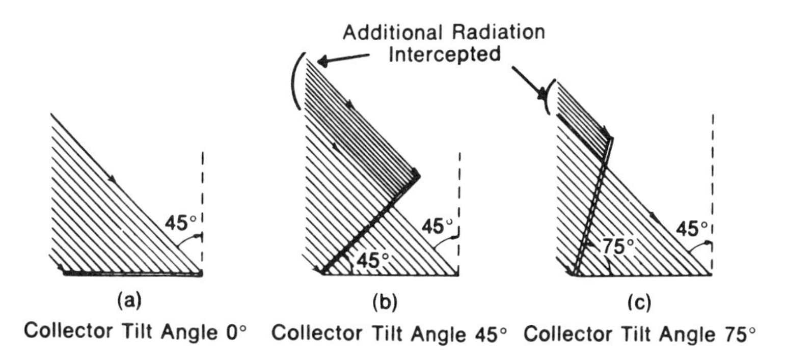


**Figure 1-13.** Effect of collector tilt on energy intercepted. (*Source. Fundamentals of Solar Heating*, D.O.E., 1978)

## REVIEW QUESTIONS

1. The sun is structured in seven layers. List them from the interior outward.

   a. ________ b. ________

   c. ________ d. ________

   e. ________ f. ________

   g. ________

2. Which part of the sun provides the radiation received by the earth?

   a. Thermosphere b. Troposphere

   c. Chromosphere d. Photosphere

   e. Mesosphere

3. Our changing seasons are caused by which of the following?

   a. Rate of the sun's movement at 15 deg per hour

   b. Hydrodynamics of the sun

   c. Earth's rotation on its axis

   d. Earth's rotational axis is tilted

   e. None of the above

4. In the southern hemisphere, the south end of the axis is tilted away from the sun during the summer.

   a. True b. False

5. When the angle of declination is at its greatest in the northern hemisphere, what is the period of time called?

   a. Spring equinox

   b. Summer solstice

   c. Fall equinox

   d. Winter solstice

6. The angle between the equator and a line from the center of the earth to its surface is known as what?

   a. Latitude

   b. Longitude

   c. Declination angle

   d. Azimuth

   e. Altitude

7. When the sun's position is true south, the azimuth is zero and the altitude is a maximum.

   a. True b. False

8. Which of the following methods is used to determine the position of the sun in relation to specific geographic locations?

   a. Direct visual observation

   b. Sun path diagram

c. Graphic projection
d. Calculative methods
e. None of the above

9. What is the sun's position in Columbus, Ohio, on February 21 at 2 P.M.?
   a. Altitude 30°, azimuth 36°
   b. Altitude 32°, azimuth 35°
   c. Altitude 29°, azimuth 35°
   d. Altitude 33°, azimuth 36°
   e. Altitude 30°, azimuth 32°

10. The fusion reaction for our sun is what type of reaction?
    a. Proton-proton
    b. Gamma
    c. Neutrino
    d. Carbon-nitrogen
    e. Electron-electron

11. In order of increasing wavelength, what are the three major energy regions of radiation from the sun that we are concerned with?
    a. ________ b. ________ c. ________

12. The solar constant in BTU/ft$^2$-hr is approximately equal to which of the following?
    a. 1337 b. 365
    c. 436 d. 1.97
    e. None of these

13. If 200 BTU/ft$^2$-hr are available on a surface, what is the equivalent value in terms of langleys per hour?
    a. 54.2 b. 101.5
    c. 17.2 d. 85.3
    e. None of the above

14. If the altitude at a certain geographic location is 36 deg at solar noon, what will the air mass be?
    a. 1.5 b. 1.2
    c. 0.81 d. 0.59
    e. 1.7

15. The total energy received at the earth's surface which is comprised of diffuse and direct beam radiation is known as which of the following?
    a. Air Mass b. Insolation
    c. Insulation d. Absorption
    e. Scattering

# Understanding the Heat Transfer

As we have learned from Chapter 1, the sun's energy is available for our use at the surface of the earth in the form of radiant energy. The key to using this solar energy for heating is to absorb this energy and transfer it to the medium that we wish to heat. This movement of heat energy between two substances at different temperatures is known as heat transfer. A knowledge of the basic principles of heat transfer is fundamental to an understanding of solar heating systems. Before we discuss specific collector systems in detail, we will briefly review the methods of heat transfer encountered in solar domestic hot water systems.

## THE CONCEPT OF HEAT

Heat is a form of energy related to the motion of molecules within a substance. It is the sum total of all the molecular energy of a body. Formally, the change in heat $\Delta Q$ of a body can be related to the change in temperature $\Delta T$ of a substance by Equation 2-1. The change in the heat (content) of a body is then positive or negative, depending on whether the temperature of the body increases or decreases.

*Equation 2-1*

$$\Delta Q = C_p M \, \Delta T$$

In this equation, $M$, is the mass of the material that has had its temperature changed by $\Delta T$ degrees. The term $C_p$ is the specific heat of a substance at constant pressure and is a function of temperature. The higher the temperature of a substance, the greater is the average kinetic energy (energy of motion) of its particles. As the average kinetic energy increases, the frequency of molecular collision increases, and therefore the amount of heat energy is increased. The amount of heat energy contained in body is measured in units called British Thermal Units (BTU) or in the metric system in units called calories. A BTU is defined as the quantity of heat required to raise the temperature of one pound of water one degree Fahrenheit. In the metric system, the amount of heat needed to raise one gram of water one degree Centigrade is a calorie. The equivalence of these units can be stated as follows:

$$1 \text{ BTU} = 252 \text{ cal}$$
$$1 \text{ calorie} = 0.003968 \text{ BTU}$$

Let us discuss each of the terms of which Equation 2-1 is comprised.

Initially, the distinction between the terms ***heat*** and ***temperature*** should be realized. ***Temperature*** is not heat energy. Temperature is a measure of the average translational energy per molecule. It is an ***indicator*** of the ***intensity*** or degree of heat stored in a body. The most commonly used measure of temperature is in degrees Fahrenheit (°F) and in degrees Celsius or Centigrade (°C). The relationship between these two common scales for temperature measurement is illustrated in Figure 2-1.

Heat is transferred between two substances at different temperatures, always from the higher to the lower temperature. Whether an object feels hot or cold depends on the direction of the heat transfer. Objects at temperatures higher than our body temperature feel hot because heat flows from the object to our body. Conversely, objects at lower than body temperatures feel cold because heat from our body is transferred to the colder object. We will illustrate the difference between heat and temperature with an example. Using Equation 2-1, we can determine what increase in heat energy is necessary to increase the temperature of one gallon of water from 70° F to the boiling point, 212° F.

$$\Delta Q = C_p M \, \Delta T$$

$$\textit{where } C_p = 1 \text{ BTU/lb-°F}$$
$$M = 8.33 \text{ lb/gal}$$
$$\Delta T = (212° \text{ F} - 70° \text{ F}) = 142° \text{ F}$$

Therefore,

$$\Delta Q = (1 \text{ BTU/lb-°F})(8.33 \text{ lb/gal})(1 \text{ gal})(142° \text{ F})$$

And

$$\Delta Q = 1183 \text{ BTU}$$

If 2 gallons of water are heated, twice as much heat energy is needed to achieve the same temperature.

Next we should consider why two objects at the same temperature may not necessarily feel like the same temperature to the touch. Different materials absorb different amounts of heat energy at a different rate of transfer in changing to the same difference in temperature. A metal object transfers heat at a faster rate than a nonmetal object and therefore feels colder to the touch due to its greater ability to absorb heat. The measure of a material's ability to absorb heat is the specific heat. The ***specific heat***, $C_p$, is the quantity of heat (in BTU) absorbed by 1 lb of a material to produce a 1° F temperature change, (BTU/lb-°F). (In the metric system it is the number of calories absorbed by 1 gram of material to rise 1° C in temperature.) The

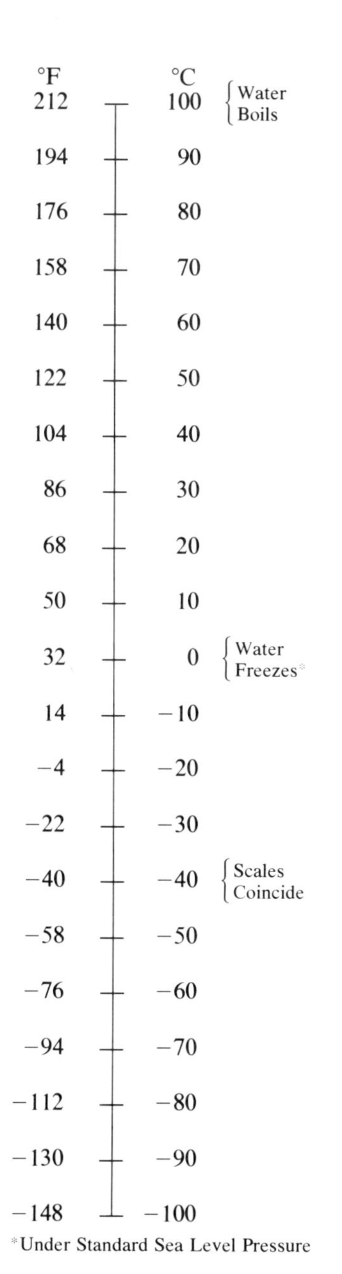


**Figure 2-1.** Temperature conversions. (Courtesy of U.S. Department of Commerce; National Climatic Center)

The standard formulas to convert °F to °C and °C to °F are shown below:

°F = 9/5 °C + 32
°C = 5/9 (°F −32)

Alternate, easy to remember conversion methods follow:

°F = 9/5 (°C + 40) −40
°C = 5/9 (°F + 40) −40

To use the alternate conversion formulas for converting from one scale to the other:

(a) add 40 to the value to be converted
(b) multiply that sum by the fraction:
(5/9 for °F to °C)
(9/5 for °C to °F)
(c) subtract 40 from the product

For example, to convert 68°F to °C:

(a) add 40: 68 + 40 = 108
(b) multiply the sum by 5/9 (°F to °C):
5/9 × 108 = 60
(c) subtract 40: 60 − 40 = 20
(d) answer: 68°F = 20°C

reason for the variation of specific heat from substance to substance lies in the different masses of the atoms. The ratio between the specific heats of two bodies of equal mass is defined by the ratio of the temperature changes experienced when the two bodies are brought into thermal contact. Therefore, if the temperature changes of two bodies $A$ and $B$ with equal mass are found to be $\Delta \dot{T}a$ and $\Delta \dot{T}b$, the ratio between the specific heats of these bodies, $(C_p)a$ and $(C_p)b$ is defined as Equation 2-2:

*Equation 2-2*

$$\frac{(C_p)a}{(C_p)b} = \frac{-\Delta Tb}{\Delta Ta}$$

Consequently, if the specific heat of water is set equal to unity as a reference, then the specific heats of other materials can be determined.

Objects of the same material may absorb different quantities of heat energy when changing the same amount of temperature. The factor affecting a difference in the quantity of heat absorbed in this case is the ***mass***. Some confusion between the concepts of weight and mass arise because the two terms are routinely interchanged in our everyday language. The weight of an object is actually the measure of the force exerted on the object (mass) due to the earth's gravity. Objects with greater mass have greater weight under the influence of gravity. Weight is a force and has the units of force (pounds). The mass of a body, on the other hand, is the quantity of inertia it possesses. Newton's second law states that $m = f/a$, where $m$ is the mass, $f$ is the force, and $a$ is the acceleration. The constant ratio of force to acceleration can be considered a property of the body called its mass. The weight of a body means the gravitational force exerted on it by the earth. For our practical purpose on earth the standard pound by definition is a body of mass 0.4535924277 kg. The pound of force is the force that gives a standard pound an acceleration equal to the standard acceleration of earth's gravity which is 32.1740 ft/sec$^2$. Now, let's once again determine what increase in heat energy is necessary to increase the temperature of one gallon of water from 70° F to the boiling point, 212° F. In units of the metric system we have:

$$\Delta Q = C_p M \Delta T$$

$$\begin{aligned} \textit{where}\ C_p &= 1 \text{ calorie/gram-°C} \\ M &= 8.33 \text{ lb} \times 0.454 \text{ kg/lb} = 3.78 \text{ kg} \\ \Delta T &= (212° \text{ F} - 70° \text{ F}) = (100° \text{ C} - 21.1° \text{ C}) \\ &= 78.9° \text{ C} \end{aligned}$$

Therefore,

$$\Delta Q = (1 \text{ calorie/gram-°C})\ (3.78 \text{ kg})\ (78.9° \text{ C})$$

and

$$\Delta Q = 298{,}084 \text{ calories}$$

Using the conversion factor of 252 calories per BTU, we find this answer equivalent to 1183 BTU. To reiterate, the quantity of heat needed depends on the parameters of specific heat, mass, and temperature difference.

## MECHANICS OF HEAT TRANSFER

There are three different physical mechanisms by which heat may be transferred from one place to another. These include conduction, convection, and radiation. In ***conduction***, heat is transferred as a result of molecular motion and the collisions between fast- and slow-moving molecules, without any gross motion of matter. ***Convection***, on the other hand, is a result of gross motion of matter, and is important in liquids and gases. Finally, heat ***radiation*** is an electromagnetic interaction between bodies, and does not require a material medium between the bodies.

***Conduction*** is the transfer of energy through a quantity of material by direct molecular interaction. The heated molecules transfer some of their vibrational energy directly to adjacent cooler molecules, resulting in a larger-scale energy transfer. Energy is lost by heat conduction through direct physical contact with objects of lower temperature. Conversely, heat is gained by direct contact with objects of higher temperatures. The ability of a material to permit the flow of heat is called its thermal conductance, $C$. ***Thermal conductance*** is the quantity of heat per unit time that will pass through the unit area of a particular material or body when a unit average temperature is established between the surfaces. The units of measure are BTU/hr-ft$^2$-°F. ***Thermal conductivity***, $K$, specifies the thermal conductance of a material for a certain thickness of material, such that $K = Ct$, where $C$ is the thermal conductance and $t$ is the thickness of the material. The units of measure are BTU – in/hr-ft$^2$-°F. The rate of heat flow, $Q$ (BTU/hour), depends on the thermal conductivity of the substance, the cross-sectional area of the conductor, its thickness, and the temperature difference between the surfaces considered. The rate of heat flow is directly proportional to the area through which the heat energy can move and is inversely proportional to the thickness of the material as depicted in the heat conduction Equation 2-3.

*Equation 2-3*

$$Q = \frac{K}{t} A (T_2 - T_1)$$

*where*

$Q$ = rate of heat transfer (BTU/hr)
$K$ = thermal conductivity (BTU-in/hr-ft$^2$-°F)
$t$ = thickness (inches)
$A$ = surface area (ft$^2$)
$(T_2 - T_1)$ = temperature difference between surfaces (°F)

For example, if a collector has a 24 ft$^2$ back with a 2-in. glass fiber insulation of thermal conductivity, $K$, of 0.23 BTU-in/hr-ft$^2$-°F, then if the absorber plate temperature is 180° F and the ambient air is 50° F, the rate of heat transfer by conduction from the back of the absorber plate to the outside environment can be found using Equation 2-3 as follows:

where

$$Q = \frac{23}{2}(24)(180 - 50)$$

and

$$Q = 358.3 \text{ BTU/hr}$$

From this example, we can see that the lower the $K$ value or greater the thickness of insulating material, the lower the rate of heat transfer by conduction. The tendency of a material to retard heat transfer is known as ***thermal resistance***, $R$. Thermal resistance of a material is the inverse of its thermal conductance such that $R = 1/C$. The units of measure are hr-ft²-°F/BTU. All materials have some resistance to heat flow. One with high thermal resistance is called ***insulation***. The concept of thermal resistance is very useful in computing the heat loss over a composite wall (made up of different materials). In many situations, the same amount of heat must flow through each insulative layer. The situation is similar to an electrical circuit with elements connected in series. Series circuits combine the overall circuit resistance as the sum of the individual resistances. For example, the simple wall illustrated in Figure 2-2 has a total overall resistance between framing members (due to conduction) of $R_T = R_1 + R_2 + R_3 + R_4 = 13.58$ hr-ft²-°F/BTU. As more thermal resistances are involved, the overall effect is simply the sum of the individual components. The ***overall coefficient of transmittance***, $U$, is the reciprocal of the total thermal resistance such that $U = 1/R_T$. The units of measure are the same as thermal conductance. By definition, therefore, Equation 2-3 can be represented as Equation 2-4.

***Equation 2-4***

$$Q = UA(\Delta T)$$

*where* $Q$ = overall rate of heat transfer (BTU/hr)
$A$ = overall area (ft²)
$U$ = overall coefficient of transmission (BTU/hr-ft²-°F)
$\Delta T$ = temperature difference between surfaces (°F)

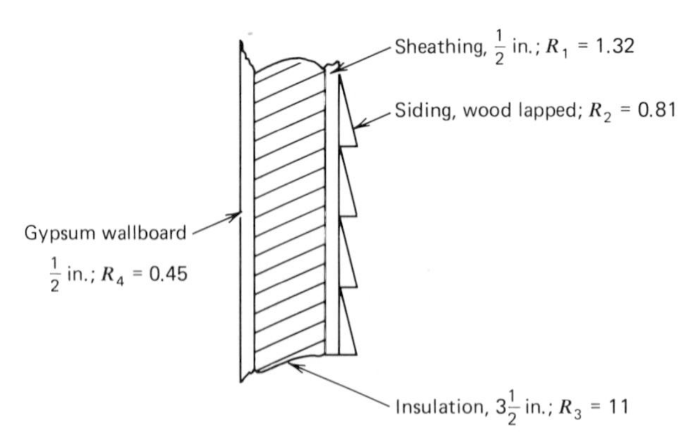


**Figure 2-2.** Conduction through a wall.

For example, assume that the inside temperature of a house was to be maintained at 68° F. If the outside ambient air was 40° F and the total wall area of Figure 2-2 was 256 ft$^2$, the total rate of heat transfer, $Q$, could be found using Equation 2-4 as follows:

$$Q = UA\,\Delta T$$
$$Q = (0.074)(256)(28)$$

and

$$Q = 530.4 \text{ BTU/hr}$$

The heat loss through the wall for this particular example does not include the convective losses.

***Convection*** involves the transfer of heat energy by the actual movement of the heated fluid in contact with solid surfaces. The air or liquid molecules exchange energy with adjacent molecules by carrying the energy via a fluid transport. There are two types of convection: (1) natural or free and (2) forced. ***Natural convection*** or free convection occurs due to the heating or cooling of any fluid when it contacts an object. As the air changes temperature, it changes density and rises or falls due to the action of gravity. ***Forced convection*** occurs when the fluid has a significant velocity relative to the object encountered. A fluid at a higher speed of travel will cause more heat transfer than one at a lower speed of travel. For example, a person will feel colder on a windy day of 10° F than on a day with the same temperature and no wind at all, because the body heat is dissipated quickly. This effect is known as the "chill factor." The rate of heat transfer due to the convection is similar to the conduction Equation 2-3. Convection is directly proportional to the temperature difference between the surface and adjacent fluid, $(T_s - T_f)$, the heat transfer area, $A$, and a film or surface coefficient, $h$, as depicted in Equation 2-5.

***Equation 2-5***

$$Q = Ah(T_s - T_f)$$

*where*
$Q$ = rate of heat transfer (BTU/hr)
$A$ = surface area (ft$^2$)
$h$ = film or surface coefficient (BTU/hr-ft$^2$-°F)
$(T_s - T_f)$ = temperature difference between surface and adjacent fluid (°F)

The film or surface coefficient, $h$, increases with fluid velocity. For instance, the free convection coefficient of air next to an inside wall (still air) is approximately 1.5 BTU/hr-ft$^2$-°F. On a windy day with 15 mph wind speeds, the free convection coefficient of air next to an outside wall is approximately 5.9 BTU/hr-ft$^2$-°F. Many examples of the coefficients can be found in the *ASHRAE Handbook of Fundamentals*, 1977. The total heat loss through the wall depicted in Figure 2-2 should now be calculated to include the convective losses that also exist. The total overall thermal resistance including conduction and convection between framing members is $R_T = R_1 + R_2 + R_3 + R_4 + R_5 + R_6 = 14.43$ hr-ft²-°F/BTU as illustrated in Figure 2-3.

The overall rate of heat transfer including both conduction and convection resistances can be determined using Equation 2-4. If the inside temperature of a house was to be maintained at 68° F, and the outside ambient air was 40° F, and the total wall area of Figure 2-3 was 256 ft², the total rate of heat transfer would be

$$Q = UA(\Delta T)$$
$$Q = (.069)(256)(28)$$

and

$$Q = 494.6 \text{ BTU/hr}$$

It can be seen from this example that the greater the velocity of the moving air, the more the resistance of the wall decreases; thus the increase in total heat loss.

As discussed in Chapter 1, ***radiation*** involves the transfer of energy from a warm body by electromagnetic waves. This form of heat transfer does not require a medium for propagation, and direct contact with the radiating source is not necessary. As previously illustrated, radiant energy exists at varying wavelengths including gamma rays, X rays, ultraviolet rays, visible rays, infrared rays, radio waves, and electric waves. When a body absorbs radiation, the body tends to restore its original state by reradiating and redistributing the extra energy. As a result of the energy redistribution, the emitted radiation may have a wavelength distribution different from that of the originally absorbed radiation. The distribution is controlled mainly by the temperature of the body. When a heated object emits a maximum amount of radiation as efficiently as possible regardless of the emitting surface, it is called a ***blackbody*** emitter. The thermal radiation properties of a material are described by its overall emissivity, $\epsilon$. By definition, emissivity, $\epsilon$, is the ratio of actual power reradiated at any wavelength to the power that

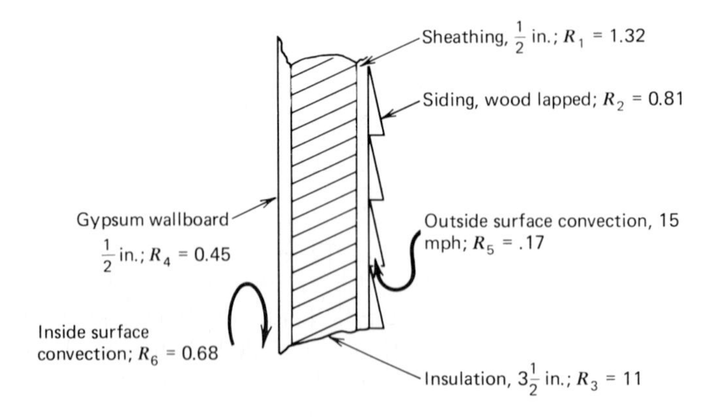


**Figure 2-3.** Conduction and convection through a wall.

would be emitted by a perfect blackbody at that wavelength. We usually find that the radiation for real bodies, is not distributed quite the same as that of a blackbody. Therefore, the body is assigned an overall emissivity, $\epsilon$, such that at a certain temperature, the body emits a fraction $\epsilon$ of the energy emitted by a blackbody at that temperature. Furthermore, the body is assigned the properties of reflectivity ($\rho$), absorptivity ($\alpha$), and transmissivity ($\tau$), which accounts for the total intensity ($I_o$) incident upon a surface. The sum of these properties is equivalent to the unity as shown in Equation 2-6.

***Equation 2-6***

$$(\rho)I_o + (\alpha)I_o + (\tau)I_o = 1$$

At any given temperature, the emitting power of a blackbody is directly proportional to its absorbing power and at any given wavelength we have $\epsilon(\lambda)$ equivalent to $\alpha(\lambda)$. Note, therefore, that the properties $\epsilon$, $\rho$, $\alpha$, and $\tau$ lie between 0 and 1 for real bodies. For a true blackbody these values would be 1, 0, 1, and 0, respectively. The ideal absorber plate has a surface with a high absorptivity to absorb as much solar radiation as possible and a low emissivity to reduce the thermal reradiative losses. Such an absorber is said to have a ***selective surface***.

## HEAT ENERGY COLLECTION AND THE SOLAR COLLECTOR

The flat plate solar collector is designed to collect both diffuse and direct beam radiation while maintaining minimum heat loss. The principal heat loss factors include: (1) conduction loss from the back of the absorber plate through the insulating material, (2) conduction losses through the sides of the collector, (3) convection losses upward through the glazing, and (4) the upward radiation loss. These heat losses can be quite large because the area for such losses is essentially equal to the area of energy collection.

Figure 2-4 illustrates the heat loss of a collector without a glazing cover. Much of the radiation absorbed by the flat black absorber plate is lost from the top surface due to convection and radiation. Convection losses can exceed radiation losses by more than a factor of 5 at wind speeds of only 10 mph. Useful heat is retained without a glazing cover only if the temperature of the absorber plate is close to the temperature of the ambient air.

Figure 2-5 illustrates the heat loss of a collector with a glazing cover. In this situation, the radiation absorbed by the flat black absorber plate is reemitted, but the glazing cover blocks much of the loss of this reemitted radiation to the outside. Energy is trapped in two ways. The temperature of the reradiated energy from the absorber surface is such that the energy distribution curve is shifted to the right so the surface only emits infrared (long-wave) radiation. A solar collector glazing is essentially opaque to longwave radiation and reradiates this energy back to the absorber plate.

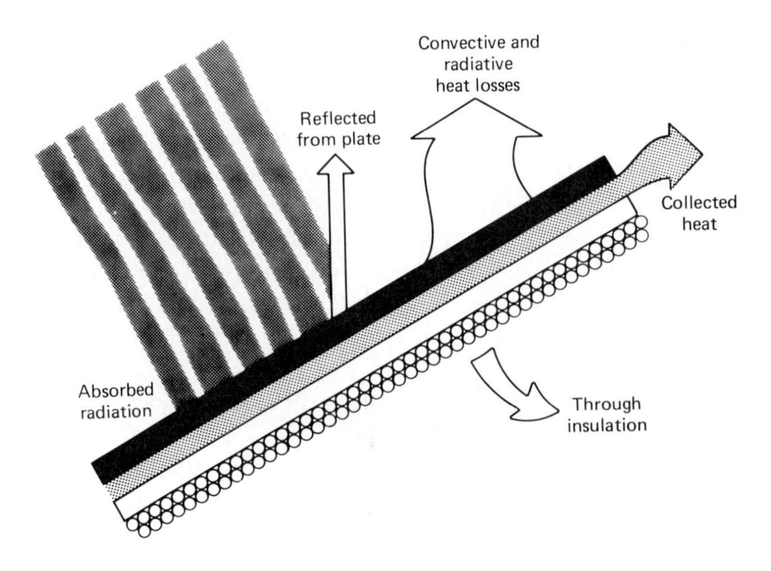


**Figure 2-4.** Flat black absorber plate without glazing. (Courtesy of Copper Development Assn., Inc., Stamford, Conn.)

The glazing also traps a layer of still air next to the absorber and reduces the convection heat loss. This combination of energy entrapment is a phenomenon known as the "greenhouse effect." There is some heat loss through conduction and convection as illustrated. A reflected energy loss is established with the addition of a glazing cover. If glass is used as a glazing cover, it can be etched by a thin film such as a fluoride-based acid bath so that the overall reflective loss is reduced. The amount of absorption in the glass can also be reduced by lowering the iron content.

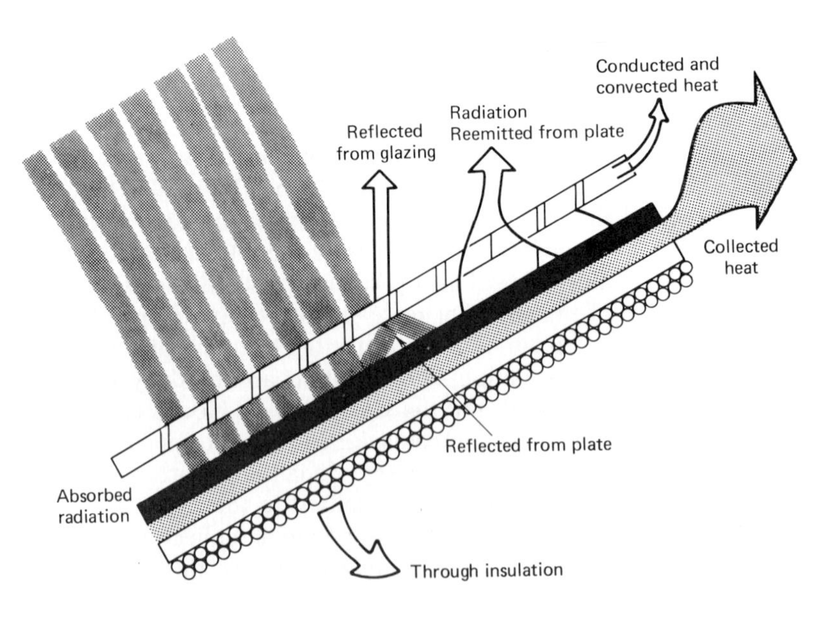


**Figure 2-5.** Flat black absorber plate with glazing cover. (Courtesy of Copper Development Assn., Inc., Stamford, Conn.)

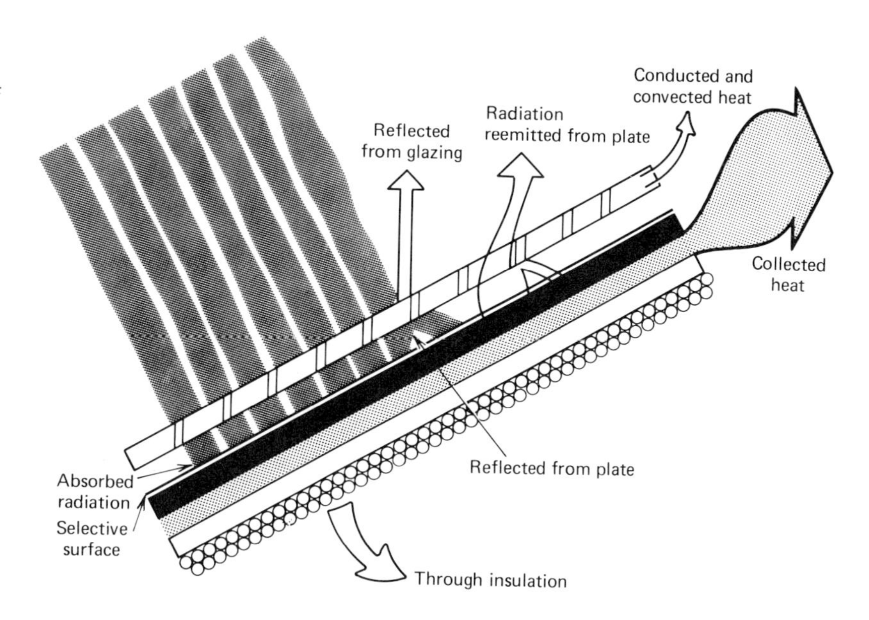


**Figure 2-6.** Selective surface absorber with glazing cover. (Courtesy of Copper Development Assn., Inc., Stamford, Conn.)

Figure 2-6 illustrates the heat loss of a collector with a glazing cover and an absorber with a selective surface. The selective surface reradiates a much smaller portion of the absorbed energy than does a flat, black nonselective surface. There is still some heat loss through conduction and convection to the outside air. This type of surface is sensitive to contamination by dust and does not retain its unique properties if exposed to the weather. We shall discuss the solar collector in further detail in Chapter 4, "The Solar Collector: Configurations, Materials, and Performance."

## REVIEW QUESTIONS

1. If 454 g of water are heated from 70° to 90° F, what is the change in the amount of heat?
   a. 9080 BTU
   b. 11 BTU
   c. 40 BTU
   d. 20 BTU
   e. 8500 BTU
2. In the metric system, what is the change in the amount of heat in review question 1?
   a. 2772 calories
   b. 10,080 calories
   c. 5040 calories
   d. $2.14 \times 10^6$ calories
   e. $2.29 \times 10^6$ calories

3. What is the value of 145° F converted to the Celsius scale?
   a. 110.5° C  b. 293.0° C
   c. 90.1° C  d. 53.4° C
   e. 62.8° C

4. If 2 gal of water are heated from 70° to 90° F, what is the change in the amount of heat?
   a. 333.2 BTU  b. 166.6 BTU
   c. 1499.4 BTU  d. 40.0 BTU
   e. 180.0 BTU

5. Assume a body is brought into thermal contact with an amount of water of equal mass. If the initial temperature of the water is 40° F, the inital temperature of the body is 20° F, and the final equilibrium temperature of both is 35° F, what is the specific heat of the body?
   a. 2.0  b. 0.5
   c. 4.0  d. 3.0
   e. 0.3

6. If a material is 2 in. thick and has a conductance of 5 BTU/hr-ft$^2$-° F, what is the conductivity, $K$, of the material?
   a. 2.5  b. 5
   c. 7.5  d. 10
   e. 12

7. If the conductivity of a material, $K$, is 0.30, the area is 1 ft$^2$ with a heat loss of 50 BTU/hr with a temperature change of 120° F, what is the thickness of the material?
   a. 0.36 in.  b. 0.72 in.
   c. 2.4 in.  d. 1.5 in.
   e. 3.6 in.

8. If the thermal resistances of a three-layer composite wall are $R_1 = 11$, $R_2 = 3.5$, and $R_3 = 1.5$, what is the overall coefficient of transmittance?
   a. 0.06  b. 0.09
   c. 0.29  d. 0.67
   e. 1.05

9. What is the overall rate of heat transfer in Review Question 8 if the temperature difference over a 100 ft$^2$ surface area is 50° F?
   a. 300 BTU/hr
   b. 5250 BTU/hr
   c. 3350 BTU/hr
   d. 1450 BTU/hr
   e. 450 BTU/hr

10. If the transmissivity of a material is 0.75 and its absorptivity is 0.12, what is its reflectivity?
    a. 0.25 b. 0.88
    c. 1.75 d. 0.13
    e. None of the above

# Generic Types of Solar Domestic Hot Water Systems

# 3

Many manufactured systems are available to the homeowner for solar domestic hot water, but the choice can be quite confusing if one is unfamiliar with the basic function of each generic type. Before making your initial investment, you should determine which system type would best serve your needs.

There are two generic classes for active solar systems in general. These include concentrating and nonconcentrating collection systems. Concentrating collectors attempt to concentrate the sun's rays to a small volume of transfer fluid to achieve a high temperature. Temperatures achieved by concentrating collectors are more suited to solar cooling than to domestic hot water (DHW) due to higher efficiency at higher operational temperatures. Reflecting and tracking concentrating collectors are normally more expensive, less durable, and effective only in very sunny regions because they do not utilize diffuse or scattered radiation. Evacuated tube-type collectors, on the other hand, which may be categorized as concentrating collectors, do offer high BTU yield under hazy, diffused sunlight conditions. These types of collectors are well suited for use in solar domestic hot water applications.

Nonconcentrating collectors, otherwise known as flat plate collectors, are used primarily in DHW and space heating applications. This chapter will familiarize the reader with the six generic types of solar hot water systems that employ the flat plate collector. These include:

1. Thermosiphon
2. Closed loop freeze-resistant
3. Drain back
4. Drain down
5. Air-to-liquid system
6. Phase-change

Each of these types is categorized as either (1) open loop (direct) or (2) closed loop (indirect) systems.

1. An ***open loop (direct) system***, illustrated in Figure 3-1*a*, is one in which the fluid heated in the collectors is the primary fluid that flows directly to the

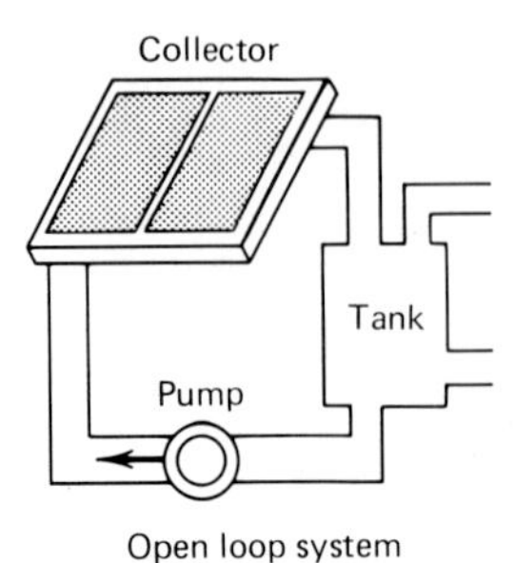


**Figure 3-1*a*.** Open loop system.

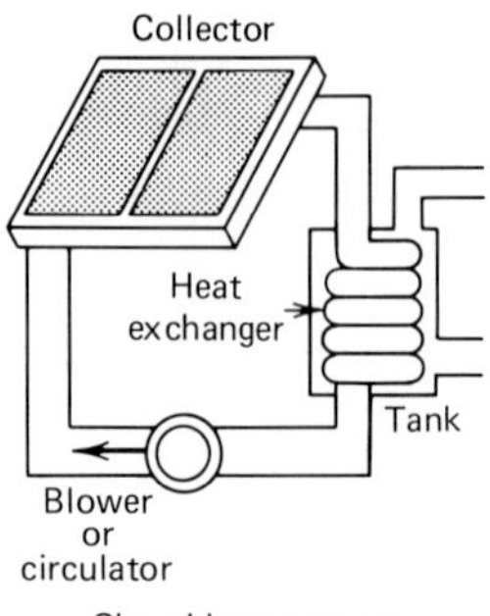


**Figure 3-1*b*.** Closed loop system.

faucet or washing machine (i.e., tap water). This system is one that can be open to atmospheric pressure. With the exception of a thermosiphon system, open loop systems may require a pump with horsepower sufficient to overcome the vertical head to the collectors from the storage tank and to overcome friction losses within the system if the collector loop is not maintained at sufficient water main pressure. The pump must be bronze or stainless steel to prevent corrosion of the water wetted surfaces.

2. A ***closed loop (indirect) system***, illustrated in Figure 3-1*b*, is one in which the fluid heated in the collectors is the secondary fluid (i.e., freeze resistant liquid or air). The heat from this secondary fluid is passed to the primary fluid (tap water) through a heat exchanger, such as a finned coil inside a storage tank. There is no exposure of the fluid to the atmosphere, and the system is pressurized. Because there are only friction losses to overcome, we can use a circulator which is a type of pump that is normally thought of in terms of less horsepower (i.e., 1/20 hp). A circulator with an iron housing and impeller can normally be used for liquid systems, depending on the transfer fluid.

## THERMOSIPHON SYSTEM

A thermosiphon system is normally an open loop system and is the simplest type of solar hot water system because of its limited number of components. It can be used in southern regions where there are no possibilities of freezing or in northern regions for seasonal hot water heating. The collectors and storage are depicted in Figure 3-2.

The bottom of the storage tank should be approximately 18 in. above the top of the collectors. As water is heated in the collectors it becomes less dense and is replaced by colder water from the bottom of the storage tank. The water in the collector, therefore, is forced into the storage tank. As long as sufficient insolation is available, this thermosiphoning action will occur and hot water will circulate into the tank. As long as the bottom of the tank is

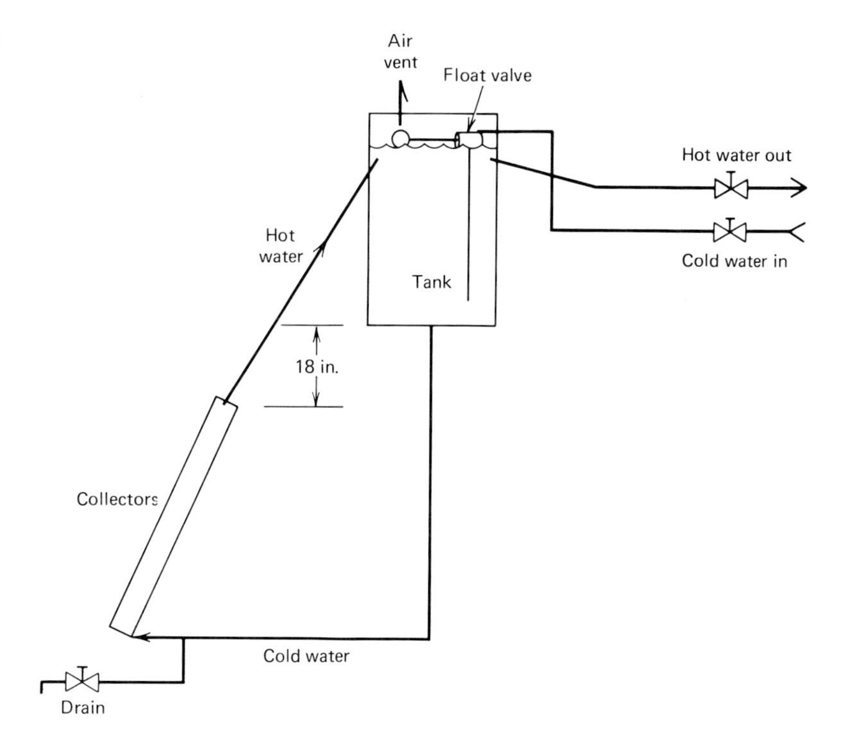


**Figure 3-2.** Thermosiphon system.

higher than the collector, reverse thermosiphon cannot occur during noninsolation periods, thus eliminating any need for a check valve. Piping should be a minimum 1 in. diameter to reduce pipe resistance to fluid. Pipe diameters and fluid resistance per length of run will be discussed in Chapter 8.

## CLOSED LOOP FREEZE-RESISTANT SYSTEM

The closed loop freeze resistant system, as illustrated in Figure 3-3, is the most common type of closed loop system used. This type of system heats a freeze resistant transfer fluid, which in turn heats the domestic water through a heat exchanger in the storage tank. A 1/20 hp (horsepower) circulator is normally used with 3/4 in. diameter piping, depending on the length of run. A nontoxic transfer fluid is circulated through this closed loop whenever sufficient insolation is available. Temperature control is established with a differential controller. This device signals the circulator to start when there is sufficient temperature gradient (15° to 20° F) between the collectors and storage tank so heat can be accumulated. It also signals the circulator to stop when the storage temperature is within 3° to 5° F of the collector fluid outlet so heat won't be lost from storage.

Components necessary for a closed loop freeze-resistant system include:

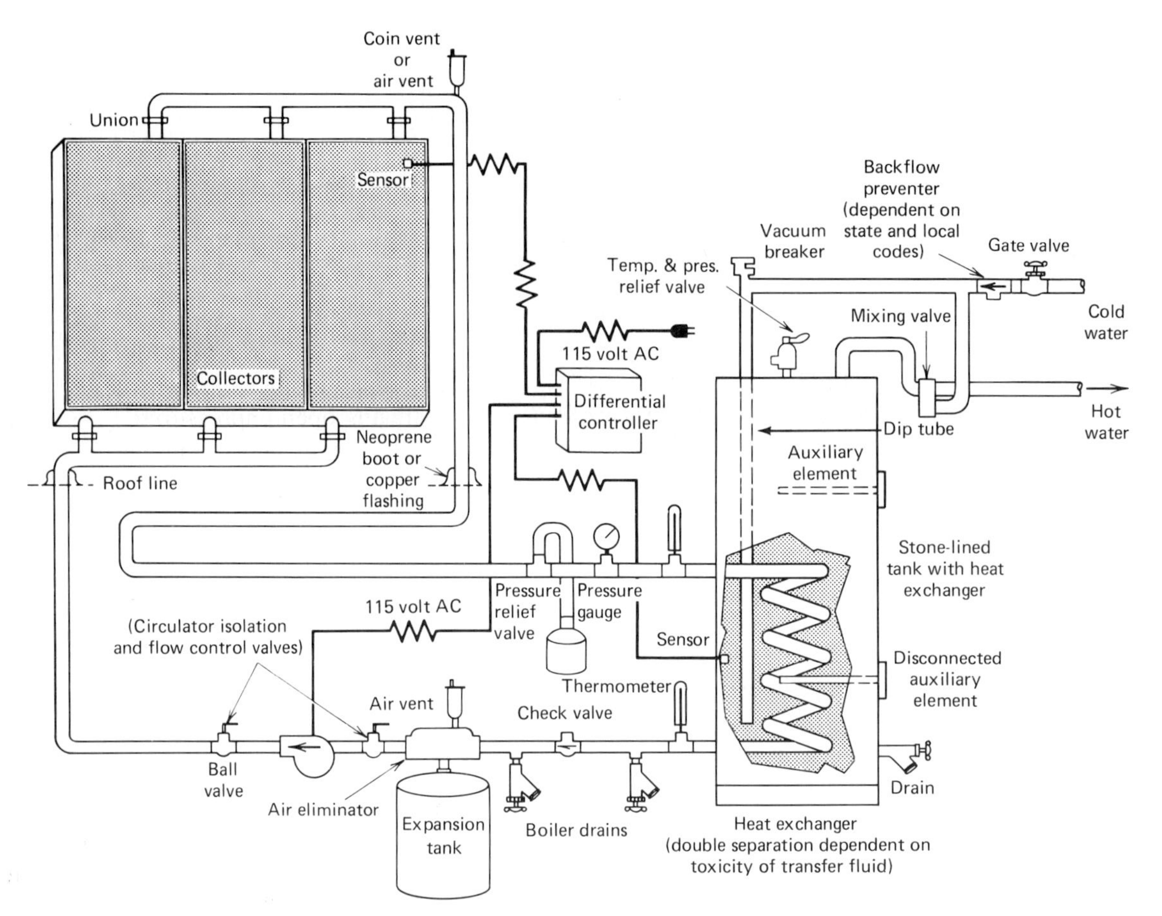


**Figure 3-3.** Closed loop freeze-resistant system. (Adapted from *Installation Guidelines for Solar DHW Systems in One and Two Family Dwellings*; H.U.D. in cooperation with D.O.E.)

Collectors
1/20 hp circulator (cast iron is suitable)
Expansion tank
Storage tank with exchanger (stone lined)
Fill drain assembly
Differential controller
Air purger with air vent
Pressure gauge (1)
Temperature gauge (2)
Air vent or coin vent
Check valve
Pressure relief valve
Vacuum relief valve

Back-flow preventer (depending on local codes)

Each of these components is discussed in Chapter 5.

## DRAIN BACK SYSTEM

The second most widely used system is the drain back system. The drain back type of system provides (1) passive freeze protection without a freeze resistant fluid, and (2) it can be used in conjunction with a glass lined tank. A drain back system differs from that of a closed loop system in the city or well water flows through a heat exchanger instead of a freeze resistant transfer fluid. The domestic water supply remains in a closed loop as illustrated in Figure 3-4*a* and the collector loop remains open/unpressurized. Water remains in the collector loop only while the pump is running. When the temperature difference is not adequate to provide heat to storage, a differential controller shuts off the pump and the water drains automatically into the storage tank by gravity. One must ensure that the internal or external manifolds of the collectors are pitched approximately 1 in. per 3 ft so proper drain back will occur. Water is used in the collector loop instead of freeze-resistant fluid because it would not be economical to fill the storage tank with a freeze-resistant fluid. Furthermore, the specific heat of water is superior to any of the transfer fluids. Storage tanks for drain back systems are typically 90, 120, or 150 gal insulated polyethelene containers depending on whether a 3, 4, or 5 panel system is needed. The storage tank is unpressurized (open loop), and supply and return lines connect the tank and collector. A pump rather than a circulator is used, and is sized to overcome the static head. A 1/7 hp pump will normally provide the 40 to 50 ft head required. System efficiency can be enhanced somewhat if a return path is established with a check valve to the cold water inlet of the storage tank, as depicted in Figure 3-4*b*. This provides a thermosiphoning arrangement between the back-up heater and storage tank and therefore increases heat storage capacity. It should be noted that the check valve must be tilted so that its gate is vertical, allowing valve operation by the small hydrostatic force caused by thermosiphoning.

Another type of drain back system which should be considered is illustrated in Figure 3-5. A smaller 6 to 10 gal polyethelene tank is used instead of a large storage tank. This smaller insulated tank is used for containment of water during drain back conditions. When there is sufficient insolation such that there is a 15° to 20° F difference between the collector outlet and the domestic hot water tank, a differential controller starts both pump and circulator. Water from the collectors and water from the domestic hot water tank are circulated through a heat exchanger in the drain back tank. Because the exchanger is outside the storage loop, however, the drain back system is not as efficient as a closed loop freeze-resistant system, thereby producing lower temperatures for equivalent collector areas.

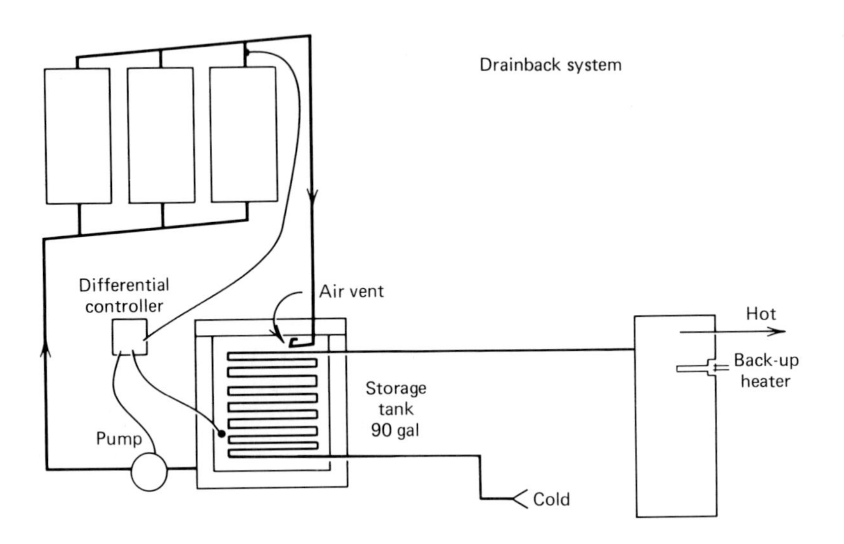


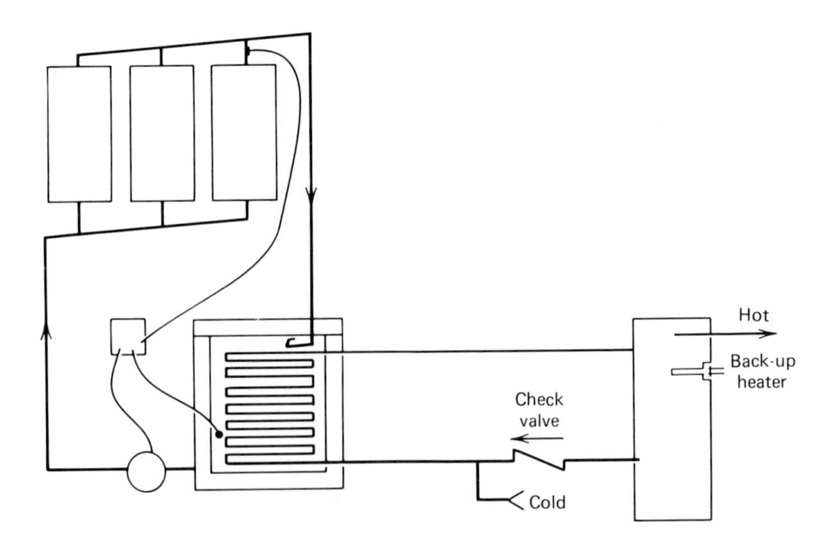


**Figure 3-4*a* and *b*.** Drain back system.

## DRAIN DOWN SYSTEM

The drain down system is a closed loop system and differs from the closed loop freeze resistant and drain back systems in that no heat exchanger is used, thereby increasing effective heat transfer. Potable water is circulated directly from the storage tank through the collector loop. Freeze protection is provided by a differential controller which de-energizes three solenoid valves to their normally open or closed conditions when the ambient temperature approaches 32° F. This system is illustrated in Figure 3-6.

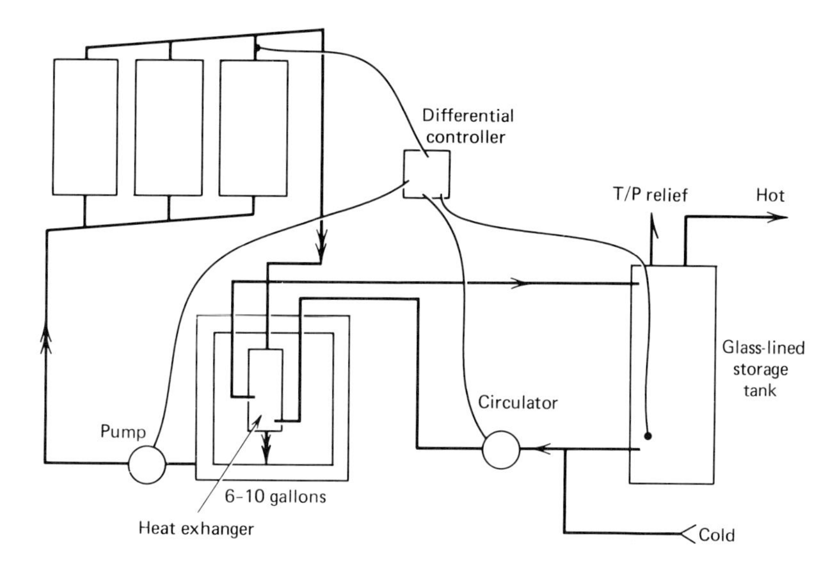


**Figure 3-5.** Drain back system.

When sufficient isolation is available, the solenoid valves are actuated, the drain down loop is closed, and the potable water loop from storage to collectors is opened. Collector manifolds and piping must be pitched so the system will automatically drain down upon solenoid deactivation.

Components necessary for a drain down system include:

Collectors
Storage tank
Pump (bronze or stainless steel)
Solenoid drain valves
Check valve
Vacuum relief (tank)
Temperature/pressure relief
Air vent
Differential controller

## Air to Liquid System

An air to liquid system is a closed loop system. A heat exchanger is used in this particular system and is located in the return path of the air duct as illustrated in Figure 3-7. The blower is normally 1/12 hp (horsepower) to 1/8 hp and is sized to provide approximately 1.5 to 3 cubic feet per minute (cfm) for every square foot of effective collector area. The storage tank is similar to the one used for a drain down system, and potable water is circulated through the loop. A 1/100 hp circulator is normally used with 1/2 in. diameter piping to the heat exchanger, depending on the length of run. There is a remote possibility that freezing can occur if the heat exchanger is placed in

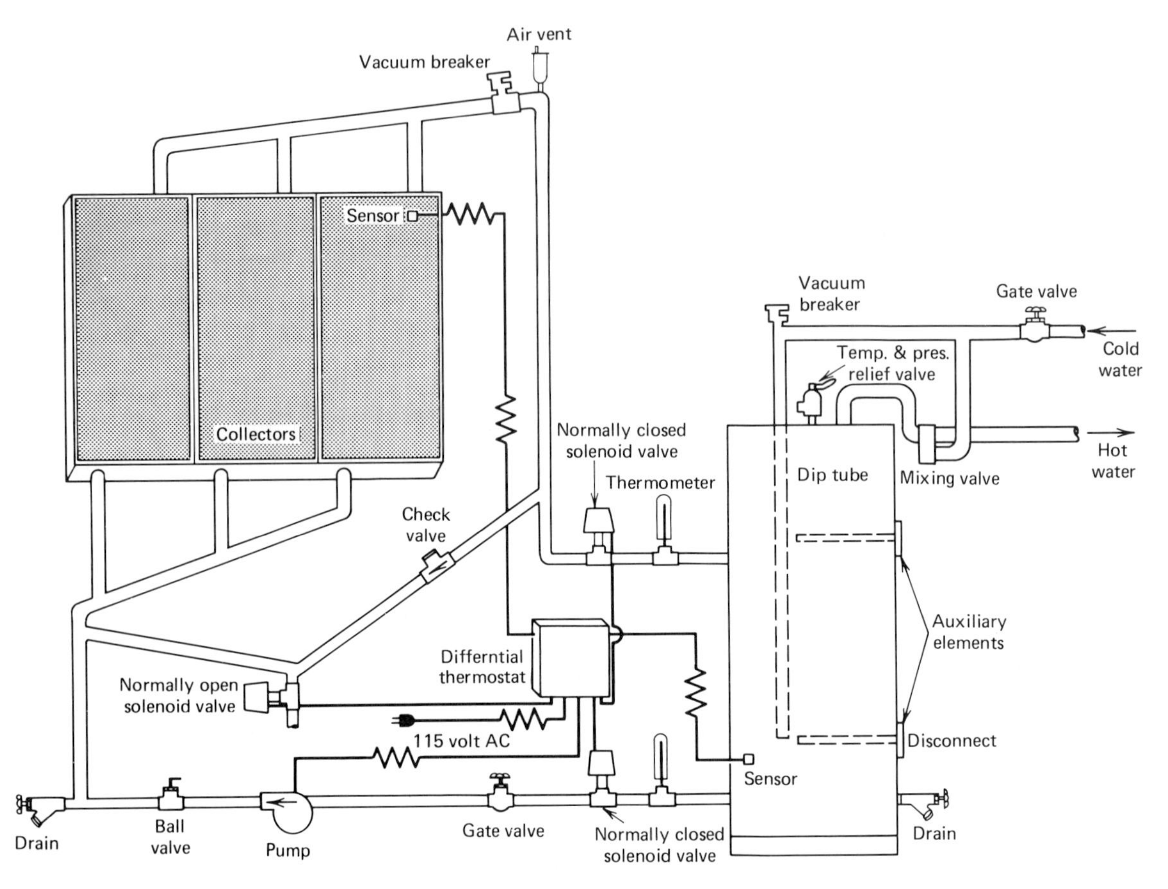


**Figure 3-6.** Drain down system. (Adapted from *Installation Guidelines for Solar DHW Systems in One and Two Family Dwellings*; H.U.D. in cooperation with D.O.E.)

unheated spaces or if cold air from the collector (on days without available insolation) is directed through the heat exchanger without the circulating pump running. A differential controller signals the blower to start at a typical value of 110° F and starts the circulator when there is a sufficient temperature gradient (20° to 25° F) between the exchanger and storage tank. The blower is typically stopped when the air temperature is reduced to 105° F and the circulator is stopped when the storage temperature is within 10° F of the exchanger outlet. These criteria may vary from one area to another.

Air has a definite advantage over other fluids because it is noncorrosive and it will not freeze or boil. It does, however, require greater installation space than piping. A typical duct work arrangement is pictured in Figures 3-8*a* and 3-8*b*. The largest drawback of this type of system is the lower efficiency of heat transfer from air to the liquid medium.

## PHASE-CHANGE SYSTEMS

A phase-change system is a closed loop system of either (1) passive or (2) subambient design. The principle operation of these designs is based upon the change of state of liquid to a gas and return to a liquid state.

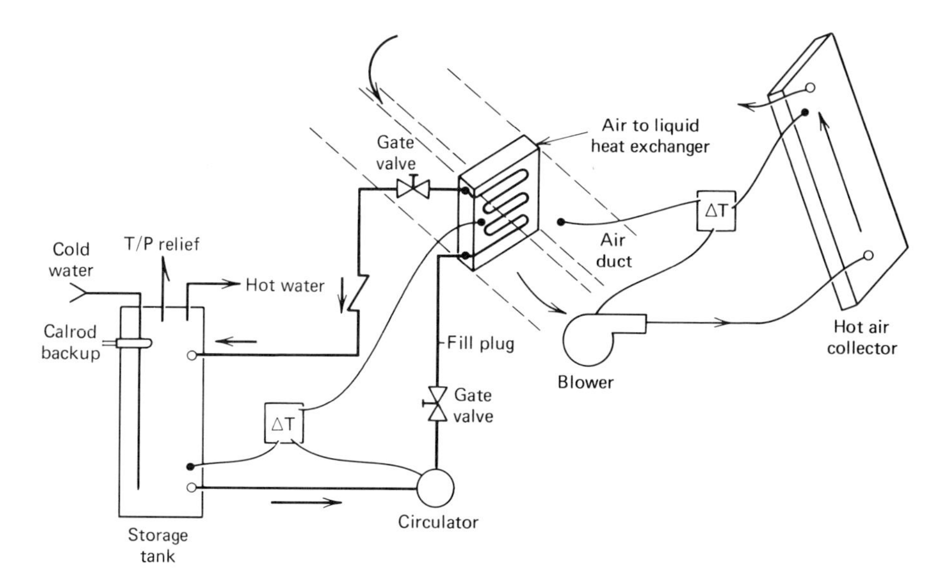


**Figure 3-7.** Air to liquid system.

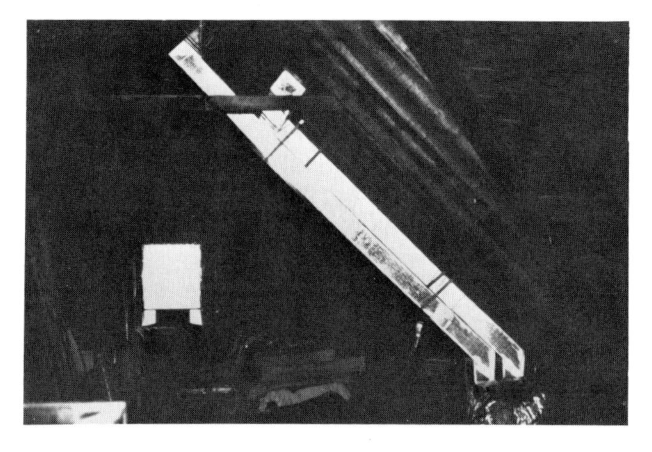

**Figure 3-8*a* and *b*.** Air to liquid system ductwork. (Courtesy of Granite State Solar Industries, Inc., Dover, N.H.)

## Passive Type

The principle transfer media or phase change material is Freon 114. Such a system is illustrated in Figure 3-9.

As the sun shines on the solar collectors that are filled with liquid Freon 114(TM), the liquid boils, changing into a vapor. (Freon 114 is a registered trademark of I. E. DuPont de Nemours Co., Inc.) The vapor rises into the heat exchanger which is inside the hot water tank and transfers the heat to the water. As the vapor gives up heat, it returns to its liquid state and is returned by gravity to the collector where the cycle continues. The system is similar in nature to a thermosiphon system except that the transfer medium changes state.

Thermostatic control is provided by setting the air pressure on the accumulator (pressure limiter). This accumulation takes the place of a differential controller. When the temperature of the coldest point in the hot water tank reaches the corresponding pressure set (i.e., 150° F with vapor pressure at 82 psi), the pressure exerted by the Freon 114 vapor will exceed the air pressure in the accumulator, stopping the heat transfer. When hot water is drawn from the tank, it is replaced by cold water, which reduces the temperature of the heat exchanger and has a corresponding effect on the vapor pressure of the Freon. The air pressure in the accumulator forces the liquid Freon back into the collectors, and heat transfer resumes until the water in storage is heated to the desired temperature as long as sufficient insolation is available.

There are no moving parts to this system, which is why it is termed *passive*, and there is minimum maintenance. The storage tank must be 12 to 18 in. higher than the collectors and, therefore, the weight of the tank when filled with water is an important consideration (i.e., 100 gal of water weigh approximately 833 lb). All fittings must be forged fittings, refrigerant grade, and joints must be silver soldered. Leaks must be detected with a halide leak

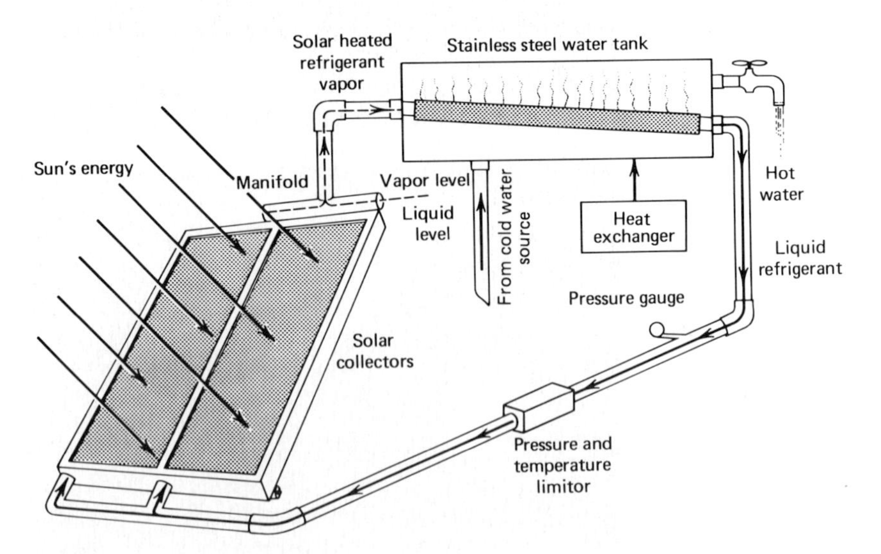


**Figure 3-9.** Passive phase-change system.

detector and collectors must be able to withstand pressure to 150 psi. The average homeowner would find this system more difficult to install than the other types of hot water systems.

## Subambient Type

The principle transfer media or phase-change material is Freon refrigerant grade R12. This type of phase change system is unique because it operates in all weather conditions (sunshine, rain, clouds, or even after sundown). It is necessary, however, to use some input energy to run the compressor in order to receive some output energy to heat hot water. This type of system is ideal for homes that are not oriented within the specifications discussed in Chapter 6, "Collector Array Sizing and Siting." A typical collector array is illustrated in Figure 3-10.

It sounds strange, doesn't it, to have a solar system that operates without sun? Well, the idea is somewhat similar to that of a heat pump, although its basic operation is quite different. For example, a refrigerator is a one-way heat pump. It can take heat out of water but cannot put it back in. It is possible, however, to change the direction of heat flow by means of reversing valves. The subambient phase change system works on a standard refrigeration cycle in which outside air is refrigerated, and the extracted heat is exhausted into water.

**Figure 3-10.** Subambient phase-change collector array. (Courtesy of Solar Specialties, Inc., Golden, Colo.)

Figure 3-11 illustrates the heating cycle of this particular system. The system starts up when the aquastat calls for hot water to enter storage. A solenoid valve is then opened, allowing liquid Freon to enter the primary thermal expansion valve. Rapid expansion and atomization of the liquid cool it, and the low temperature, low pressure, liquid/vapor Freon enters the collectors. Energy transferred from the environment and available insolation causes the refrigerant to change state from a liquid to a vapor. There is no glazing on the collectors as can be seen in Figure 3-10, because any air movement across them will warm the collectors, increasing the energy intake of the system. This is unlike other flat plate collectors in which glazing is an important parameter in determining collector efficiency.

The Freon in the vapor state is then pumped to the compressor by way of the suction accumulator. The accumulator allows any remaining liquid to boil off and ensures that only gas enters the compressor. Gas that gets too hot in the collectors could damage the compressor. Its temperature is therefore measured just prior to the compressor, and if it is too hot, a desuperheater valve is allowed to open and to add cold Freon to the line at the accumulator. When enough Freon has entered the system, it raises the pressure in the lines, and the compressor is activated by a high pressure control switch. The compressor increases the pressure resulting in a corresponding temperature rise. The high pressure, high temperature vapor is then cooled in a tube-in-tube condenser, which changes the vapor state to a liquid

**Figure 3-11.** Subambient phase change system. (Courtesy of Yankee Resources, Brunswick, Maine)

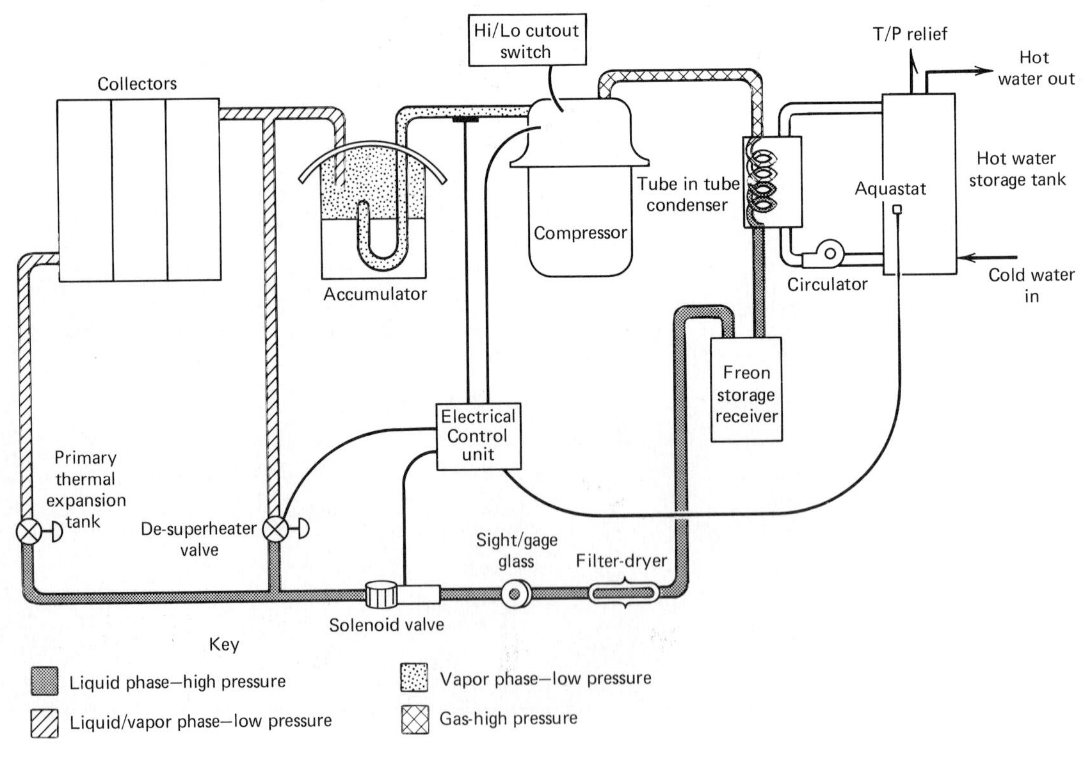

state. This change of state releases the energy absorbed by the refrigerant in the collector in the form of heat. Water is circulated through the condenser coil absorbing the heat released and supplying that heat to storage. The refrigerant leaves the condenser as a liquid and passes through a valve that expands the liquid into the liquid/vapor mixture which enters the collectors, thus completing the loop.

Basic differences between this subambient system and a heat pump should be distinguished.

1. The transfer medium in a heat pump is R22 vice R12.
2. When insolation is available, the collectors become superambient and system output may be doubled or tripled. Heat pumps do not have collection areas that are capable of increasing system output.
3. The heat pump evaporator must be defrosted regularly by pumping heat back outside to melt the ice. The subambient system does not need to be defrosted.
4. This system uses natural convection for heat transfer, whereas heat pumps use high horsepower fans to force convection air across the coil.

We have now discussed the six generic types of available solar hot water systems. Each type has its own distinctive features, which are summarized in Table 3-1. For regions that need freeze protection, the closed loop freeze resistant, drainback, and air to liquid systems are the most reliable systems to install.

**Table 3-1**
**Solar Domestic Hot Water Systems**

| *Type* | *Main Features* | *Advantages* | *Disadvantages* |
|---|---|---|---|
| Thermosiphon system | Flat plate liquid collectors<br>Normally open loop but no pump or external power (passive DHW system)<br>Storage tank higher than collector | No external power<br>Few components<br>High performance | Seasonal; dependent on freezing locations (if water is collector fluid)<br>Needs structural support for high storage tank |
| Closed loop freeze resistant system | Flat plate liquid collectors<br>Closed loop of piping from collectors to storage tank<br>Uses external energy (circulator and differential controller)<br>Uses nonfreezing collector fluid<br>Pressurized stonelined storage tank | Can be used in coldest climates<br>More and better established competition<br>Circulator; small consumption of external energy<br>High performance | Liquid; service/ maintenance required<br>More components required |

**Table 3-1** *(continued)*

| *Type* | *Main Features* | *Advantages* | *Disadvantages* |
|---|---|---|---|
| Drain back system | Flat plate liquid collectors<br>Water is collector fluid (open loop)<br>Potable water circulates through the heat exchanger in the storage tank (not through the collectors)<br>Large heat exchanger<br>Pitched headers | Can be used in the coldest climates<br>No antifreeze used<br>Most simple of active flat plate systems (no valves) | Larger pump; larger consumption of external energy<br>System must drain thoroughly<br>Use of corrosion inhibitor recommended |
| Drain down system | Flat plate liquid collectors<br>Potable water circulated through collectors<br>Line pressure feeds collectors (open loop)<br>Has automatic drainage valves<br>Pitched headers | No heat exchanger or extra storage tank needed<br>High performance | In some instances a larger pump; larger consumption of external energy<br>System must drain thoroughly<br>No corrosion inhibitor possible<br>Freeze danger with valve failure |
| Air-to-liquid system | Flat plate air collectors<br>Air-to-water heat exchanger<br>Ductwork and blower<br>Pipes and circulator<br>Larger collector area than liquid system | Won't freeze (dependent on exchanger location)<br>Air leaks won't cause damage<br>Integrates well with space heating | Hard to detect leaks<br>More space required for ducts<br>Blower and circulator required<br>More carpentry involved<br>Less efficient than other systems |
| Phase change system | Flat plate liquid collectors<br>Freon 114 or R12<br>Storage tank higher than collectors (passive type)<br>Closed loop refrigerant grade piping from collectors to storage tank (passive type) or to condenser (subambient type) | No external power (passive type)<br>Can be mounted at any location (subambient type) | Very hard to detect leaks<br>Special equipment to install<br>More components required (subambient type) |

In the final year of the solar domestic hot water systems tests conducted by the New England Electric Company, the following results were obtained for several types of installations as shown in Table 3-2.

**Table 3-2**
**System Performance Comparisons**

| *Type of System* | *Units Installed* | *Average Energy Savings (%)* |
|---|---|---|
| Water drain down | 12 | 38 |
| Water drain back | 4 | 41 |
| Closed loop freeze-resistant | | |
| Glycol | 78 | 40 |
| Silicone | 2 | 57 |
| Air to liquid | 4 | 27 |

Although these numbers do not reflect a balanced statistical comparison, they do illustrate system function in general. Phase change and thermosiphon systems were not represented in the systems installed. The type of system to be used for solar domestic hot water is one of individual preference. One thing that is assured, however, is that any of these generic types will save money in fuel costs.

## REVIEW QUESTIONS

1. There are six generic types of solar hot water systems that employ the flat plate collector. Name them.
   a. ________ b. ________
   c. ________ d. ________
   e. ________ f. ________
2. An open loop or direct system is one in which the fluid heated in the collectors is the secondary fluid and not the primary fluid (i.e., tap water).
   a. True b. False
3. In which two generic systems must the storage tank be located above the collector array?
   a. ________ b. ________
4. In which two generic systems must the manifolds of the collectors be pitched?
   a. ________ b. ________
5. Which generic system requires the use of automatic solenoid valves for freeze protection purpose?
6. Freon is the principal transfer media of which generic system?
7. Larger horsepower pumps are required in open loop systems than are required in closed loop systems.
   a. True b. False

8. In which generic system does a leak not cause an immediate problem?
9. Of the six generic systems, which one would be the most difficult for the homeowner to install?
10. In which generic system is the heat exchanger in the primary loop?

# The Solar Collector: Configurations, Materials, and Performance

# 4

Now that we have discussed basic concepts of heat transfer, the sun and its factor of insolation, and what types of systems are available for solar domestic hot water heating, we should now discuss the device that collects solar energy. This device used to transfer solar energy to heat is appropriately called a ***solar collector***. In the case of solar domestic hot water (DHW) systems, this heat is transferred from its collection point to storage and/or direct use via a transfer medium. Depending on the generic type of DHW system, this medium could be air, water, Freon, or a special freeze-resistant heat transfer fluid. (Heat transfer fluids will be discussed in Chapter 5, "Description and Explanation of System Components.")

There are three basic types of solar collectors available. These include: (1) tracking concentrators (Figure 4-1), (2) nontracking evacuated tube concentrators (Figure 4-2), and (3) flat plate collectors (Figure 4-3).

## TRACKING CONCENTRATOR COLLECTOR

As we briefly mentioned previously in Chapter 3, we do not need a ***tracking concentrating collector*** to heat domestic hot water. As in any system, the fewer moving parts there are, the fewer things that can go wrong. The relatively low temperatures needed to heat domestic hot water do not warrant the use of the more sophisticated tracking collector. A cross-sectional view and tracking operation are illustrated in Figure 4-1.

## NONTRACKING CONCENTRATING COLLECTOR

The **non-tracking evacuated tube concentrator** (Figure 4-2*a*) is a multipurpose collector that achieves high temperatures as a result of a vacuum between the absorbing surface and outer glass window, and through the use of a reflector. Specific applications include space cooling, process and domestic water heating, and space heating.

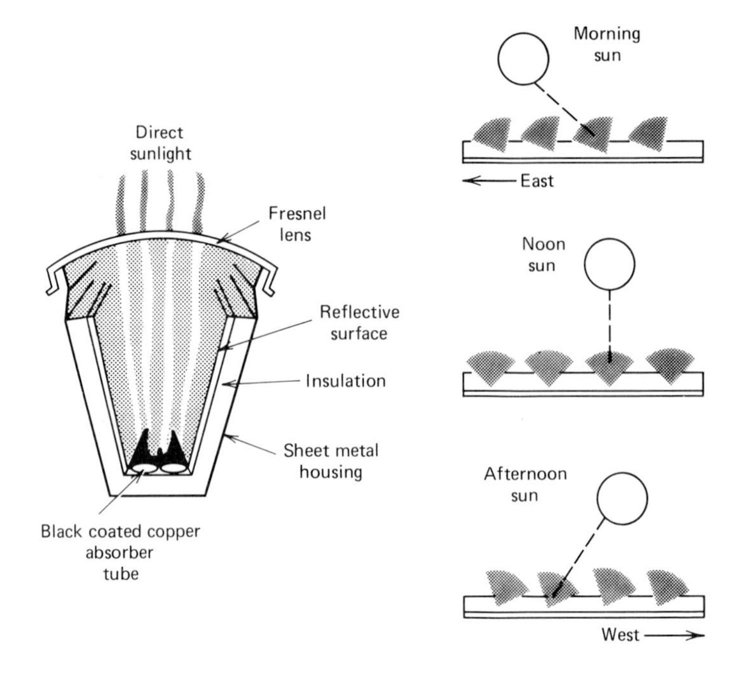


**Figure 4-1.** The tracking concentrating collector.

**Figure 4-2*a*.** General Electric vacuum tube solar collector. (Courtesy of General Electric Co., Solar Heating and Cooling Division, Philadelphia, Pa.)

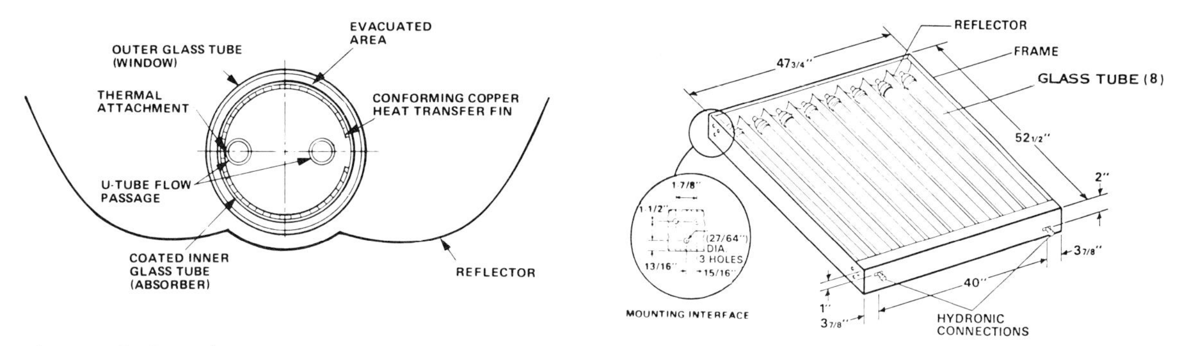


**Figure 4-2*b*.** General Electric vacuum tube solar collector. (Courtesy of General Electric Co., Solar Heating and Cooling Division, Philadelphia, Pa.)

The principal component in such a collector is the evacuated glass enclosure, as illustrated in Figure 4-2*b*, which consists of two concentric glass cylinders. The outer cylinder serves as a solar window. The inner tube is coated on the outer diameter with a selective coating and thus serves as the absorber. The space between the two cylinders is evacuated to minimize thermal losses. The tube is nested in a crosslike reflector which improves performance by concentrating available insolation on the absorber. This type of concentrating collector utilizes both diffuse and direct insolation. The energy absorbed is conducted along the fin to the working fluid flowing through a U-tube flow passage which has one leg attached to the fin with the other leg free floating to allow for thermal expansion. Because temperatures in excess of 200° F are frequent, minimum fluid pressure of 45 psi should be maintained at the collector inlet to avoid fluid boiling in the flow passages.

## FLAT PLATE COLLECTOR

The most common and most simple type of solar collector for DHW consists of a flat black plate encased with an insulated back and sides, and glazed with low iron glass or fiberglass reinforced polyester (FRP). A typical flat plate collector and its cross-sectional view are illustrated in Figures 4-3*a* and 4-3*b*, respectively. ***Flat plate collectors*** are categorized as either air or liquid. Most DHW applications use a liquid collector plate because heat exchanger transfer from liquid to liquid is more efficient than from air to liquid. This type of energy transfer device has only a few parts, but one must be able to decide what those parts should consist of and for what application each is intended. What type of glazing material should be used? What types of heat transfer fluids are available? These types of questions will lead us to answers about the performance and durability of the solar collector panel. In the remainder of our discussion we will focus on the flat plate collector.

## COLLECTOR COMPOSITION

The flat plate solar collector is composed basically of four parts. These parts include (1) the housing or shell of the collector, (2) insulation, (3) glazing,

**Figure 4-3a.** A flat plate collector. (Courtesy of Columbia Chase Corp., Holbrook, Mass.)

and (4) absorber plate. It is important to remember that collectors must withstand large thermal gradients, and the performance of the materials that comprise the collector is a prime factor in overall collector and system operation. Resistance to weather, moisture and corrosion, durability, economics of maintenance, and thermal efficiency are among the many factors to consider.

## Collector Housing or Shell

First we should consider the exterior box which integrates the other components which make up the collector. There are four common types of enclosures.

### Steel

This metal has the advantage of being exceptionally strong and resistant to damage by temperature variations. The shell should be galvanized and covered with a baked enamel finish. It should use one piece for the back and sides and the corners should be neatly united and welded. The steel shell

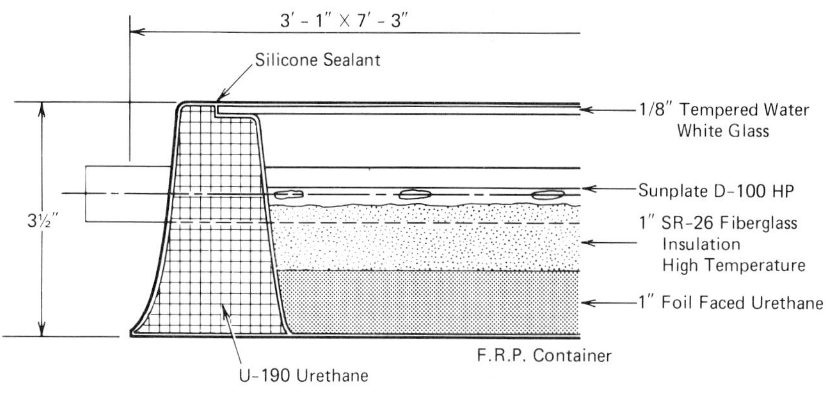


*Collector Case*
Fiberglass reinforced polyester monolithic seamless enclosure, nonheat conducting, corrosion resistant.

*Mounting Option*
1 in. integral full perimeter mounting flange.

*FRP Properties*

| | |
|---|---|
| Flexural strength | 28 psi × $10^3$ |
| Flexural modulus | 12 psi × $10^5$ |
| Tensile strength | 18 psi × $10^3$ |
| K-value | 1.2 BTU/HR/FT$^2$/°F/IN |
| Density | .058 lbs./in.$^3$ |
| Specific heat | .33 BTU/LB/°F |
| Specific gravity | 1.5 |
| Continuous heat resistance | 350° F |
| Thermal coefficient of expansion | 16 in./IN/°F × $10^{-6}$ |
| Section modulus of side rail | 7.007 |
| Impact strength izod | 12ft-lb/in. of notch |
| Wall thickness | Avg. 1/8 in. |

**Figure 4-3*b*.** Cross-sectional view of a flat plate collector. (Courtesy of Columbia Chase Corp., Holbrook, Mass.)

should also not come into contact with the absorber plate if the metals are dissimilar because of the possibility of galvanic corrosion. The principal disadvantage is one of weight. A heavier panel is harder to install and provides more stress to the roof framing than an aluminum or fiberglass housing.

### Aluminum

This metal is lower in weight and has excellent durability. It can also be anodized in color to blend in with the roof. The aluminum must be anodized or otherwise it will pit and corrode. Aluminum is more difficult to weld, so seams may not be as tight as the steel counterpart. Aluminum housings are typically made with extruded sides and a flat back plate. One should look for loose seams and ensure that joined backs and corners are tight. This material is ideal because of its resistance to weather and low weight factor.

### Monolithic fiberglass

This material is the lightest of the commonly used materials and also offers freedom from corrosion. Precision molding tends to make this shell a little more expensive than most. Because the housing is a complete mold, one has less seams to worry about in leakage. Upon installation, this shell should be handled more carefully than its metal counterparts because it is more subject to damage. This material is ideal because of its resistance to weather and its low weight factor.

### Wood

This material is relatively less expensive and more easily worked. It is generally not a good material to use in solar panels, however, because of the extreme temperature gradients encountered. Wood must be protected against the weather and must be treated every few years, thus requiring more maintenance. It is more economical to spend the money initially for a better frame than to have the maintenance and replacement problems associated with wood housings.

## Insulation

Insulation in a solar collector affects the heat loss to the environment from the absorber plate. The collector must be insulated on both the ***sides*** and ***back*** to minimize heat loss. Some collectors may not have insulated sides and this should be checked. The insulation used should not have the tendency to absorb moisture or outgas upon heating, and should have the ability to withstand stagnation temperatures of 350° F.

Outgassing is considered to be a function of time and temperature. It occurs at high temperatures within the solar collector, where vapors are expelled by organic binders in glass fiber insulation. Generally these vapors condense on the inside surface of the outermost glazing, reducing the incident solar energy on the absorber plate. The purpose of the organic binder is to give the insulation resiliency to prevent settling of the material and therefore preventing a reduction in the thermal conductivity, $K$, (BTU-in./hr-ft$^2$-° F). If a binder-free fiberglass is used, it should be used in conjunction with an insulation containing a binder thus providing resiliency with minimum outgassing.

Most insulation materials will absorb moisture, causing a degradation of thermal properties. Glass fiber insulations are more prone to this problem than are foam insulations. Table 4-1 shows the thermal conductivity, $K$, for insulation materials used in solar collectors. The manufacturer's specification sheets should be reviewed and the dealers queried as to these facts. One must be cautious of insulation such as "beadboard," which has a tendency to break down under high temperatures.

## Glazing

Two types of glazing common to the solar market are low iron glass and fiberglass reinforced polyester (FRP). The purpose of the glazing is to reduce the heat loss due to the external air temperature by insulating the absorber plate from the outside environment, and to trap the infrared radia-

**Table 4-1**
**Thermal Conductivity of Insulation Materials for Solar Collectors (BTU-in./hr-ft²-° F)**

| | Mean Temperature | | | |
|---|---|---|---|---|
| *Type* | *75° F* | *150° F* | *250° F* | *350° F* |
| Glass fiber building insulation | 0.30 | — | — | — |
| Glass fiber high density board | 0.22 | 0.27 | 0.33 | 0.41 |
| Glass fiber low binder | 0.231 | 0.275 | 0.345 | — |
| Glass fiber no binder | 0.23 | 0.28 | 0.36 | 0.41 |
| Polyisocyanurate foam | 0.17 | — | — | — |
| Urethane foam | 0.27 | — | — | — |

tion from the absorber surface. A comparison of an unglazed and glazed collector will be discussed under collector performance later in this chapter. The parameters to consider are coefficient of expansion, softening temperature, energy transmission, and tensile strength. These factors are depicted in Table 4-2 for various cover plate materials.

Solar energy transmission is approximately 87 percent for most FRPs and anywhere from 86 to 92 percent for low iron glass. An FRP material is approximately eight times lighter than glass, and will not shatter. National Bureau of Standards testing shows glass to be weakest at the corners of collectors when subjected to their hail gun test. Glass will break, whereas fiberglass will star. This starring is an internal fracture of glass fibers.

Prolonged exposure to high temperatures (>319° F at 66 psi) may cause some discoloration to an FRP, whereas glass will have no discoloration. An FRP must have some type of acrylic-modified gel or polyvinyl fluoride film (Tedlar®) to protect the fiberglass against ultraviolet degradation and erosive atmosphere; glass does not.

Which material should a collector use? Either one is satisfactory; both have their advantages and both have their disadvantages. One must look at the overall collector and determine which one has the preferred material compatibility.

## Absorber Plate

### Material

Absorber plates for DHW use are normally composed of aluminum or copper. ***Aluminum*** is less costly but has only half the thermal conductivity of copper. In a direct water system, the dissimilarity between the aluminum plate and copper piping can result in a galvanic corrosion problem. In a freeze-resistant closed loop system, this problem can be lessened by controlling the amount of fluid acidity (pH) of the transfer media. ***Copper*** is the

**Table 4-2**
**Properties of Typical Cover Plate Materials[1]**

| Material / Property | Poly(vinyl fluoride) | Poly(ethylene terephthalate) | Polycarbonate | Fiberglass Reinforced Plastics | Poly(methyl methacrylate) | Fluorinated ethylene-propylene | Clear Lime Glass (Float) | Sheet Lime Glass | Water White Glass |
|---|---|---|---|---|---|---|---|---|---|
| % Solar Transmittance (for thickness listed below) | 92-94 | 85 | 82-89 | 77-90 | 89 | 97 | 83-85 | 84-87 | 85-91 |
| Maximum Operating Temperature (°F) | 227° | 220° | 230-270° | 200° | 180-190° | 248° | 400° | 400° | 400° |
| Tensile Strength (psi) (ASTM D-638) | 13000 | 24000 | 9500 | 15000-17000 | 10500 | 2700-3100 | 4000 annealed 10000 tempered | 4000 annealed 10000 tempered | 4000 annealed 10000 tempered |
| Thermal Expansion Coefficient (in/in/°F x $10^{-6}$) | 24 | 15 | 37.5 | 18-22 | 41.0 | 8.3-10.5 | 4.8 | 5.0 | 4.7-8.6 |
| Elastic Modulus (psi x $10^6$) (D-638) | .26 | .55 | .345 | 1.1 | .45 | .5 | 10.5 | 10.5 | 10.5 |
| Thickness (in) | .004 | .001 | .125 | .040 | .125 | .002 | .125 | .125 | .125 |
| Weight (lb/ft²) For above thickness | .028 | .007 | .77 | .30 | .75 | .002 | 1.63 | 1.63 | 1.65 |
| Refractive Index | 1.45 | 1.64 | 1.59 | - | 1.49 | 1.34 | 1.52 | 1.52 | 1.52 |

1/ These values were obtained from the following references:

Grimmer, D. P., Moore, S. W., "Practical Aspects of Solar Heating: A Review of Materials Used in Solar Heating Applications." LA-UR-75-1952, paper presented at SAMPE Meeting, October 14-16, 1975, Hilton Inn.

Kobayashi, T., Sargent, L., "A Survey of Breakage-Resistant Materials for Flat-Plate Solar Collector Covers," paper presented at U.S. Section-ISES Meeting, Ft. Collins, Colorado, August 20-23, 1974.

Scoville, A. E., "An Alternate Cover Material for Solar Collectors,' paper presented at ISES Congress and Exposition, Los Angeles, California, July, 1975.

Clarkson, C. W., Herbert, J. S., "Transparent Glazing Media for Solar Energy Collectors," paper presented at U.S. Section-ISES Meeting, Ft. Collins, Colorado, August 21-23, 1974.

Modern Plastics Encyclopedia, 1975-1976, McGraw-Hill Publishing Company.

Toenjes, R. B., "Integrated Solar Energy Collector Final Summary Report," LA-6143-MS, Los Alamos Scientific Laboratory, Los Alamos, New Mexico, November, 1975.

*Source.* Intermediate Minimum Property Standards Supplement, 1977 Edition, U.S. Dept. of Housing and Urban Development.

preferred type of plate to use for the following reasons. Generally, corrosion of copper will decrease with a decreasing amount of oxygen in the transfer media. In a closed system, where oxygen is expelled, the corrosion rate will be negligible. One must remember though that this *does not include* the consideration of the ***decomposition*** of the ***transfer media***.

Corrosion of aluminum can continue, however, even in the absence of oxygen because alternative corrosion can occur. If the pH of the transfer media is not maintained above 8, anodic dissolution of aluminum is accompanied by the evolution of molecular hydrogen. Material properties are shown in Table 4-3.

### Construction

There are four basic methods used in the construction of absorber plates. One method consists of bonding together two sheets of preformed channelized sheets of copper creating integral tubes as depicted in Figure 4-4.

**Table 4-3**
**Properties of Typical Absorber Substrate Materials[1]**

| Material / Property | Aluminum | Copper | Mild Carbon Steel | Stainless Steel |
|---|---|---|---|---|
| Elastic Modulus, Tension psi x $10^6$ | 10 | 19 | 29 | 28 |
| Density lbs/cu.in. | 0.098 | 0.323 | 0.283 | 0.280 |
| Expansion Coefficient (68-212°F) in/in/°F x $10^{-6}$ | 13.1 | 9.83 | 8.4 | 5.5 |
| Thermal Conductivity (77-212°F) Btu/hr·ft$^2$·°F·ft | 128 | 218 | 27 | 12 |
| Specific Heat (212 °F) Btu/lb·°F | 0.22 | 0.09 | 0.11 | 0.11 |

1/ Typical valves: standard specifications or manufacturer's literature should be consulted for specific types or alloys.

*Source.* Intermediate Minimum Property Standards Supplement, 1977 Edition, U.S. Dept. of Housing and Urban Development.

The second method involves bonding the tubes to the plate by silver-brazing as depicted in Figure 4-5*a*, or by some other type of metallurgical bond as illustrated in Figure 4-5*b*.

The third method provides a mechanical lock around the tube as shown in Figure 4-6. The absorber plate uses a 360 deg roll-formed lock seamed mechanical bond between the absorber fins and the fluid tubes.

The fourth method consists of bonding two sheets together and then expanding certain areas to form channels for fluid paths. These channels are formed during the process of fusing the two sheets. This type of absorber is illustrated in Figures 4-7*a* and 4-7*b*.

**Figure 4-4.** Integral tubes.

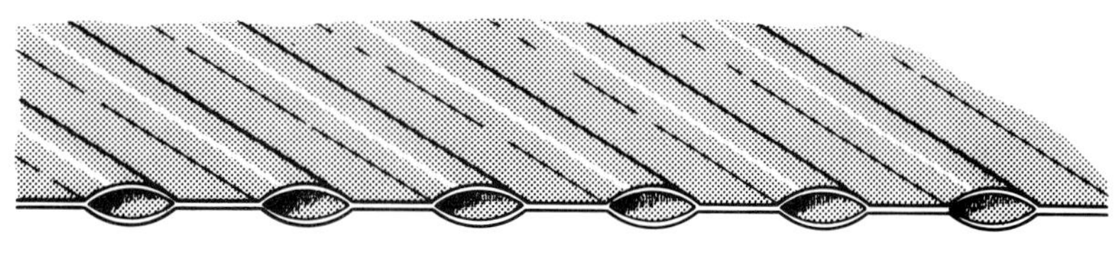

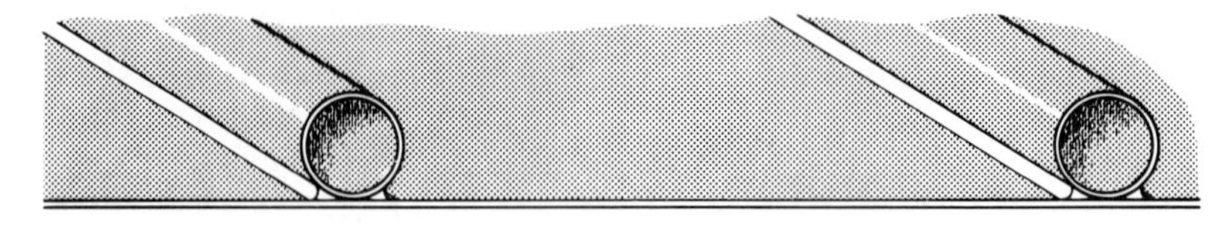


**Figure 4-5*a*.** Separate tubes.

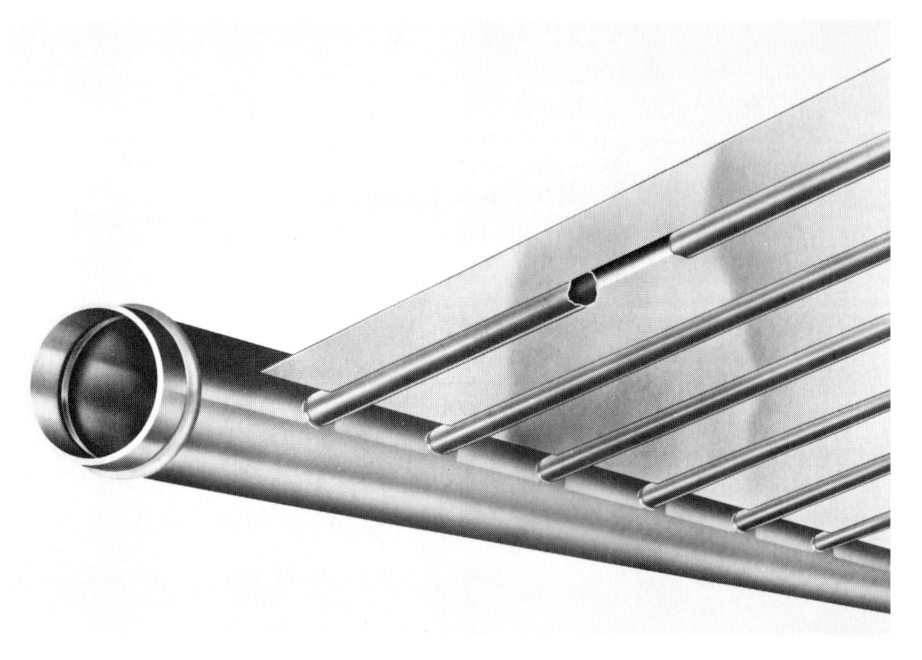

**Figure 4-5*b*.** Metallurgical bond. (Courtesy of Terra-Light, Inc., Billerica, Mass.)

The absorber plate is the "heart" of the solar collector, and thus its design and construction are critical. Design criteria include type and thickness of the metal and the size and spacing of the tube in relation to fin area. The metal sheet to which the tube is connected conducts the heat to the tube which is transferred to the passing liquid. A very long and very thin fin would have very low efficiency because most of the heat would be lost to the surroundings rather than to the tube. If the fin is long and thick, too much heat may be collected per tube, so that the tube becomes much warmer than the water flowing inside of it, resulting in heat loss to the surroundings.

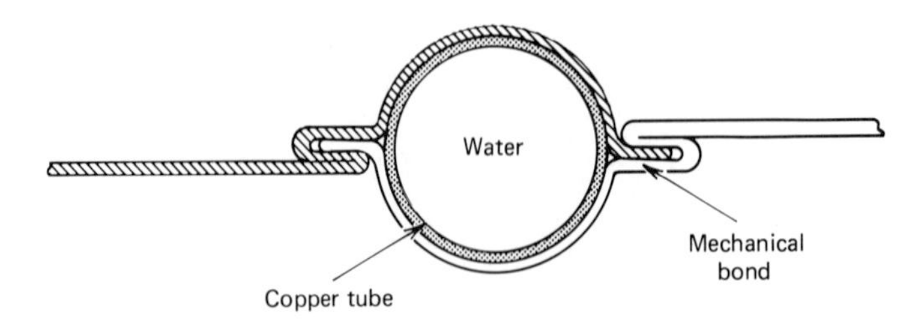


**Figure 4-6.** Mechanical bond absorber plate (Courtesy of Phelps Dodge Industries, Inc., New York, N.Y.)

**Figure 4-7*a*.** Tube expansion absorber plate. (Courtesy of Yankee Resources, Brunswick, Maine)

**Figure 4-7*b*.** Tube expansion. (Courtesy of Olin Brass Corp., East Alton, Ill.)

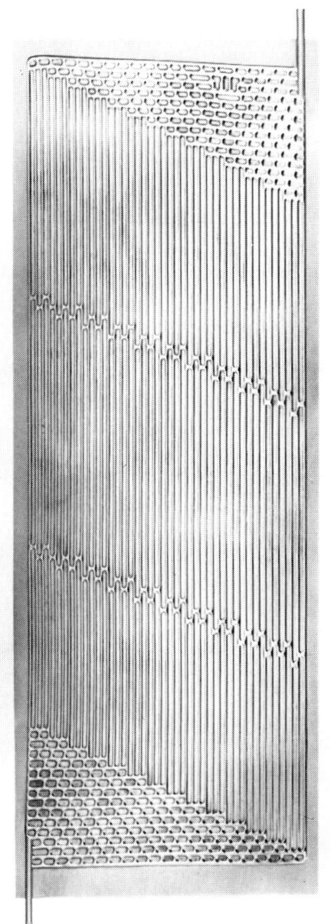

Tube area is also very important. The smaller the tube diameter, the faster the transfer fluid must flow to remove the same amount of heat. The faster the flow of the transfer fluid, the more susceptible is the tubing to corrosion. At this point we can understand that fluid flow and its associated pressure drop, tube size, fin area, relative distance between fluid passages, and material are all very important parameters interrelated in the construction of the absorber plate. It is this complexity of variable relationships that should dissuade the "do-it-yourselfer" from constructing his or her own absorber plate. If "do-it-yourselfers" decide to build a solar collector, they should, at least purchase the absorber plate from one of the several manufactured plates available.

A typical flow rate through each absorber plate for a solar DHW system is 0.5 gal per minute per plate. The tube diameters are nominally 1/2 in. and the headers at the top and bottom of the absorber are usually 3/4 to 1 in. in diameter. Absorber plates are normally designed to withstand pressures from 25 to 150 psi. Typical closed loop freeze-resistant systems will operate at 15 to 30 psi and open loop systems will usually operate anywhere from atmospheric pressure for a drain back system to 90 psi for a drain down system depending on city or well water pressure.

Absorber plates are manufactured with either internal manifolds for pressurized systems or with an added nipple for external manifold connection for unpressurized or pitched systems (i.e., drain down or drain back). These two types of plates are illustrated in Figures 4-8 and 4-9, respectively. In the case of internal manifolds, the absorber header is used as the distribu-

**Figure 4-8.** Internal manifold nonpitched header.

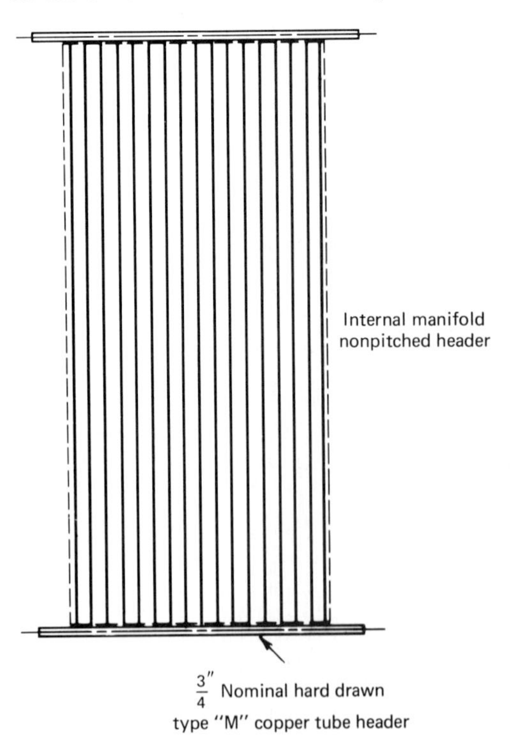

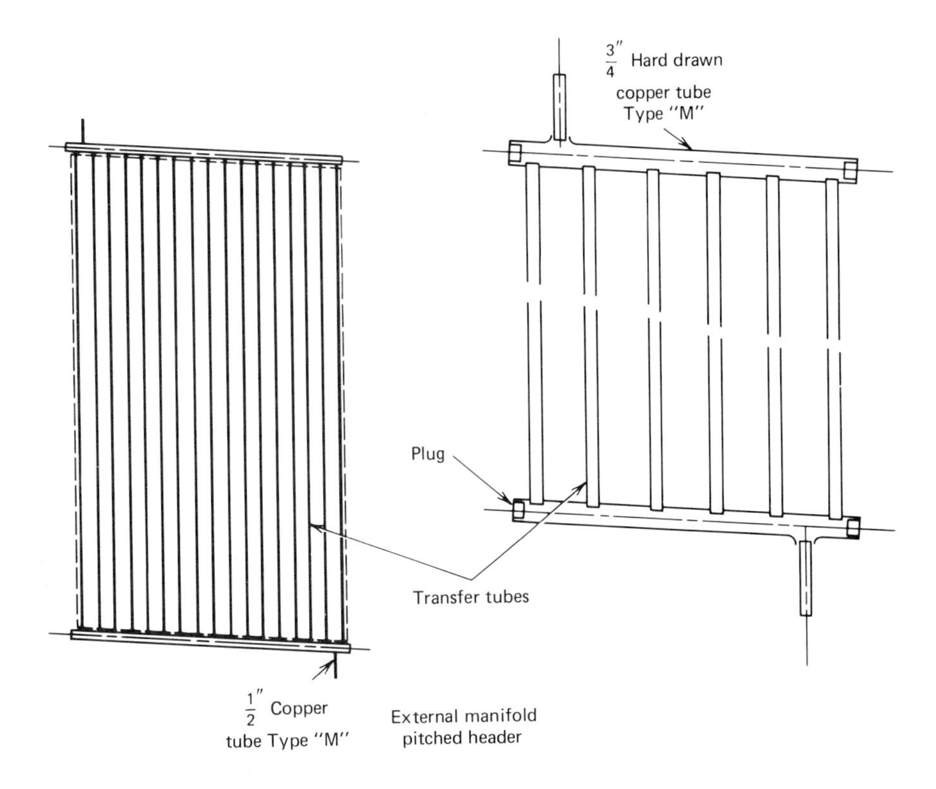


**Figure 4-9.** External manifold pitched header.

tion manifold and is usually connected in parallel flow arrangements which we will soon discuss. Internal manifolds within the collector housing will save the cost of pipe insulation. In the case of external manifolds, a distribution manifold external to the collectors is used. External manifolds involve extra piping as well as extra insulation. It is, therefore, more economical to purchase collectors with internal manifolds if the collectors need not be pitched or require something other than a parallel flow arrangement. Figures 4-10 and 4-11 illustrate a closed loop freeze-resistant system with and without external manifolds, respectively.

Finally, the absorber plate should "float" within the collector housing and should not be fastened directly to the sides. Because a typical stagnation temperature of 350° F can be reached, allowance should be made for thermal expansion in the collector plate. A rigidly fastened plate could result in a bent plate because the housing and plate expand at a different rate causing uneven heating and leakage or both. The plate should be thermally isolated from the outside shell.

## Coatings

Absorptive coatings are of two types: selective and nonselective. A selective surface is a special coating that impedes the reradiation of infrared energy from the hot absorber plate, therefore retaining more heat to be transferred to the liquid media. The absorbtivity is high and emissivity is low. A

**Figure 4-10.** Closed loop freeze-resistant system with external manifolds. (Courtesy of Applied Technologies, Inc., Kittery, Maine)

nonselective coating, on the other hand, exhibits high absorbtivity and high emissivity. The higher the heating requirement, the more a selective surface is needed. Costs and an unknown life factor are items to consider. Systems that require 200° F and higher need selective coatings for efficient operation. Examples of absorptive coatings are illustrated in Table 4-4.

**Figure 4-11.** Closed loop freeze-resistant system with internal manifolds. (Courtesy of Applied Technologies, Inc., Kittery, Maine)

**Table 4-4**
**Characteristics of Absorptive Coatings**

| Property / Material | Absorptance [1/] α | Emittance ε | α/ε | Breakdown Temperature °F (°C) | Comments |
|---|---|---|---|---|---|
| Black Chrome | .87-.93 | .1 | . ∿9 | | |
| Alkyd Enamel | .9 | .9 | 1 | | Durability Limited at High Temperatures |
| Black Acrylic Paint | .92-.97 | .84-.90 | ∿1 | | |
| Black Inorganic Paint | .89-.96 | .86-.93 | ∿1 | | |
| Black Silicone Paint | .86-.94 | .83-.89 | ∿1 | | Silicone Binder |
| PbS/Silicone Paint | .94 | .4 | 2.5 | 662 (350) | Has a High Emittance for Thicknesses >10μm |
| Flat Black Paint | .95-.98 | .89-.97 | ∿1 | | |
| Ceramic Enamel | .9 | .5 | 1.8 | | Stable at High Temperatures |
| Black Zinc | .9 | .1 | 9 | | |
| Copper Oxide over Aluminum | .93 | .11 | 8.5 | 392 (200) | |
| Black Copper over Copper | .85-.90 | .08-.12 | 7-11 | 842 (450) | Patinates with Moisture |
| Black Chrome over Nickel | .92-.94 | .07-.12 | 8-13 | 842 (450) | Stable at High Temperatures |
| Black Nickel over Nickel | .93 | .06 | 15 | 842 (450) | May be Influenced by Moisture at Elevated Temperatures |
| Ni-Zn-S over Nickel | .96 | .97 | 14 | 536 (280) | |
| Black Iron over Steel | .90 | .10 | 9 | | |

1/ Dependent on thickness and vehicle to binder ratio.

G. E. McDonald, "Survey of Coatings for Solar Collectors", NASA TMX-71730, paper presented at Workshop on Solar Collectors for Heating and Cooling of Buildings, November 21-23, 1974, New York City.

G. E. McDonald, "Variation of Solar-Selective Properties of Black Chrome with Plating Time", NASA TMX-71731, May 1975.

S. W. Moore, J. D. Balcomb, J. C. Hedstrom, "Design and Testing of a Structurally Integrated Steel Solar Collector Unit Based on Expanded Flat Metal Plates", LA-UR-74-1093, paper presented at U. S. Section-ISES Meeting, Ft. Collins, Colorado, August 19-23, 1974.

D. P. Grimmer, S. W. Moore, "Practical Aspects of Solar Heating: A Review of Materials Use in Solar Heating Applications", paper presented at SAMPE Meeting, October 14-16, 1975, Hilton Inn.

R. B. Toenjes, "Integrated Solar Energy Collector Final Summary Report", LA-6143-MS, Los Alamos Scientific Laboratory, Los Alamos, New Mexico, November 1975.

G. L. Merrill, "Solar Heating Proof-of-Concept Experiment for a Public School Building", Honeywell Inc., Minneapolis, Minnesota National Science Foundation Contract No. C-870.

D. L. Kirkpatrick, "Solar Collector Design and Performance Experience", for the Grover Cleveland School, Boston, Massachusetts, paper presented at Workshop on Solar Collectors for Heating and Cooling of Buildings, November 21-23, 1974, New York City.

*Source.* Intermediate Minimum Property Standards Supplement, 1977 Edition, U.S. Dept. of Housing and Urban Development.

To document and quantify the advantage of a selective surface, Olin Brass conducted a series of tests to compare collector performance, using black chrome versus black paint. The tests were performed by an independent laboratory, Desert Sunshine Exposure Testing (DSET), using American Society of Heating, Refrigeration and Air Conditioning Engineers (ASHRAE) 93-77 procedures. Results of these tests included the following:

1. The black chrome-plated absorber showed higher collector efficiency as the temperature difference between absorber and environment increased. This comparison is illustrated in Figure 4-12.

2. Temperature differences between the fluid in the collector and the environment of 30° to 80° F are typical of summertime operation of solar domestic hot water systems. When the insolation is perpendicular to the collector surface, the advantage of the black chrome selective surface ranged from 4 to 10 percent for 300 BTU/ft$^2$-hr, as depicted in Figure 4-13. As the angle of incident insolation increased relative to the collector surface, (i.e., morning and afternoon), the advantage of the selective surface increased from 8 to 35 percent or 200 BTU/ft$^2$-hr.

Typical winter temperature differences between the fluid in the collector and the environment are still greater and may range from 75° to 100° F above the ambient temperature. The midday advantage of black chrome then becomes 22 to 46 percent for 300 BTU/ft$^2$-hr. The morning and afternoon advantage then become 74 percent to well over 100 percent improvement.

3. The single glazed collector using black chrome-plated absorber was found to be effectively equivalent to the double glazed collector with black painted absorber at lower temperature differences but was found to be significantly better at higher temperature differences. This difference is shown in Figure 4-14 and illustrates that systems requiring higher temperature outputs (i.e., 200° F) need selective coatings for efficient operation.

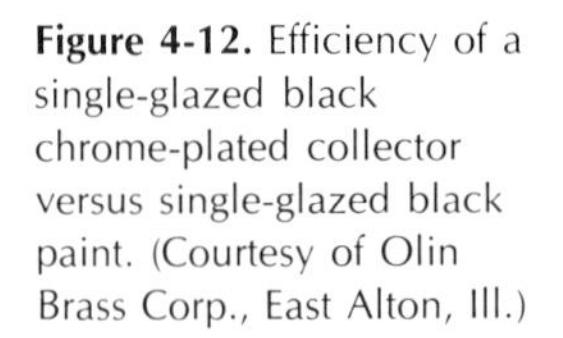

**Figure 4-12.** Efficiency of a single-glazed black chrome-plated collector versus single-glazed black paint. (Courtesy of Olin Brass Corp., East Alton, Ill.)

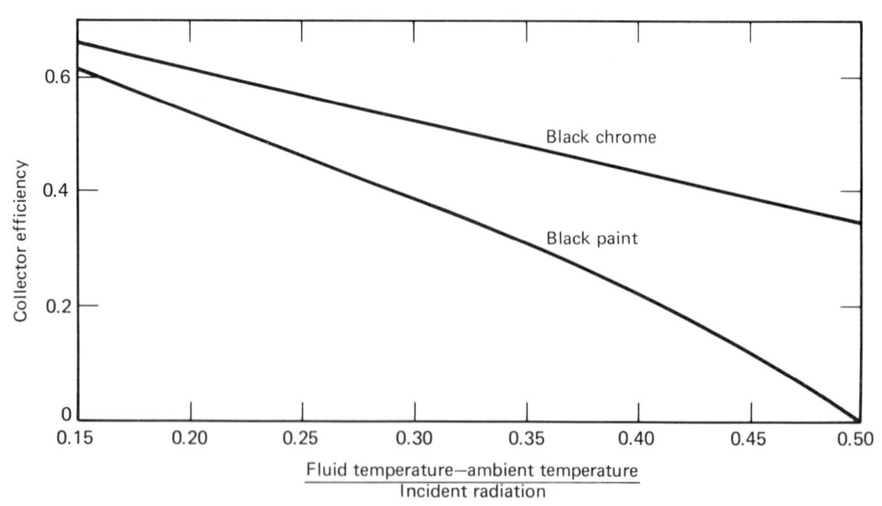

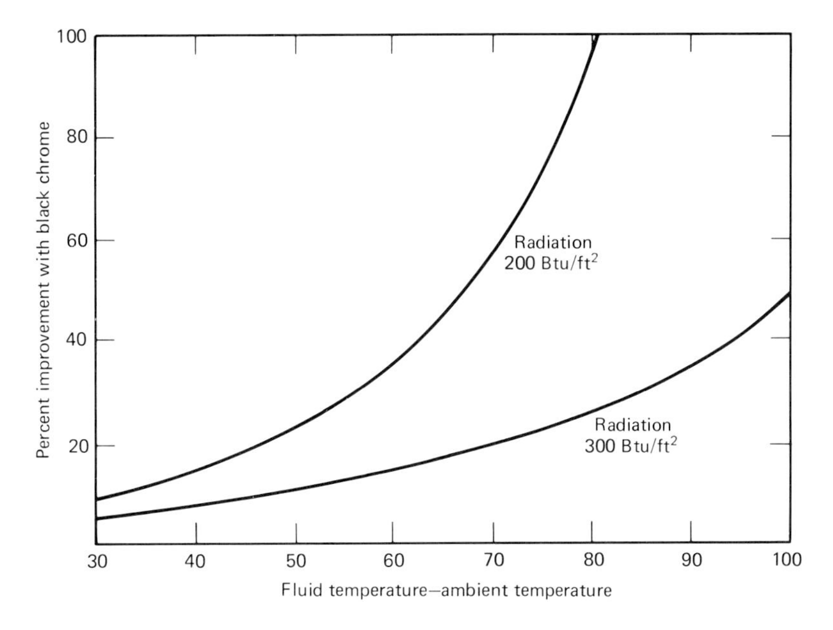


**Figure 4-13.** Relative performance improvement of a black chrome-plated absorber compared to black paint. (Courtesy of Olin Brass Co., East Alton, Ill.)

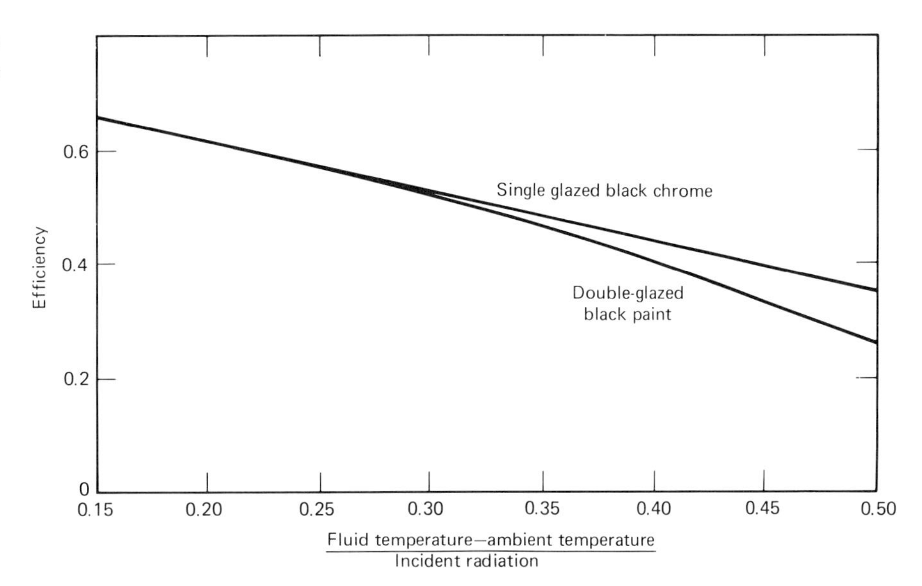


**Figure 4-14.** Efficiency of a single-glazed black chrome plated collector versus double-glazed black paint. (Courtesy of Olin Brass Co., East Alton, Ill.)

## COLLECTOR ARRANGEMENT

The solar collector area is composed of individual collectors arranged to operate as a single system. The arrangement and relationship of one collector to another is extremely important for effective solar collection and efficient system operation. Three basic multiple collector arrays are illustrated as follows.

## Parallel Flow—Direct Return

A direct return piping arrangement circulates the transfer medium from the bottom of the collector to a return manifold at the top. This arrangement may cause severe operating problems by allowing wide temperature variations from collector to collector due to flow imbalance. Although the pressure drops across each collector are essentially the same and at the same flow rate, high pressure drops occurring along the supply/return header or manifold will cause flow imbalance. This problem can be reduced by installing manual balancing valves as shown in Figure 4-15. Provisions must be made to measure the pressure drop or temperature in order to adjust the flow rate to prevent collectors closer to the circulating pump from exceeding design flow rates, and those farther away from receiving less.

## Parallel Flow—Reverse Return

Reverse return piping systems are considered preferable to direct return for their ease of balancing. Because the total length of supply piping and return piping serving each collector is the same and the pressure drops across each collector is equal, the pressure drop across each manifold is also theoretically equal. The major advantage of reverse return piping is that balancing is seldom required because flow through each collector is the same. Provisions for flow balancing may still be required in some reverse return piping systems depending on the overall size of the collector array and the type of collector. External and internal manifold arrangements are shown in Figure 4-16.

## Series Flow

Series flow is often used in large arrays to reduce the amount of piping required, by allowing several collector assemblies to be served by the same supply/return headers or manifolds. Series flow can also be employed to increase the output temperature of the collector system. Either direct or reverse return distribution circuits can be employed, but unless each collector branch has the same number of collectors, the reverse return system has no advantage over direct return; each would require flow balancing. This system is shown in Figure 4-17.

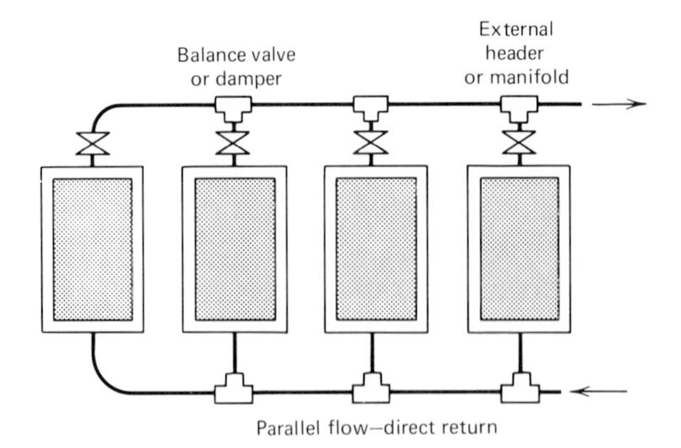


**Figure 4-15.** Parallel flow-direct return.

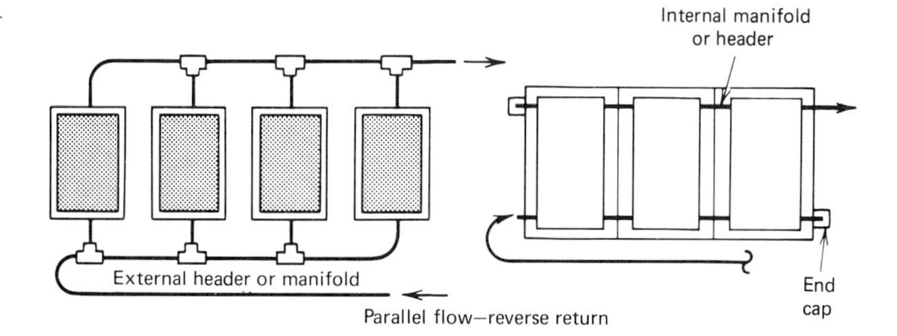


**Figure 4-16.** Parallel flow-reverse return.

## COLLECTOR PERFORMANCE

Once a person has decided which materials are desired, whether for noncorrosive abilities, weight, or aesthetics, there must then be a means of comparing collectors to determine which collector would work best under different circumstances. Such a method does exist and is used for performance testing flat plate collectors with ASHRAE test method 93-77. This method provides efficiency versus operating conditions to construct a normalized curve for insolation, $I_o$, and temperature difference between the heat transfer fluid collector inlet temperature, $Ti$, and ambient air, $Ta$. A typical collector efficiency curve is depicted in Figure 4-18.

This type of graph should be available for all manufacturer's data. Because all data should be derived under the same testing requirements, all graphs should be comparable with one another. Although this graphical illustration may look a bit foreboding at first, it is important to understand how this curve is derived and what it actually means to you. For those who are mathematically oriented or who wish to rejuvenate their algebra, the following derivation of the collector efficiency curve will be of great interest. If you are not so inclined, however, you can proceed to the examples at the end of this chapter which illustrate the use of this curve. Remember, this curve depicts collector performance only and does not depict system daily performance.

The simplest type of collector consists of a flat black plate without any glazing. The heat to be extracted from this plate will be less than the solar

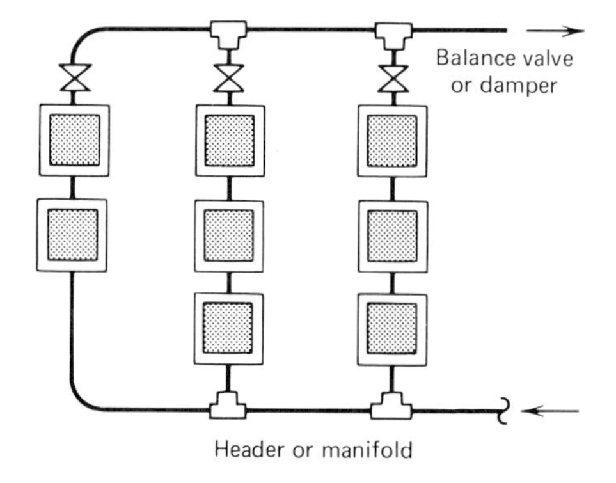


**Figure 4-17.** Series flow.

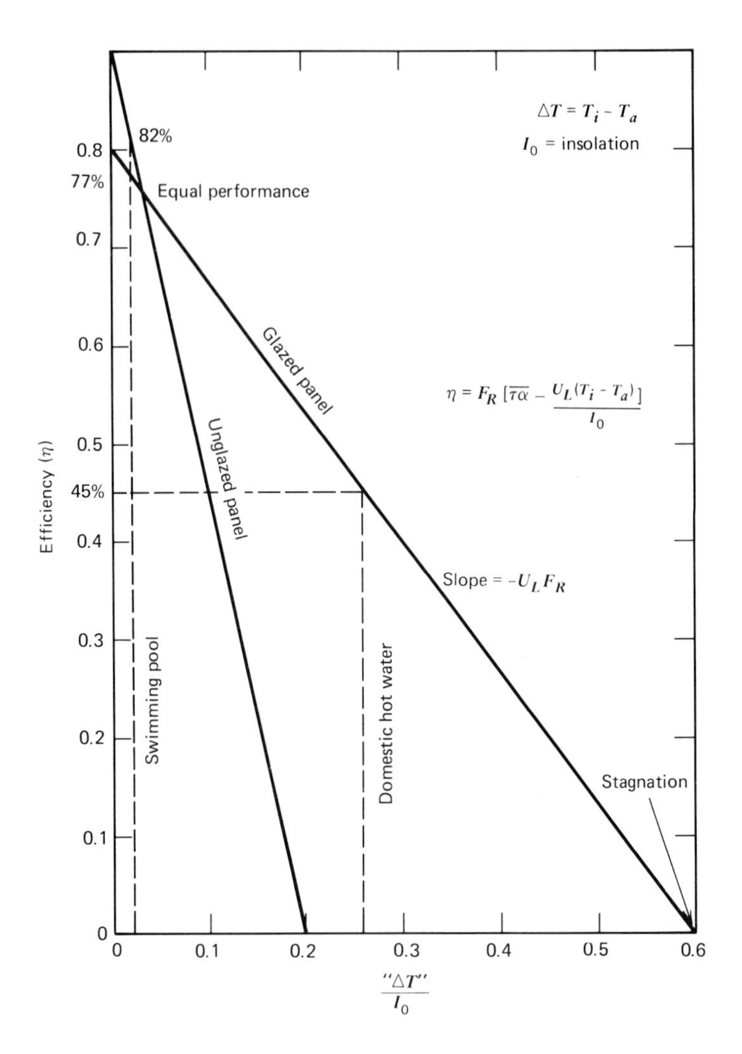


**Figure 4-18.** Collector efficiency curve.

energy collected by the plate. This is due to the amount of heat loss due to convection/conduction and radiation loss of the collector to the surroundings.

The amount of heat loss is a function of the temperature difference between the surface and its surroundings and emissivity of the absorber plate. Efficiency $\eta$ of the collector therefore will be defined as output divided by input. Output is the amount of energy absorbed less losses and input is the amount of insolation $I_o$ available incident to the collector surface. Efficiency is depicted in Equation 4-1.

***Equation 4-1***

$$\eta = \frac{\text{output}}{\text{input}} = \frac{\text{absorbed energy} - \text{losses}}{\text{insolation}}$$

And algebraically separating these terms:

$$\eta = \frac{\text{absorbed energy}}{\text{insolation}} - \frac{\text{losses}}{\text{insolation}}$$

The maximum efficiency that can be obtained by the collector would result if there were no losses as shown in Equation 4-2.

***Equation 4-2***

$$\eta = \frac{\text{absorbed energy}}{\text{insolation}} = \alpha \quad \text{where losses} = 0$$

This ratio is designated as solar absorptance $\alpha$, which is characteristic of the collection surface. Now we know the efficiency won't exist without losses to the absorbed heat. To keep it algebraically simple, these losses are approximated as being proportional to the temperature difference between the absorber plate surface temperature $Tp$ and the ambient air temperature $Ta$. The convection/conduction and radiation losses are combined into one proportional constant, $U_L$, which we will call the heat loss coefficient.

Therefore, we have Equation 4-3, which depicts the losses as the product of the heat loss coefficient and the temperature difference between absorber plate and the ambient air.

***Equation 4-3***

$$\text{Losses} = U_L(Tp - Ta)$$

And combining equations 4-1, 4-2, and 4-3, we have Equation 4-4:

***Equation 4-4***

$$\eta = \alpha - U_L \frac{(Tp - Ta)}{I_o}$$

Remember, we are still discussing the flat black plate only; *no glazing material. Note in Equation 4-4 that the losses are reduced as the collector operates near the ambient temperature.* When, or if, these losses equal the absorbed energy, we note that the efficiency is zero, or

$$\eta = 0 \qquad \text{when} \qquad \alpha = U_L \frac{(Tp - Ta)}{I_o}$$

Once the plate temperature $Tp$ can no longer increase, stagnation occurs. This is one of the most severe conditions for a solar collector. It can occur in the heat of summer under no flow conditions, or on a cold, windy, clear day where convection and radiation losses are large and the heat loss coefficient $U_L$ approaches the magnitude of the energy absorbed.

The simple flat black collector without a glazing has good low-temperature applications such as in swimming pools in the summer. For swimming pools where the maximum pool temperature is 90° F with an average of 80° F, the unglazed collector will collect more solar energy. An example of this will be illustrated later. When the difference in plate temperature and air temperature must be in excess of 100° F as in heating hot water in the winter, the value of $U_L$, which is the proportional constant for heat losses, becomes very important. On a cold, windy, clear day convection and radiation losses could be large such that the collector would reach stagnation supplying no useful amount of heat. A change in the collector can be affected by either $\alpha$ or $U_L$. Because $\alpha$ is typically 0.8 to 1.0, its effective change is less adaptable than is $U_L$. By decreasing the radiative and convective losses while still maintaining a high solar absorptance, a collector can attain a greater useful amount of heat. A transparent cover or glazing is therefore used to decrease $U_L$. In adding a glazing, the solar absorptance is also decreased but only slightly. This is due to reduced transmitted energy to the plate. That transmitted fraction is designated $\tau$, where

$$\tau = \frac{\text{transmitted energy}}{\text{incident energy}}$$

When the plate absorbs this transmitted solar energy, it converts it to heat and a portion is reradiated toward the glazing at infrared (IR) wavelengths. The *IR* transmission factor of most glazings is very low and therefore very little energy is reradiated to the sky. Radiation losses are thereby reduced by trapping the radiation; convective/conduction losses are reduced upon the creation of a dead air layer between the glazing and absorber plate.

Because the solar absorptance is now slightly decreased due to transmission properties of the glazing, we now have Equation 4-5. Note the difference from Equation 4-4.

***Equation 4-5***

$$\eta = \tau\alpha - U_L \frac{(Tp - Ta)}{I_o}$$

And because a portion of the energy not absorbed by the plate is reflected back to the plate by the glazing, the solar absorptance factor is an effective transmission absorptance product yielding $\overline{\tau\alpha}$. Equation 4-5 is now written as

***Equation 4-6***

$$\eta = \overline{\tau\alpha} - U_L \frac{(Tp - Ta)}{I_o}$$

As $Ta \rightarrow Tp$ for maximum efficiency,

$$\eta = \overline{\tau\alpha}$$

The plate temperature $Tp$ varies continuously over the surface of the absorber plate. From a measurement standpoint, therefore, it is more convenient to measure the collector inlet, $Ti$, temperature than to measure the internal varying plate temperature. One might also think that a better represented average would be $(Ti + To)/2$. However, because we are more concerned with the storage outlet temperature to the collectors, the value of $Ti$ would be more representative of collection efficiency.

In order for the efficiency to represent the entire collector and allowing the substitution of $Ti$ for $Tp$, a heat removal factor $FR$ must be used to account for the fluid flow rate, collector to fluid interface, and inherent properties of the fluid itself. Therefore, multiplying both the transmission absorbtance product and the proportionate heat loss by this multiplying factor $FR$, we have Equation 4-7

***Equation 4-7***

$$\eta = FR\left[\overline{\tau\alpha} - U_L \frac{(Ti - Ta)}{I_o}\right]$$

*where* $\eta$ = collector efficiency
$FR$ = heat removal factor
$\overline{\tau\alpha}$ = effective transmissivity − absorbtivity product
$U_L$ = overall heat loss coefficient
$Ti$ = transfer fluid temperature at the collector inlet
$Ta$ = ambient air temperature
$I_o$ = instantaneous level of solar radiation

This equation represents collector efficiency and provides us with a comparable means of evaluating collector performance. We can see that this equation is in the same format as the equation of a straight line.

***Equation 4-8***

$$Y = MX + b$$

*where* $b$ = $Y$ axis intercept (ordinate)
$M$ = slope

From Equations 4-7 and 4-8 it can be seen that a straight line will result, as illustrated in Figure 4-18, if a plot of efficiency, $\eta$, versus the quantity $(Ti - Ta)/I_o$ is made, assuming slope and intercept functions are constant. The slope of the line is a function of the overall heat loss coefficient, where

$$M = -FRU_L$$

The intercept of the line is a function of the transmissivity of the cover plate(s) and the absorbtivity of the absorber plate(s), where

$$b = FR\overline{\tau\alpha}$$

In reality, however, $U_L$ is not constant under the ASHRAE 93-77 test conditions because it varies with the temperature of the collector and ambient air. A second order curve is therefore used to describe the thermal performance of the collector, where

$$Y = c + bx + ax^2$$

The intercept is still related to $\overline{\tau\alpha}$ and the slope at any point on the curve is proportional to the heat loss rate for that value of $(Ti - Ta)/I_o$.

From the graph of Figure 4-18, we can see that at a rate of $FRU_L$, the efficiency decreases as $(Ti - Ta)/I_o$ increases and once the total absorbed energy equals the total losses,

$$FR\overline{\tau\alpha} = FR\left[U_L \quad \frac{(Ti - Ta)}{I_o}\right]$$

then $\eta = 0$ and stagnation occurs.

The efficiency curve does not give absolute values for the overall heat loss coefficient ($U_L$) and effective transmissivity-absorbtivity product, $\overline{\tau\alpha}$, because both are multiplied by the heat removal factor $FR$. However, the plot does indicate relative values for these two quantities that can be used for comparing collectors. Determination of the absolute values of $\overline{\tau\alpha}$ and $U_L$ would require additional measurements beyond the normal tests performed for collector performance evaluation.

Now, how do we make decisions about the collector based on these efficiency curves? Which collector curve is better? And for what application?

Let's discuss two examples comparing an unglazed collector with a glazed collector. The first discussion will concern the use of solar radiation to heat an outdoor swimming pool. The second discussion will concern the solar application to domestic hot water.

1. Let us say that we want to heat a swimming pool. Assuming a maximum insolation of 250 BTU/ft²-hr, an inlet water temperature, $Ti$, of 80° F, and an ambient average temperature of 75° F, then

$$\frac{Ti - Ta}{I_o} = \frac{80 - 75}{250} = \frac{5}{250} = 0.02$$

Efficiency $\eta$ of the glazed collector is 77 percent and the efficiency of the unglazed collector is 82 percent. The difference in efficiency of these two collectors for this application is (0.82 – 0.77)0.05, which is equivalent to 12.5 BTU/ft²-hr more energy collected for the unglazed panel than the glazed panel.

2. Now, let us say that we want to heat domestic hot water. Again assuming a maximum insolation of 250 BTU/ft²-hr, an inlet water temperatue $Ti$ from storage of 140° F and an ambient average temperature of 75° F, then

$$\frac{Ti - Ta}{I_o} = \frac{140° - 75°}{250} = \frac{65}{250} = 0.26$$

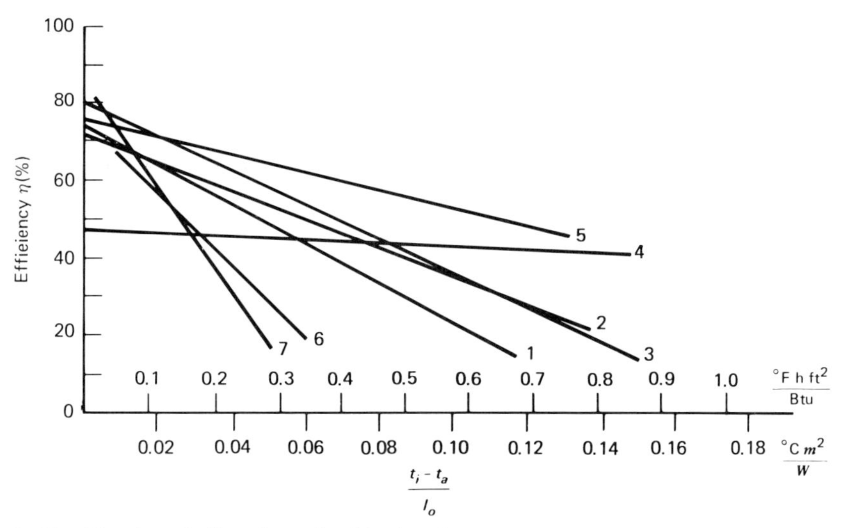


**Figure 4-19.** Typical thermal efficiency curves for liquid collectors based on collector aperture area. (*Source. Intermediate Minimum Property Standards Supplement,* 1977 Edition; U.S. Dept. of Housing & Urban Development)

1. Double glazed, flat plate, flat black paint
2. Antireflective double glazed, flat plate, black chrome selective surface
3. Single glazed, flat plate, black chrome selective surface
4. Single glazed, evacuated tube, concentric selective absorber, no rear reflector
5. Linear Frennel lens, tracking-concentrating, black chrome selective surface
6. Single glazed, flat plate, flat black paint
7. Single glazed, trickle type

Efficiency $\eta$ of the glazed collector is 45 percent, whereas an unglazed collector simply would not work in this condition, as depicted in Figure 4-18.

Examples of general efficiency curves for various combination of materials are illustrated in Figure 4-19.

## REVIEW QUESTIONS

1. The flat plate solar collector is composed of four basic parts. Name them.

   a. ________ b. ________

   c. ________ d. ________

2. Which material is best used as an absorber plate?

   a. Aluminum b. Steel

   c. Copper d. Magnesium

   e. Brass

3. If collectors need not be pitched or do not require something other than a parallel flow arrangement, a collector with internal manifolds would be more economical to use than a collector with external manifolds.

   a. True b. False

4. Absorptive coatings for collector plates consist of two types. What are they?

   a. ________ b. ________

5. Which of the three basic multiple collector array configurations can cause severe operating problems by allowing wide temperature variations from collector to collector as a result of flow imbalance?

6. For solar DHW purposes, tests indicate that a single glazed collector with a selective absorber surface is effectively equivalent to a double glazed collector with a black painted nonselective absorber surface.

   a. True   b. False

7. The efficiency of a collector can be defined by which of the following relationships?

   a. Absorbed energy divided by insolation available
   b. Net energy absorbed divided by insolation available
   c. Losses divided by insolation available
   d. Absorbed energy divided by total losses
   e. None of the above

8. When losses in a collector are equal to the energy absorbed, what is the efficiency of the collector?

   a. 100 percent
   b. 0 percent
   c. 50 percent
   d. Indeterminant
   e. None of the above

9. Assume a collector efficiency curve intersects the ordinate at 0.6 and the abscissa at 0.8. What is the overall heat loss coefficient?

   a. −0.75   b. −1.25
   c. +0.75   d. +1.25
   e. None of the above

10. Assuming a maximum insolation of 200 BTU/ft$^2$-hr, an inlet water temperature from storage of 120° F, and an ambient average temperature of 60° F, what is the efficiency of a collector that has the same curve represented in review question 9?

    a. 0.60   b. 0.24
    c. 0.30   d. 0.75
    e. 0.37

# Description and Explanation of System Components

There are many components in addition to the collector essential to the operation of each generic type of solar domestic hot water system.

A majority of the components required will now be discussed in alphabetical order under the following headings:

- Differential Controllers
- Heat Exchangers
- Heat Transfer Fluids
- Piping and Miscellaneous Hardware
- Pumps and Blowers
- Storage Tanks

We shall physically describe all components and explain their function to provide a better understanding of their application.

## DIFFERENTIAL CONTROLLERS

The varying nature of solar energy dictates the use of a differential type rather than a fixed type of temperature controller for all generic types of solar hot water systems. Only the thermosiphon and passive phase-change systems do not require a controller. A differential controller (Figures 5-1*a* and 5-1*b*) constantly monitors collector temperature and storage temperature and compares the temperature difference. When the collectors are hotter than the water in the storage tank by a preselected difference (normally 15° to 20° F), a circulator, pump, or blower is activated to transfer the collected energy via a transfer medium through the open or closed loop system to storage. Once the storage has accumulated a sufficient amount of energy such that the temperature difference is within a preselected difference (normally 3° to 5° F), the circulator, pump, or blower is deactivated so that the stored heat is not expended from storage back through the collectors, thus assuring an overall net energy gain from collectors to storage.

The sensors most commonly used to measure the temperature at the

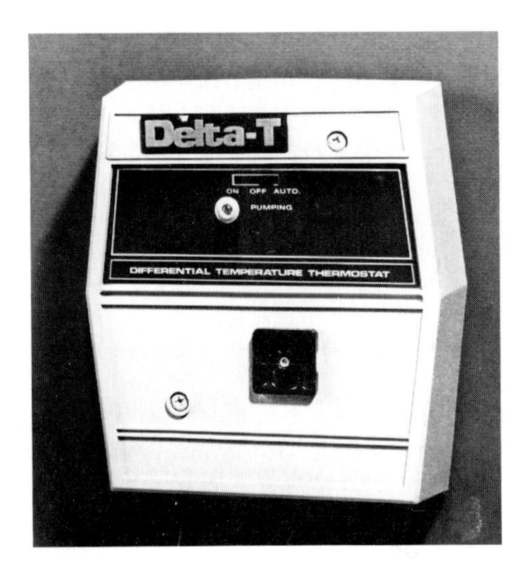


**Figure 5-1*a*.** Typical differential controller. (Courtesy of Heliotrope General, Inc., Spring Valley, Calif.)

**Figure 5-1*b*.** Typical differential controller. (Courtesy of Independent Energy, Inc., E. Greenwich, R.I.)

collectors and storage are thermistors. This device is a semiconductor material that is nonlinear and decreases in resistance with increasing temperature. It is normally encapsulated in a pipe plug or a copper lug as shown in Figure 5-2.

When connected to the electrical circuit of a differential controller, current flow varies in proportion to changes in resistance of the thermistor, which is exposed to changing air or liquid temperatures. Through the use of solid state electronics the current change is amplified to close or open a relay and in turn operate a circulator, pump, or blower as necessary. Thermistors are usually specified by a resistance value at 25° C. The 10,000-ohm and

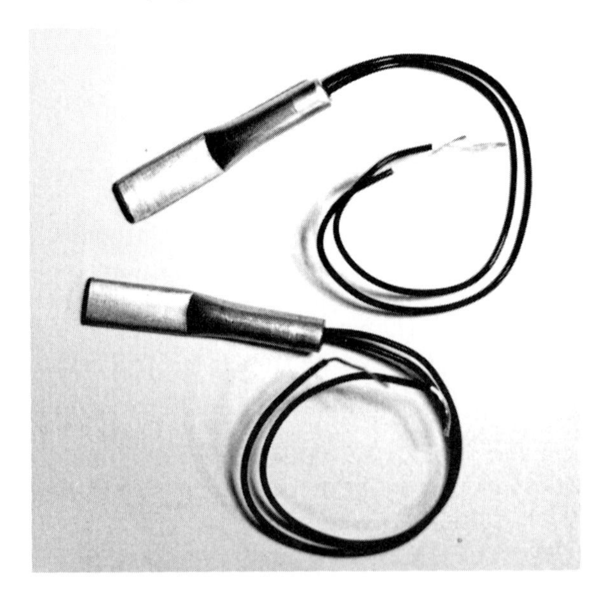

**Figure 5-2.** Typical strap on thermistor sensors. (Courtesy of Applied Technologies, Inc., Kittery, Maine)

3000-ohm thermistors are the most commonly used. Other sensor applications may include high limit turn-off to limit the temperature of the storage tank and freeze recirculation or drain down to protect the collectors from freezing.

Most differential controllers are the on-off type, where the circulator, pump, or blower is either fully on or off. Proportional controllers are also available which can vary motor speeds as a function of the temperature differential between collector outlet and storage. Such variable motor speeds can maximize energy collection. For example, on clear, sunny days, if a circulator ran only at a low speed, the energy process would be inefficient. On cloudy days, if a circulator ran only at high speeds, it would continually cycle on and off. Proportional controllers should not, however, be used for drain down or drain back systems unless special interfacing circuitry is used, because initially a pump may not receive enough power to fill the system.

A typical daily operating cycle of an on-off type of differential controller is illustrated in Figure 5-3. We shall assume the system is a closed loop with a circulator.

The initial temperature difference between the collector and storage needed to energize the circulator is $\Delta T$ on (i.e., 20° F) and the temperature difference needed to deenergize is $\Delta T$ off (i.e., 5° F). As the morning progresses and more insolation is available, the collector temperature rises to $\Delta T$ on at point 1. The heated transfer fluid is then circulated through the heat exchanger in the storage tank and returned to the collector, causing an initial drop in the collector outlet temperature to point 2. At this point a hysteresis circuit inherent to the design of the controller provides a time delay, preventing the circulator from cycling on and off while the collector temperature is increasing. The collector temperature will continue to rise until solar noon depending on ambient temperature and available insolation. Once the temperature between the collector output and storage is equivalent to $\Delta T$ off, the circulator is de-energized at point 3. Because the circulator is off at this point but we still have some available insolation, a stagnation condition exists, and the collector temperature increases slightly to point 4.

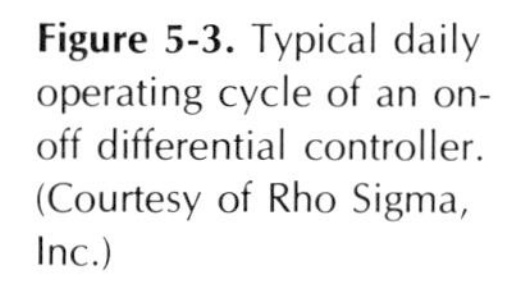
**Figure 5-3.** Typical daily operating cycle of an on-off differential controller. (Courtesy of Rho Sigma, Inc.)

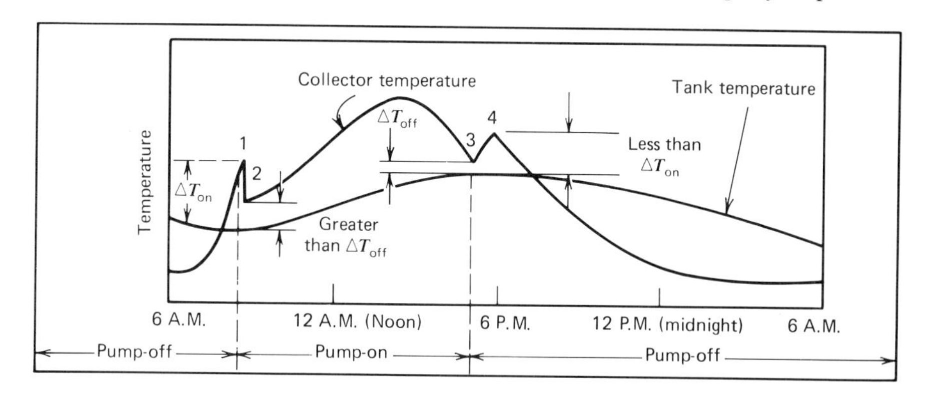

At this point the hysteresis circuit provides a time delay, preventing the circulator from cycling on and off while the collector temperature is decreasing. As the sun sets, the circulator would remain off because the collector temperature would be lower than the storage tank temperature.

## HEAT EXCHANGERS

### Description

A heat exchanger is a device used to transfer heat from one medium to another without violating the integrity of either medium. Heat exchangers separate the heat transfer fluid (either air or liquid) in the collector loop from the domestic water supply in the storage tank. Solar domestic hot water systems using heat transfer fluids which are toxic are required to have double-walled heat exchangers. Heat transfer fluids that are nontoxic require only single-walled heat exchangers, depending on local codes. Of the six generic types of solar domestic hot water systems, neither the thermosiphon or drain down systems require the use of a heat exchanger because the domestic water supply itself is circulated through the collectors. Heat exchangers which are categorized as liquid-to-liquid and air-to-liquid are associated with the four remaining generic systems. These two categories are broken down into specific types of heat exchangers that are normally found for each of the following systems shown in Table 5-1.

#### Coil-in-Tank Heat Exchanger

The coil in tank is the most effective type of heat exchanger because it is located directly in the storage tank. The exchanger is normally a finned coil (shown in Figure 5-4) in the bottom of the water tank. The rule of thumb for

**Table 5-1**
**Heat Exchangers and Associated Systems**

| *Generic System* | *Heat Exchanger Category* | *Definitive Type* |
|---|---|---|
| Closed loop | Liquid-to-liquid | |
| Freeze resistant | Coil in tank | Finned coil |
| Drainback | Liquid-to-liquid | |
| | Counterflow | Tube in tube |
| | Mixed flow | Shell and tube |
| Air to liquid | Air-to-liquid | Cross-flow |
| Phase change | Liquid-to-liquid | |
| Passive | Coil in tank | Finned coil |
| Subambient | Counterflow | Tube in tube |

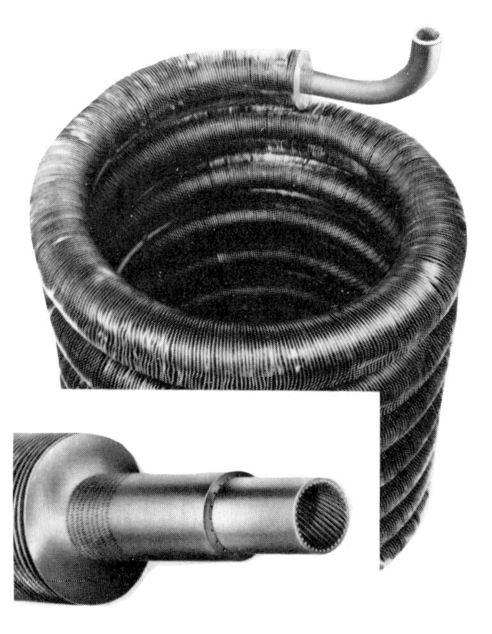

**Figure 5-4.** Typical finned coil heat exchanger. (Courtesy of Forge Fin Division, Noranda Metal Industries, Inc., Newton, Conn.)

calculating maximum fin height is 65 percent of the tube diameter for most materials.

The tank water is coldest at the bottom of the tank and as heat from the transfer fluid is exchanged to the storage tank, it rises to the top of the tank by convection. The amount of fluid contained in typical coil-in-tank heat exchangers is depicted in Table 5-2. This table will be addressed again in Chapter 9 in the discussion of filling the system.

### Counterflow Heat Exchanger

The counterflow is the second most effective. One such device is the tube in tube exchanger shown in Figure 5-5. This device is self-contained in that it contains two mediums flowing in opposite directions. The coldest fluid on the return path to the collectors always thermally contacts the coldest fluid from the storage tank to be heated, which aids collector efficiency.

**Table 5-2**
**Typical Heat Exchanger Capacities**

| *Tank Size (gal)* | *Heat Exchanger Surface Area (sq ft)* | *Fluid Capacity (gal)* |
|---|---|---|
| 65 | 15 | 0.2 |
| 80 | 20 | 0.25 |
| 120 | 40 | 0.5 |

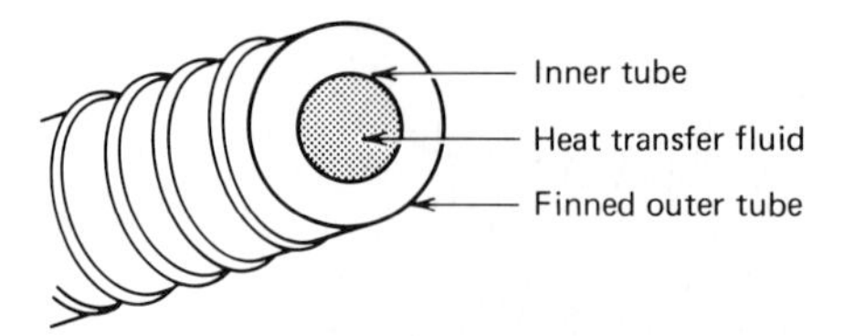


**Figure 5-5.** Tube in tube heat exchanger.

### Mixed Flow Heat Exchanger

This heat exchanger is one in which one fluid thermally contacts the average temperature of another fluid. A shell and tube exchanger (as shown in Figure 5-6) consists of an outer casing or shell surrounding a bundle of tubes. The water to be heated is normally circulated in the tubes and the hot liquid is circulated in the shell. This design is less effective than the counterflow design.

### Air to Liquid Heat Exchanger

Air to liquid exchangers consist of a continuous finned coil as illustrated in Figure 5-7. The heat transfer coefficent on the air side is much lower than on the liquid side so the exchanger surface area is placed directly in the hot air collector outlet ductwork. Because water cannot practically be made to flow countercurrent to the air stream due to space restriction, a cross-flow design configuration is normally used, as illustrated in Figure 5-8. To permit increased area without excessive duct size, two cross-flow exchangers can be used in series.

## Design Elements

Ideally, the collector inlet temperature should be equal to the lowest temperature in the storage tank. The addition of a heat exchanger will compromise this ideal situation. There are many parameters involved in the selection of a heat exchanger design. These variables include inlet/outlet dimensions, flow rates, exchange surface areas, compatibility of materials, pressure and temperature limitations, and thermal heat transfer coefficients.

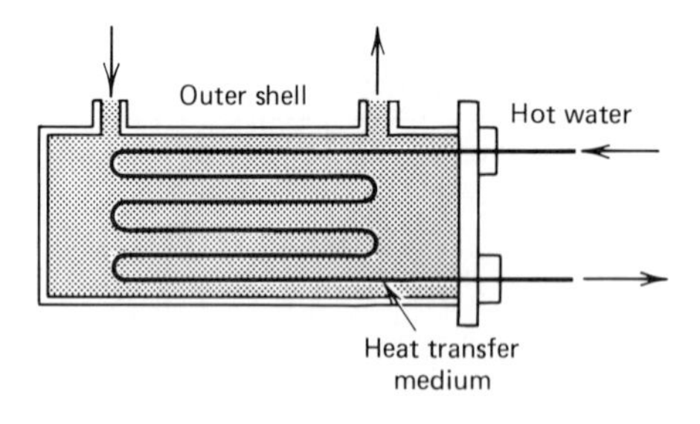


**Figure 5-6.** Shell and tube heat exchanger.

**Figure 5-7.** Air to liquid heat exchanger. (Courtesy of Granite State Solar Industries, Inc., Dover, N.H.)

Because of the number of interrelationships that exist, we shall not pursue the design criteria of the various types of heat exchangers in detail. In general, for a coil in tank heat exchanger, manufacturer's data indicate an average heat exchanger surface area of 0.25 $ft^2$ per square foot of collector area.

From physics, the second law of thermodynamics states that heat is transferred from a hot body to a cooler one until an equilibrium is reached. Mathematically, we can represent this exchange of heat from the heat exchanger to storage as Equation 5-1:

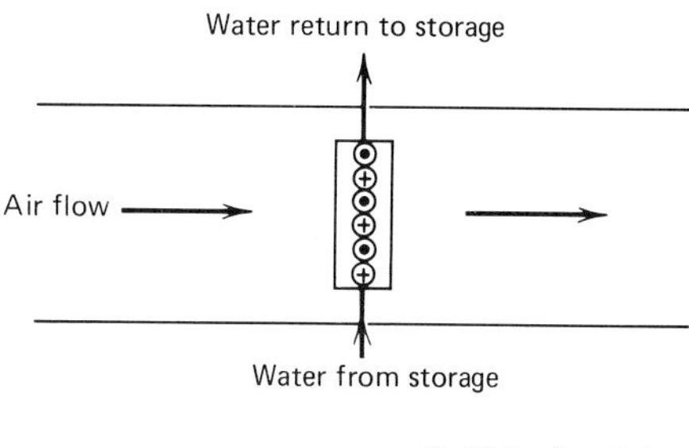


**Figure 5-8.** Cross-flow heat exchanger.

***Equation 5-1***

$$Q = m \int_{T_1}^{T_2} C_P(T)\, dT$$

*where*
$Q$ = change in heat (BTU/hr)
$m$ = mass flow rate (lb/hr)
$C_p$ = specific heat (BTU/lb-°F)
$T_2—T_1$ = temperature change (inlet—outlet) = $\Delta T$(°F)

We have an integral sign in Equation 5-1 because the specific heats of all substances vary somewhat with temperature. If we assume no heat losses and the specific heat is constant for a specific temperature range, then we can reduce this equation to an algebraic relationship as shown in Equation 5-2.

***Equation 5-2***

$$\Delta Q = mC_p \Delta T$$

For example, if we have a three collector system with a flow rate of 2.1 gal per minute, a synthetic hydrocarbon heat transfer fluid with a specific heat $C_p$ of 0.56, and a temperature difference $\Delta T$ from inlet to outlet of the heat exchanger of 30° F (i.e., 160° F inlet and 130° F outlet), then we can find the ***maximum*** amount of heat $\Delta Q$ transferred to storage. From Equation 5-2 we have

$$Q = mC_p \Delta T$$

*where* $m$ = (2.1 gal/min) × (7 lb/gal) = 14.7 lb/min
$C_p$ = 0.56 BTU/lb-°F
$\Delta T$ = 30° F

and $\Delta Q$ = (14.7 lb/min)(0.56 BTU/lb-°F)(30° F)
$\Delta Q$ = 247 BTU/min

or $\Delta Q$ = 14,818 BTU/hr

Remember that the rate of heat exchange will vary throughout the course of a day. In actuality, the addition of a heat exchanger will elevate the collector temperature, leading to increased system heat losses and a decrease in overall system performance. The final result is a heat exchanger factor, $E$, which must be introduced to Equation 5-2 to determine the ***actual*** amount of heat transferred to storage. The heat exchange actually achieved can then be expressed as a proportion $E$ of the maximum, where $E$ is called the ***effectiveness*** of the heat exchanger as depicted in Equation 5-3.

***Equation 5-3***

$$\Delta Q = EmC_p \Delta T$$

The effectiveness $E$ is independent of the temperature level, and depends only on the exchanger heat transfer properties and the fluid flow. The derivations and final equations of this effectiveness factor $E$ will vary from one type of heat exchanger to another.

## HEAT TRANSFER FLUIDS

### Fluid Characteristics

Fluids used in the transfer of heat from the solar collectors to the storage are either aqueous or nonaqueous fluids. In the drain back, drain down, or thermosiphon systems, the fluid is normally water (aqueous). In the phase change systems, the fluid is Freon (nonaqueous) and in the air to liquid system, the fluid is air (nonaqueous). It is the closed loop freeze-resistant systems that normally use a variety of aqueous or nonaqueous fluids.

In some instances, the type of generic system chosen will determine the type of heat transfer fluid to be used. In other instances, other considerations must be examined, including geographic location of the system (must it be protected from freezing or boiling), the potential of stagnation (fluid deterioration), component compatibility (some fluids will corrode or adversely affect aluminum, copper, and seal and gasket material), and environmental aspects (some fluids are toxic, others have unpleasant odors when spilled).

The selection of the transfer fluid to be used should not be a casual afterthought. These fluids are important to the life and operation of the solar domestic hot water system. Each type of heat transfer fluid has its own unique properties, as illustrated in Table 5-3. Variations in parameters such as viscosity, specific heat, coefficient of expansion, freezing point, boiling point, and flash point of heat transfer fluids will determine the size and compatibility of many system components.

Temperatures attained by transfer fluids under operating and stagnant conditions are recommended by the HUD Intermediate Property Standards *not* to exceed 100° F below the flash point of the fluid.

Ideally the transfer fluid should have thermal stability and should not degrade at service temperatures or cause corrosive deterioration of system components.

***Galvanic corrosion*** is normally associated with aqueous liquids; however, galvanic corrosion can occur with nonaqueous liquids if:

1. The liquid is initially aggressive to the containment material.
2. Water is present in the liquid or contaminates the liquid.
3. Liquid decomposition generates corrosive products.

**Table 5-3**
**Typical Heat Transfer Liquids**[a]

| | | *Glycols* | | | *Hydrocarbons* | | | |
|---|---|---|---|---|---|---|---|---|
| | *Water* | *50% Ethylene Glycol/Water* | *50% Propylene Glycol/Water* | *Silicone* | *Aromatics* | *Paraffinic Oil* | *Synthetic* | *Air* |
| Freezing point, °F (°C) | 32 (0) | −33 (−36) | −28 (−33) | −58 (−50) | −100 to −25 (−73 to −32) | +15 (−9) | −40 (−40) | None |
| Boiling point, °F (°C) (at atm. pressure) | 212 (100) | 230 (110) | — | None | 300–400 149–204 | 600 (316) | 625 (329) | None |
| Fluid stability | Requires pH or inhibitor monitoring | Requires pH or inhibitor monitoring | Requires pH or inhibitor monitoring | Excellent | Good | Good | Excellent | Excellent |
| Flash[b] point °F (°C) | None | None | 600 (315) | 450 (232) | 145–300 (63–149) | 455 (235) | 345 (174) | None |
| Specific heat at 100°F [BTU/(lb-°F] | 1.00 | 0.80 | 0.85 | 0.39 | 0.36–0.42 | 0.46 | 0.56 | 0.25 |
| Viscosity (cstk at 77° F) | 0.9 | 21 | 5 | 20 | 1–100 | — | 20 | 0.5 |
| Toxicity | No | Yes | No | No | Yes | Yes | No | No |
| Coefficient of of expansion (% per °F) | 0.025 | 0.035 | 0.015–0.025 | 0.059 | — | 0.04 | 0.04 | — |

[a]These data are extracted from manufacturers' literature to illustrate the properties of a few types of liquid that have been used as a transfer fluid. Data are average values.

[b]It is important to identify the conditions of tests for measuring flash point. Because the manufacturers' literature does not always specify the test, these values may not be directly comparable.

***Pitting corrosion*** can also be associated with both types of transfer liquids in that ions from one type of metal can be transported by the fluid and deposited on another type of metal in the system, causing pitting. Corrosion can be caused by many parameters including the composition of metals in the water which varies from one locale to another, flow rate, the presence of additives, or decomposition of the liquid at elevated temperatures. Because of intermixed variables the user should monitor the *pH* or acid content periodically to prevent any acid formed from corroding the metal parts of the system. For copper and aluminum this *pH* factor should be greater than ($>$) 8.

We shall now discuss each of the types of transfer fluids listed in Table 5-3.

### Water

Untreated water is the least expensive and most easily available fluid. The four chemical compounds in water chemistry that cause problems in solar energy systems include calcium carbonate, magnesium hydroxide, calcium silicate, and calcium sulfate. This is due to their decreased solubility with increasing temperatures. If freezing or boiling is not a problem, if replacement water is readily at hand, and if there are no mineral hardness problems, water is an excellent transfer fluid. Because its specific heat is greater than any of the other fluids it can deliver more BTU to storage. Whatever the local water conditions, distilled or deionized water is recommended for drain back systems with a heat exchanger. One must remember to use a pump of bronze or stainless steel to prevent corrosion of the water-wetted surfaces.

### Gylcols—Propylene and Ethylene

To overcome the deficiencies characteristic to water, propylene or ethylene glycols and corrosion inhibitors can be added to water normally at a 50/50 or 60/40 glycol to water ratio. This addition of glycols to water will solve many of the problems associated with water as a transfer fluid, with the exception of the vapor pressure problem. At 400° F stagnation temperatures, pressures of over 150 psi may be generated in a closed system.

It is extremely important to note that ethylene glycol is ***toxic*** and requires the use of a double-walled heat exchanger in a closed loop to preclude the danger of contaminating the potable water. Food grade propylene glycol U.S.P./water mixtures, when certified nontoxic, can be used with a single-walled heat exchanger if no toxic dyes or inhibitors have been added to the mixture.

Glycol solutions should not be used with zinc galvanized plumbing because the required corrosion inhibitors react with zinc. Glycols may also damage certain materials such as butyl rubber membranes in some types of expansion tanks. During stagnation, glycols can decompose rapidly at approximately 280° F, forming sludge and organic acids. The higher the temperature, the more rapid is the degradation. Ethylene and propylene glycol also break down in use. The buffers added to glycols are intended to prevent the *pH* from dropping. Their effect is not permanent and if the glycol degradation is allowed to continue, the buffers will be depleted and the solution will in turn become acidic. Because of this eventual acidic state, the solution should be monitored and a regular maintenance schedule should be maintained. The fluid should be changed at least once every two years. It is the frequency of fluid change and service that increase the cost of using this transfer fluid over the lifetime of the system. An initial investment using synthetic hydrocarbons or silicones is more economical because of less frequency of fluid change and service.

### Hydrocarbons

Hydrocarbon heat transfer fluids are typically categorized as synthetic hydrocarbons, paraffinic, or aromatic refined mineral oils.

Paraffinic mineral oils are petroleum-based heat transfer fluids. Their temperature range between boiling and freezing is greater than that of water. They are also nonconducting and may have a higher viscosity than water. This type of transfer fluid is considered ***toxic*** and requires the use of a double-walled heat exchanger in a closed loop to preclude the danger of contaminating the potable water. Paraffinic-based mineral oils freeze at relatively high temperatures.

Aromatic oils have lower viscosities than paraffins, allowing the use of smaller pumps. Because they have lower flash points however, they are not as safe to use. Aromatics will also dissolve roofing tar and most elastomer seals. Viton® or neoprene seals should be used in pumps whenever paraffinic or aromatic hydrocarbons are used.

Synthetic hydrocarbons are not water miscible and therefore do not attack metals or elastomers as do aqueous solutions. They also do not develop excessive vapor pressure at normal operating temperatures as do water-based fluids. This type of heat transfer fluid will normally remain stable between five to ten years, thereby decreasing maintenance. Like the aromatic oils, if this transfer fluid is spilled onto asphalt shingles or asphalt tile floors, it should be washed up quickly to prevent decomposition. Synthetic hydrocarbons are typically *nontoxic*.

### Silicones

Silicone heat transfer fluids are essentially inert, *virtually nontoxic*, will not freeze or boil, have no odor, and will cause neither galvanic corrosion nor degradation of roofing materials. These fluids also exhibit a very high flash point. Although the initial cost of these fluids is higher than other transfer fluids, there is no need to replace or dispose of the fluid periodically, and they have a life expectancy of 20 years or more.

There are, however, a few disadvantages to the fluid. Silicones have a lower heat capacity as can be observed in Table 5-3, as well as a higher viscosity requiring twice the flow rate of most systems and thus more pump horsepower. Silicones are also incompatible with expansion tanks fitted with neoprene or butyl rubber diaphragms. EPDM (Ethylene-Propylene-Diamine) rubber or Viton materials should be used. Silicon fluids will also readily leak through the smallest pipe joint soldering flaws which normally would retain water or other fluids. Teflon tape, when used alone or with pipe dope, is unacceptable for threaded joints. All threaded connections must be sealed with either Loctite® Pipe Sealant with Teflon or Dow Corning® 730 fluorosilicone sealant to prevent leakage.

### Air

Air is a heat transfer fluid that is used in conjunction with an air to liquid heat exchanger. Its advantages are it cannot freeze or boil, it is noncorrosive, and it is free.

It's disadvantages are the higher cost of duct work and installation in comparison with copper pipe and a very low specific heat of 0.018 BTU/ft$^3$-°F (at sea level). Resulting system efficiencies are therefore less than in other hot water systems.

## Design Elements

The specific heat, flow rate, and density constitute the solar energy collection capability of the transfer fluid. All of the heat transfer fluids depicted in Table 5-3 will transfer heat at different capacities. This heat transfer is a function of the specific heat multiplied by the weight transferred in a given period of time.

For example, if it takes a flow rate of one pound per minute of water to transfer one BTU, then by comparison it would take a flow rate of 1.20 lb/min of water/glycol, 1.79 lb/min of a hydrocarbon, 2.56 lb/min of silicone, and 3.95 lb/min (56 ft$^3$/min) of air to transfer the same amount of heat due to the lower specific heat.

The density (mass per unit volume) of the transfer media, or the weight per unit, also has a major effect on the energy collection. The densities of various fluids are depicted in Table 5-4. The weight per unit of liquid transferred is a function of volume and density. Therefore, pumping rates will vary for each heat transfer fluid in order to transfer an equivalent amount of heat. Table 5-4 illustrates the relative amount of heat supplied by one gallon of a fluid at a temperature change of 1°. When flow is adjusted or normalized to the relative flow rates shown in Table 5-5, each fluid will deliver the same amount of heat.

**Table 5-4**
**BTU of Heat Supplied by One Gallon at 1° F Temperature Rise**

| *Heat Transfer Fluid* | *Specific Heat (BTU/lb-°F)* | *Density (lb/gal)* | *No. of BTU Supplied* |
|---|---|---|---|
| Water | 1.00 | 8.33 | 8.33 |
| Water/glycol | 0.83 | 8.3 | 6.89 |
| Hydrocarbon | 0.56 | 7.0 | 3.92 |
| Silicone | 0.39 | 8.0 | 3.12 |
| Air | 0.25 | 0.0075 | 0.0019 |

## PIPING AND MISCELLANEOUS HARDWARE

### Piping

Copper pipe should be used throughout the solar domestic hot water system. Thermoplastic pipe (PVC and CPVC) is not recommended for use because CPVC, itself, has a tendency to sag at approximately 160° F. Even polybutylene pipe will begin to sag at approximately 200° F. Rigid copper pipe Type "L" is normally used for indoor plumbing and rigid copper Type "M" is normally used for heating systems. Type "M" copper pipe has a thinner wall than Type "L" and therefore may be more subject to mechanical damage through abuse. Neither the Type "L" or Type "M" pipe used in solar hot water systems are subject to pressure failure due to wall thickness. Because Type "M" copper pipe is less expensive than Type "L," it can be used throughout the collector loop where local building codes permit. Table 5-6 gives dimensions of Type "L" and Type "M" copper tubing for ½, ¾, 1 and 1¼ in. diameters.

The resistance to flow, known as the friction head loss, is made up of the resistance in straight lengths of pipe, elbows, tees, couplings, various hand valves for control and isolation, the solar collectors, heat exchanger, and any other device through which the transfer fluid must be pumped or circulated. Collector manufacturers usually provide friction head loss imposed by the collector for various flow rates (average flow rates are normally approximately 0.02 gal/min/ft$^2$). Friction head loss in a collector loop will be discussed during the discussion of pumps in this chapter. Table 5-7 provides

**Table 5-5**
**Flow Rates for Equivalent Heat Delivery**

| *Heat Transfer Fluid* | *Flow Rates Necessary to Deliver the Same Amount of Heat* |
|---|---|
| Water | 1 |
| Water/glycol | 1.21 |
| Hydrocarbon | 2.13 |
| Silicone | 2.67 |
| Air | 4,384.21 |

**Table 5-6**
**Copper Tube Dimensions (inches)**

| *Nominal Pipe Size* | *Outside Diameter* | *Inside Diameter* | *Wall Thickness* | *Inside Cross-Sectional Area* |
|---|---|---|---|---|
| *Type "L"* | | | | |
| ½ | 0.625 | 0.545 | 0.040 | 0.233 |
| ¾ | 0.875 | 0.785 | 0.045 | 0.484 |
| 1 | 1.125 | 1.025 | 0.050 | 0.825 |
| 1¼ | 1.375 | 1.265 | 0.055 | 1.257 |
| *Type "M"* | | | | |
| ½ | 0.625 | 0.569 | 0.028 | 0.254 |
| ¾ | 0.875 | 0.811 | 0.032 | 0.517 |
| 1 | 1.125 | 1.055 | 0.035 | 0.874 |
| 1¼ | 1.375 | 1.291 | 0.042 | 1.831 |

a quick reference to simplify the determination of the proper pipe size to use in a closed loop piping system for varying lengths of pipe, flow rates, and storage tank and heat exchanger sizes. For example, if a solar DHW system incorporates an 80 gal tank with a 20 $ft^2$ single wall heat exchanger, a flow rate of 2 gals/min, and has a supply and return pipe run of 80 ft, then a ¾ in. copper pipe is recommended.

Every effort should be made to minimize friction head loss, and this can be accomplished by minimizing the length of pipe run and by minimizing the number of bends. In an open loop system the return pipe should be larger in diameter than the supply to allow for a quick disposal of the transfer fluid from the collectors. The total amount of fluid necessary to fill the piping for various lengths of supply and return runs is illustrated in Table 5-8. This table will be mentioned again in Chapter 9 during the discussion of filling the system.

There are many types and sizes of standard fittings and hardware that are required in plumbing the system. Figure 5-9 illustrates only a few of these items. Item (1) is a typical 6-in. "Milford" pipe hanger, item (2) is a 90° elbow (ell) ¾-in. sweat fitting, item (3) is ¾-in. × ¾-in. female adapter, item (4) is a pipe hanger, item (5) is a ¾-in. × ½-in. reducing coupler, and item (6) is a ¾-in. × ¾-in. male adapter.

Pipe runs from roof mounted collectors should be installed through wall partitions where possible and through closets where it is not possible. If pipes are routed through rooms, they should be close to the walls while maintaining at least a 2-in. gap between the supply and return lines for insulation. Piping should be supported so that it will expand and contract with temperature changes. Pipe run vertically should be supported with hangers at 10-ft intervals and pipe run horizontally should be supported at 6-ft intervals.

**Table 5-7**
**Pipe Sizing**
*65 Gal Tank 15 Ft² Heat Exchanger*

| | | Supply and Return Pipe Run (length in feet) | | | | |
|---|---|---|---|---|---|---|
| *GPM* | *COLL* | *40 ft* | *60 ft* | *80 ft* | *100 ft* | *120 ft* |
| 1.0 | 2 | ½(S) | ½(S) | ½(S) | ½(S) | ½(S) |
| 1.5 | 3 | ½(S) | ½(S) | ½(S) | ½(S) | ½(S) |
| 2.0 | 4 | ½(S)<br>¾(D) | ¾(S) | ¾(S) | ¾(S) | ¾(S) |
| 2.5 | 5 | ¾(S)<br>1(D) | ¾(S)<br>1(D) | ¾(S)<br>1(D) | ¾(S)<br>1(D) | ¾(S)<br>1(D) |
| 2.5 | 6 | ¾(S)<br>X(D) | 1(S)<br>X(D) | 1(S)<br>X(D) | 1(S)<br>X(D) | 1(S)<br>X(D) |

*80 Gal Tank, 20 Ft² Heat Exchanger*

| | | Supply and Return Pipe Run (length in feet) | | | | |
|---|---|---|---|---|---|---|
| *GPM* | *COLL* | *40 ft* | *60 ft* | *80 ft* | *100 ft* | *120 ft* |
| 1.0 | 2 | ½(S) | ½(S) | ½(S) | ½(S) | ½(S) |
| 1.5 | 3 | ½(S) | ½(S)<br>¾(D) | ½(S)<br>¾(D) | ½(S)<br>¾(D) | ½(S)<br>¾(D) |
| 2.0 | 4 | ¾(S) | ¾(S)<br>1(D) | ¾(S)<br>1(D) | ¾(S)<br>1(D) | ¾(S)<br>1(D) |
| 2.5 | 5 | ¾(S)<br>1(D) | 1(S) | 1(S) | 1(S) | 1(S) |
| 2.5 | 6 | X | X | X | X | X |

*120 Gal Tank, 40 Ft² Heat Exchanger*

| | | Supply and Return Pipe Run (length in feet) | | | | |
|---|---|---|---|---|---|---|
| *GPM* | *COLL* | *40 ft* | *60 ft* | *80 ft* | *100 ft* | *120 ft* |
| 1.0 | 2 | ½(S) | ½(S) | ½(S) | ½(S) | ½(S) |
| 1.5 | 3 | ½(S) | ½(S) | ½(S) | ½(S) | ½(S) |
| 2.0 | 4 | ½(S) | ½(S)<br>¾(D) | ¾(S) | ¾(S) | ¾(S) |
| 2.5 | 5 | ¾(S) | ¾(S)<br>1(D) | ¾(S)<br>1(D) | ¾(S)<br>1(D) | ¾(S)<br>1(D) |
| 3.0/<br>2.5 | 6 | ¾(S) | ¾(S)<br>1(D) | ¾(S)<br>1(D) | ¾(S)<br>1(D) | 1(S) |

Key: Single wall (S).
Double wall (D).
X, heat exchanger not of sufficient size for collector area.

**Table 5-8**
**Fluid Capacity of Total Pipe Run**

| *Pipe Parameters* | | | *Typical Heat Transfer Fluid Approximate Capacity (gals) at 72° F* | | | |
|---|---|---|---|---|---|---|
| *Pipe Size (in.)* | *Total Pipe Run (Ft)* | *Volume ($Ft^3$)* | *Water* | *Glycol* | *Hydrocarbon* | *Silicone* |
| | 40 | 0.218 | 1.63 | 1.82 | 1.75 | 1.62 |
| | 60 | 0.327 | 2.45 | 2.73 | 2.62 | 2.42 |
| ½ | 80 | 0.436 | 3.27 | 3.64 | 3.50 | 3.23 |
| | 100 | 0.545 | 4.08 | 4.54 | 4.37 | 4.04 |
| | 120 | 0.654 | 4.90 | 5.45 | 5.25 | 4.85 |
| | 40 | 0.491 | 3.68 | 4.09 | 3.94 | 3.64 |
| | 60 | 0.736 | 5.51 | 6.14 | 5.90 | 5.45 |
| ¾ | 80 | 0.982 | 7.36 | 8.19 | 7.88 | 7.28 |
| | 100 | 1.227 | 9.19 | 10.23 | 9.84 | 9.09 |
| | 120 | 1.472 | 11.03 | 12.27 | 11.81 | 10.91 |
| | 40 | 0.873 | 6.55 | 7.28 | 7.00 | 6.47 |
| | 60 | 1.309 | 9.81 | 10.91 | 10.50 | 9.70 |
| 1 | 80 | 1.746 | 13.08 | 14.56 | 14.01 | 12.94 |
| | 100 | 2.182 | 16.35 | 18.19 | 17.51 | 16.17 |
| | 120 | 2.618 | 19.61 | 21.83 | 21.00 | 19.40 |
| | 40 | 1.364 | 10.22 | 11.37 | 10.94 | 10.11 |
| | 60 | 2.045 | 15.32 | 17.05 | 16.41 | 15.15 |
| 1¼ | 80 | 2.727 | 20.43 | 22.74 | 21.88 | 20.21 |
| | 100 | 3.409 | 25.54 | 28.42 | 27.35 | 25.26 |
| | 120 | 4.091 | 30.65 | 34.11 | 32.82 | 30.31 |

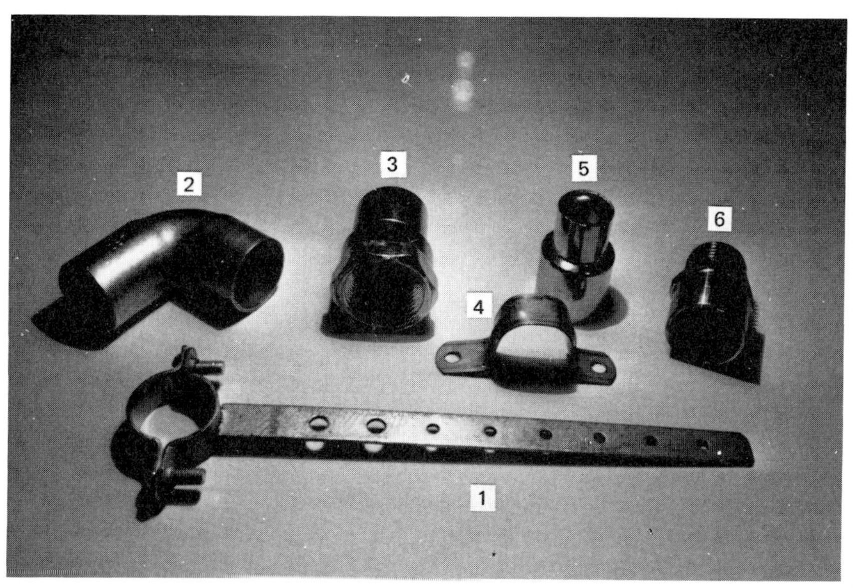


**Figure 5-9.** Typical plumbing hardware.

Typical plumbing hardware location and integration with respect to a closed loop system is illustrated in Figure 5-10.

## Miscellaneous Hardware

We shall now discuss plumbing components and materials which are used in the generic types of systems described previously in Chapter 3. Reference to Chapter 3 and Figure 5-10 is suggested while studying this section in order to determine where each item is used. The following items are discussed in alphabetical order:

1. Air purger
2. Air vent/coin vent
3. Backflow preventer
4. Expansion tank

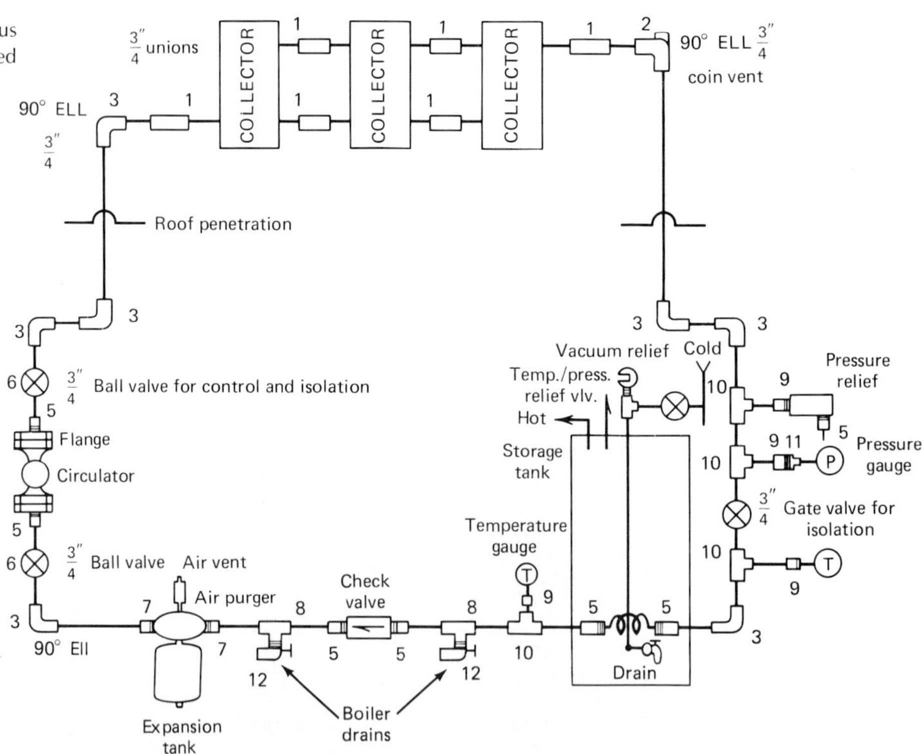


**Figure 5-10.** Miscellaneous plumbing hardware (closed loop freeze-resistant system).

*Parts Description and Quantity*

1. ¾″ Unions (6)
2. 90″ Ell ¾″ coin vent (1) or air vent
3. 90″ ell ¾″ (7)
4. Gate valve ¾″ (1)
5. ¾″ × ¾″ male adapter (7)
6. Ball valve ¾″ (2)
7. ¾″ × 1″ male adapter (2)
8. ¾″ × ¾″ tee × ¾″ female adapter (2)
9. ¾″ × ¾″ female adapter (4)
10. ¾″ × ¾″ × ¾″ tee (4)
11. ¾″ male × ½″ female adapter (1)
12. Boiler drain ¾″ threaded (2)

5. Pipe insulation
6. Pressure gauge
7. Roof penetrations
8. Solder
9. Temperature gauge
10. Unions
11. Valves

**Air Purger (Figure 5-11)**

The heat transfer fluid on filling the system will contain dissolved air. The air purger or air eliminator is a one-piece cast-iron chamber that has internal contours and baffles designed for low flow resistance characteristics and efficient separation of the air from the transfer fluid. As the fluid is circulated in a closed system, the denser portion flows through the lower portion of the air purger directly to system piping. The less dense fluid, containing dissolved air, moves into the upper portion of the air purger. This upper section of the air purger is designed to free the air for accumulation at the venting port. The fluid that is separated from its air content rejoins the main flow. The air purger must be installed so that fluid flow is in the direction of the arrow.

**Air Vent/Coin Vent (Figures 5-12 and 5-13)**

Air vents alleviate air bubbles from a system. The automatic float type of air vent is the most commonly used. As air enters the vent chambers of this device, the fluid level is lower, causing a float to fall, thus opening an air valve. As the air is released, the water level rises, raising the float and thus closing the air valve. The cap of the automatic float-type vent should be opened two turns for normal venting. The device is mounted vertically and is located at the highest point in the collector array. In a closed loop system the float air vent is attached to the air purger. Either the float air vent type or the manual air vent (coin vent) is also attached at the highest point in the collector array to facilitate the filling procedure.

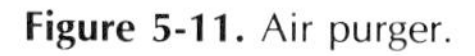
Figure 5-11. Air purger.

Figure 5-12. Float type air vent with air purger.

**Figure 5-13.** Float type air vent and coin vent.

### Blackflow Preventer (Figure 5-14)

If a toxic, non-potable heat transfer fluid is used in a closed loop system, a backflow preventer may be required by some local codes to ensure that none of the toxic liquid is able to infiltrate town or city water supplies. The backflow preventer is simply a spring loaded device that will open when pressure is in one direction and that will close and discharge the fluid when under pressure from the opposite direction. A residential type backflow preventer may discharge in any direction and should not be installed near electrical components or items that might be damaged by liquids.

### Expansion Tank (Figures 5-15 and 5-16)

An expansion tank is a necessary component of a solar domestic hot water system that has a closed circulation in the collector-to-storage loop. The expansion tank is required to relieve the pressure created when the heat transfer fluid expands in volume upon increasing temperature. This tank contains a flexible diaphragm normally charged with 12 psi air on one side and pressurized with the heat transfer fluid on the other side. As the fluid changes temperature and volume, the diaphragm compresses the air charge and makes room for the expanding fluid.

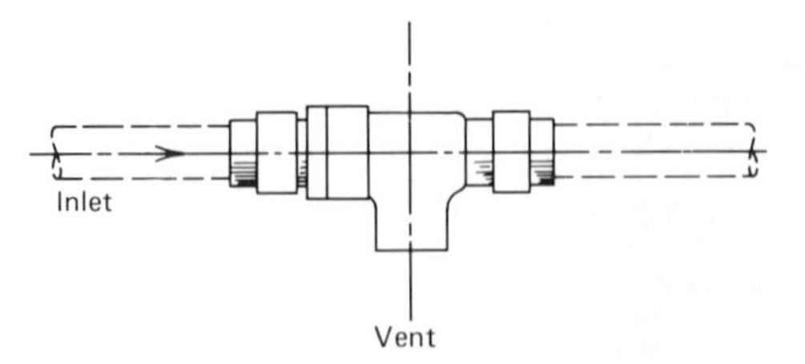


**Figure 5-14.** Backflow preventer.

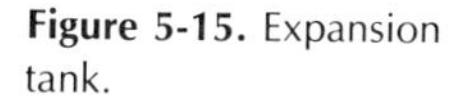

**Figure 5-15.** Expansion tank.

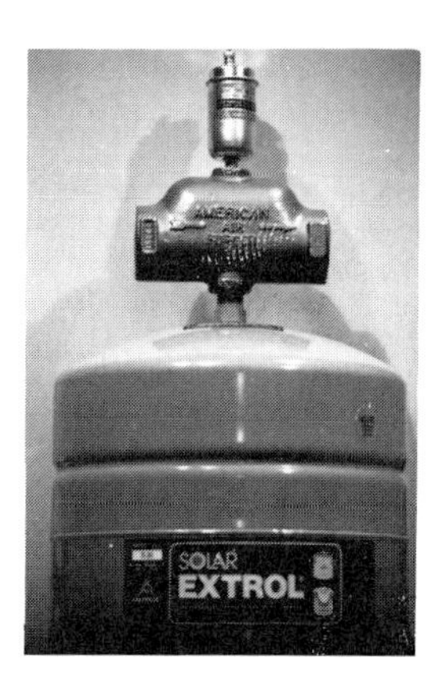


**Figure 5-16.** Expansion tank with air purger and air vent.

For example, from Table 5-3 we find that water has a coefficient of expansion of 0.025 percent per degree fahrenheit. A 5 gal volume of water at 50° F will therefore expand to 5.19 gal at 200° F. An expansion tank will compensate for this variation in volume. Expansion tanks for solar domestic hot water systems normally have capacities of 1.5 and 3.0 gal. Table 5-3 provides coefficients of expansion for various heat transfer fluids to determine the actual expansion capacity needed. As a rule of thumb, expansion tanks can be sized at 10 percent of the total fluid in the closed loop. Nonpressurized open loop systems do not require a separate expansion tank because volume variations can be handled by the storage tanks.

The expansion tank should be installed on the suction side of the circulator and attached to the air purger in a downward position to prevent heat loss. Most solar expansion tanks are pressurized by the manufacturer to 12 psi. Most closed loop systems are pressurized and operate at 15 psi and are air pressure tested at 30 psi prior to fill. If the collector system is pressure tested much beyond 30 psi, then the expansion tank may be damaged. This situation can be avoided by either removing the expansion tank from the system prior to pressure testing or by increasing the internal preset pressure of the expansion tank using the threaded valve on the tank bottom.

### Pipe Insulation (Figures 5-17, 5-18, and 5-19)

Insulating the pipes is the last step in the installation process and is discussed in Chapter 9. After pressure testing and system fill, the collector system should be checked for fluid leakage and then the piping should be insulated. Various types of pipe insulating materials are available as listed in Table 5-9 together with their respective resistance values.

Elastomers (i.e., Armaflex®, Rubatex®, etc.) are normally black in color and should be used for the indoor collector return to storage only. (In some situations where the heat transfer fluid supply line from storage to the collectors is short and little heat is lost, it may not be economical to use pipe insulation. Remember that the collector operates at a higher efficiency with

**Figure 5-17.** Elastomer and isocyanurate pipe insulation.

**Figure 5-18.** Elastomer pipe insulation.

**Figure 5-19.** Sealing isocyanurate pipe insulation.

a cooler inlet fluid.) Elastomers are not protected from ultraviolet degradation and if used outside, they will turn brittle and fall apart within a one- to two-year period unless treated by the manufacturers' recommended procedure. Isocyanurates, however, are available with a jacket composed of Tedlar® PVF film, impregnated fiberglass scim—reinforced asbestos felt, and Mylar® Polyester film, which is resistant to ultraviolet degradation and the natural environment. This type of pipe insulation should be used on all exterior piping. Similar products are also available in fiberglass and urethane variations.

**Table 5-9**
**Pipe Insulations**[a]

| | |
|---|---|
| Fiberglass | R-3 |
| Elastomers | R-3.5 |
| Urethane | R-6.5 |
| Isocyanurates | R-7 |

[a]Thermal resistance ($R$); hr-ft$^2$-°F/BTU

Elastomers can be slid over piping prior to soldering joints or it can be slit lengthwise and installed after soldering joints. If the elastomer is slid over piping, it can be held away from the joint to be soldered by using clamps or Vise Grips® until after the system has been pressure tested. Normally this type of insulation is available in 6-ft lengths. The manufacturer's recommended adhesive should be used to seal joints.

Jacketed insulation is installed on outside piping by sealing the wraparound jacket with an adhesive spray glue. All seams can be protected using adhesive-backed Mylar® tape. This type of pipe insulation is normally available in 3-ft lengths.

### Pressure Gauge (Figures 5-20 and 5-21)

A pressure gauge is necessary for use during closed loop pressure testing as well as for providing a means to monitor the system for leakage resulting in a loss of system pressure. Most pressure gauges have a ¼-in. male thread fitting. In a normal ¾-in. collector to storage loop the gauge can be installed using a ¾-in. by ½-in. female adapter and a ½-in. male by ¼-in. female adapter. This can be located in a tee in a section of the heat transfer loop between an isolation valve and the collectors as illustrated in Figure 5-10.

Pressure is defined as the ratio of force per unit area (pounds per square inch; psi) and is measured as either gauge or absolute pressure. Gauge pressure is the effective pressure for doing work because it measures the difference between pressure in a containment and that of atmosphere (14.74

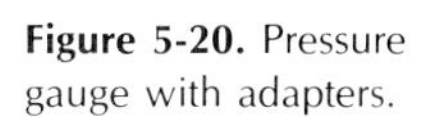

**Figure 5-20.** Pressure gauge with adapters.

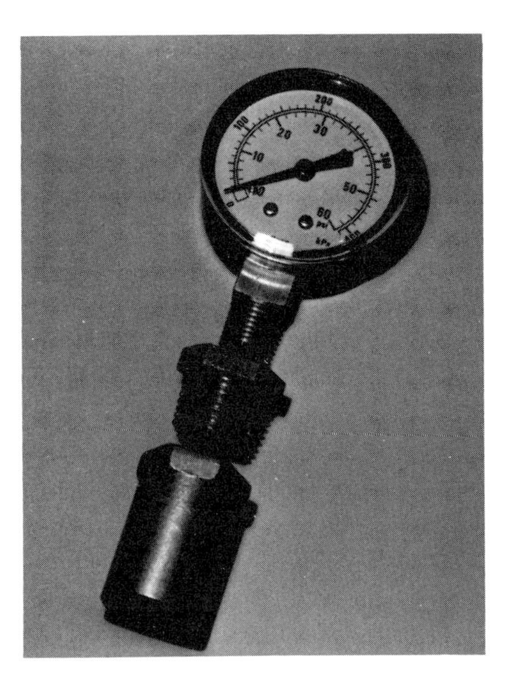

**Figure 5-21.** Adapters with in-line tee.

psi). Gauge pressure can therefore be expressed in terms of absolute pressure by simply adding 14.74 psi to the gauge pressure reading.

### Roof Penetrations (Figures 5-22 and 5-23)

Two roof penetrations are necessary to accommodate the supply and return pipe runs to the roof-mounted collectors. Standard neoprene boots used in similar circumstances for vent pipes or soil stacks can be modified to channel the nominal ¾-in. copper pipes. Pipe flanges designed expressly for solar installations are also available, whereby the ¾-in. pipe (1) passes through a ¾-in. by 1¼-in. modified coupler (2) over an insulating hose for thermal break (3) attached to a 1-in. copper nipple (4), which is in turn brazed onto a 6 in. by 9 in. copper sheet (5).

A 1-in. hole should be drilled through the roof for ¾-in. pipe. The top of the flange should be completely under the shingles and the bottom of the flange over the shingles. Silicone sealant should be used around the hole and between the shingles and flange. Under no circumstances should the pipe insulation pass through the roof. Pipe insulation should butt up against the penetration flange and the edge around the base sealed with silicone. If pipe insulation were passed through the opening of a neoprene boot, there is a good possibility of leakage through the roof if the insulation is damaged in any way.

### Solder

Soldering provides a means of forming joints between metallic surfaces using a flexible alloy (solder) whose melting point is lower than that of the metals being joined The two classes of solders are soft and hard.

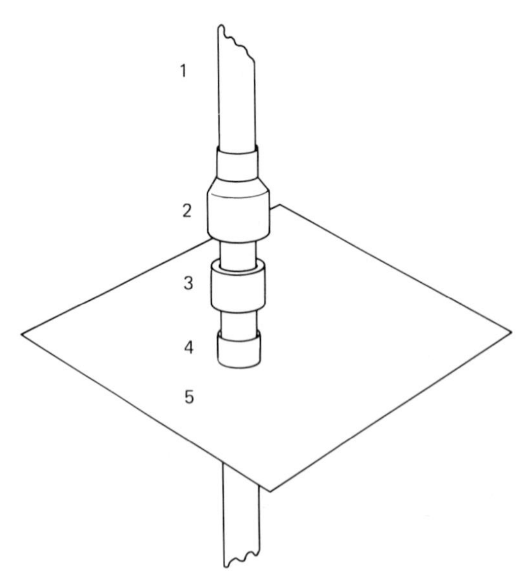


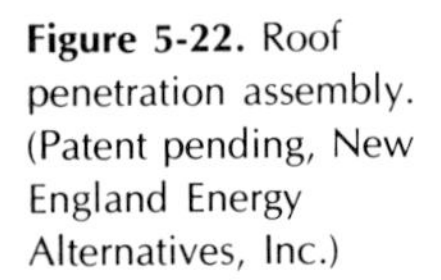

**Figure 5-22.** Roof penetration assembly. (Patent pending, New England Energy Alternatives, Inc.)

**Figure 5-23.** Roof penetration.

***Soft*** solders are normally referred to as those alloys that have melting temperatures below 600° F. Typical examples of proportions and the associated melting points are

50 percent tin (Sn)-50 percent lead (Pb), 421° F
95 percent tin (Sn)-5 percent antimony (Sb), 464° F

Collector header connections should be soldered with 95-5 (Tin-Antimony) AWS (American Welding Society) class alloy. All other joints can be soldered with 50-50 (Tin-Lead) for flat plate collectors. Evacuated tube type collector systems should be soldered with 95-5 solder throughout the system.

***Hard*** solders normally refer to those alloys that have melting temperatures greater than 600° F. Most of these alloys are silver solders melting at temperatures in excess of 1100° F.

### Temperature Gauge (Figures 5-24 and 5-25)

Temperature gauges should be installed at the inlet to the storage tank and at the outlet from the storage tank in the collector loop. Temperature gauges installed at these points will provide an indication of how the system is performing. Normally, these thermometers have a ¾-in. to 1-in. male threaded base and can be installed in conjunction with a sweat fitting to female adapter. The stem of the gauge should not be introduced to impede the flow of transfer fluid.

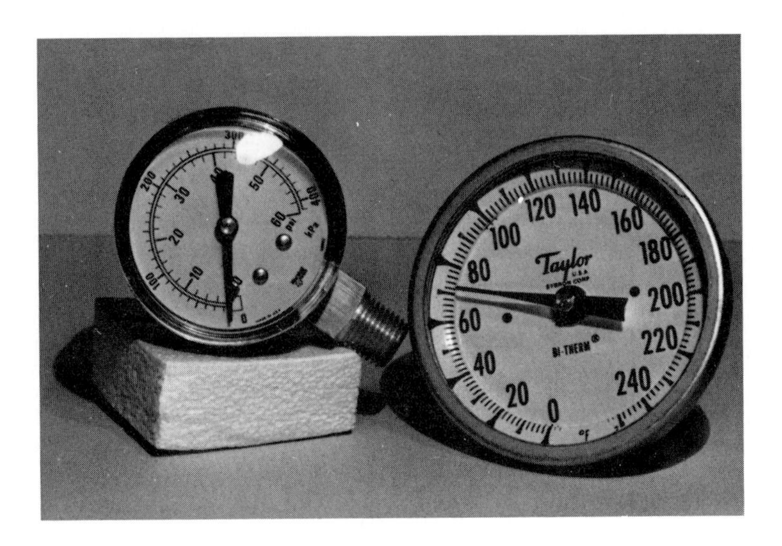


Figure 5-24. Pressure and temperature gauges.

**Unions (Figures 5-26 and 5-27)**

Unions consist of three pieces including a nut, a male adapter, and pressure fitting. Although more costly than a straight coupling between collectors, the unions provide quick and convenient assembly of one collector to another and require less time in soldering connections in a roof-mounted situation. If one collector develops a problem, it can quickly be disconnected /reconnected from the array without any soldering.

**Valves**

Special attention has been given to the selection and placement of valves in solar DHW systems. Most of the plumbing is no different from that of a

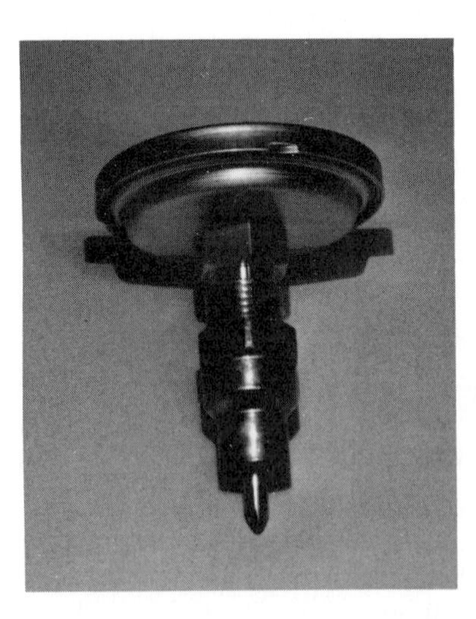

Figure 5-25. Temperature gauge with threaded adapter.

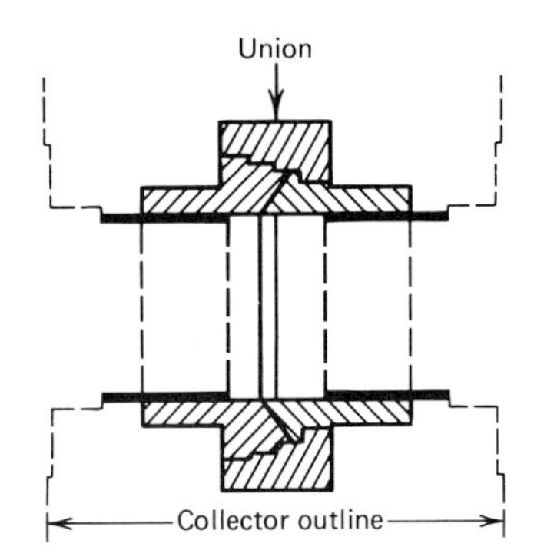


**Figure 5-26.** Unions.

conventional DHW system, but the solar DHW system will be more likely to have a collector loop separate from the regular plumbing and may use toxic or nonpotable heat transfer fluids. A variety of safety valves and control valves are necessary for the operation of a solar domestic hot water system. The types of valves used may vary slightly from one generic system to another. We shall discuss 10 of these valves. Alphabetically they include: (1) ball valve, (2) boiler drain valve, (3) check valve, (4) gate valve, (5) mixing valve, (6) pressure reducing valve, (7) pressure relief valve, (8) temperature and pressure relief valve, (9) solenoid and motorized valve, and (10) vacuum relief valve.

**(1) Ball Valve (Figures 5-28 and 5-29).** A ball valve can be used as a balancing valve to adjust the flow of heat transfer fluids through collectors that do not

**Figure 5-27.** Typical union connection between collectors.

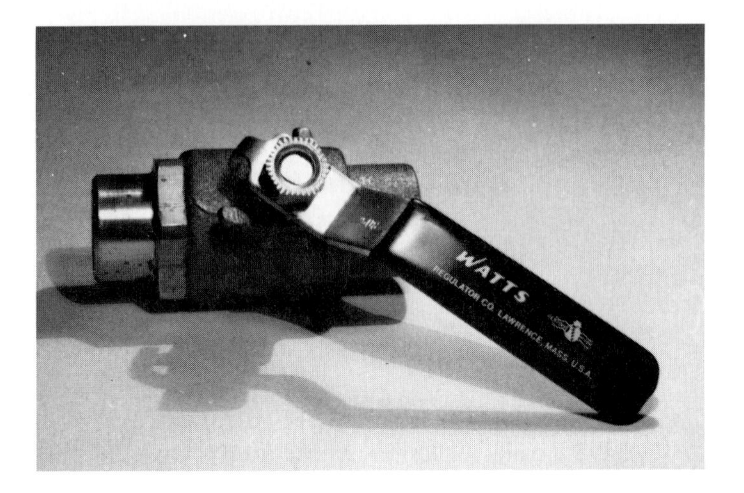


**Figure 5-28.** Typical ball valve.

have a reverse return piping arrangement, or it can be used as a control and isolation valve to "fine tune" the flow through the circulator or pump. This type of valve can be used to throttle fluid flow without causing unnecessary restrictions.

**(2) Boiler Drain Valve (Figure 5-30).** In the normal domestic hot water system, there should be one boiler drain valve at the bottom of the storage tank for draining. In the collector loop for a solar hot water system there should be two boiler drain valves, as illustrated in Figure 5-10. These two valves are used to charge and fill a closed loop system using an external pump.

**Figure 5-29.** Ball valve in half-open position.

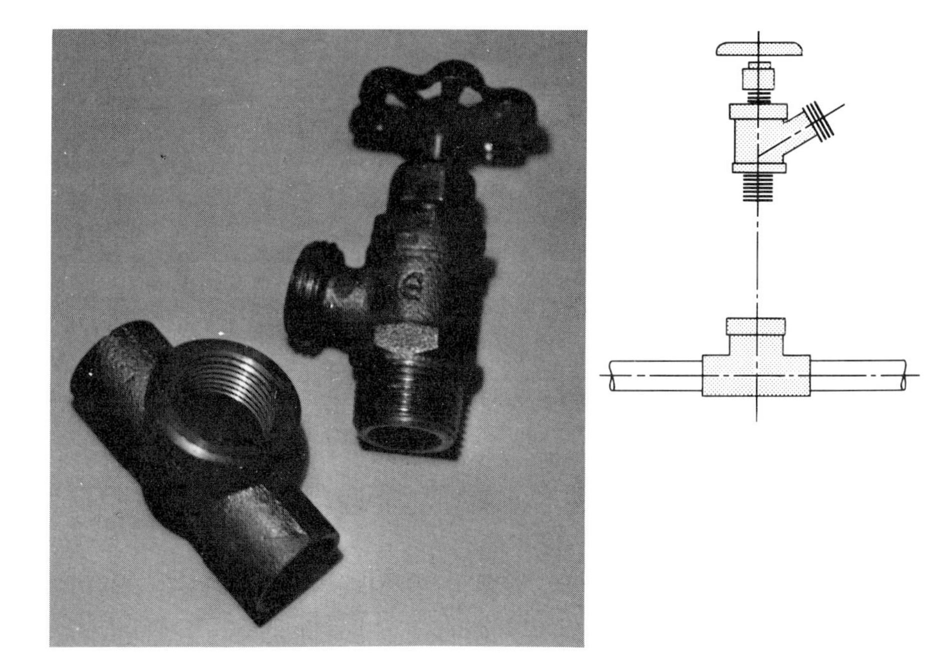

**Figure 5-30.** Boiler drain valve and tee.

**(3) Check Valve (Figures 5-31 and 5-32).** A check valve is designed to permit the flow of liquid in one direction only. They are installed in the various generic types of systems as shown in Chapter 3. The can prevent reverse thermosiphoning of heated water from storage into the collector array. A number of designs are available for horizontal and vertical piping installations; however, horizontal mounting is normally the procedure used. On installation, one must ensure that the arrow indication on the valve is pointed in the desired direction of flow. Check valves are used in drain down systems to direct the water through the collectors properly during the system operation.

**Figure 5-31.** Check valve with male adapters.

**Figure 5-32.** Fill and drain assembly.

The check valve with its male adapters, and boiler drains with tees are known as the "fill and drain assembly", which will be discussed in Chapter 9.

**(4) Gate Valve (Figure 5-33).** In a gate valve, the flow of liquid is controlled by a sliding gate operated by means of a screw spindle to move the gate to the open and closed positions. When opened this valve permits a full and unrestricted passage for fluid because there are not tortuous bends as in the

**Figure 5-33.** Example of gate valves used to isolate a pump for service and repair.

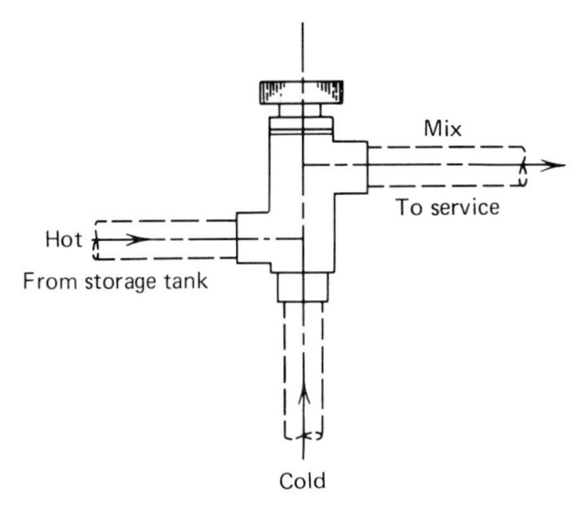


**Figure 5-34.** Mixing valve.

case of a globe-type valve. The gate valve is therefore an excellent valve to use for isolation purposes.

**(5) Mixing Valve (Figure 5-34).** This valve is used to temper the hot water from the storage with cold water. Although the normal household faucet can be used as a mixing valve, it is recommended that one be installed if there are small children in the household. This valve can be preset to 140° F so a person will not be scalded accidentally. The valve should be installed 12 in. below the hot water outlet with the cold water entering from the bottom.

**(6) Pressure Reducing Valve (Figure 5-35).** Pressure reducing valves are often used in conventional water supply systems to reduce incoming water pressure and to prevent damage to some components. These valves are usually installed when the incoming pressure is greater than the working pressure of any component, such as parts of a drain down system. Pressure reducing valves should be preceded by a strainer and isolated by shutoffs for cleaning.

**(7) Pressure Relief Valve (Figure 5-36).** Pressure relief valves are designed to allow transfer fluids to escape from a closed loop prior to the working

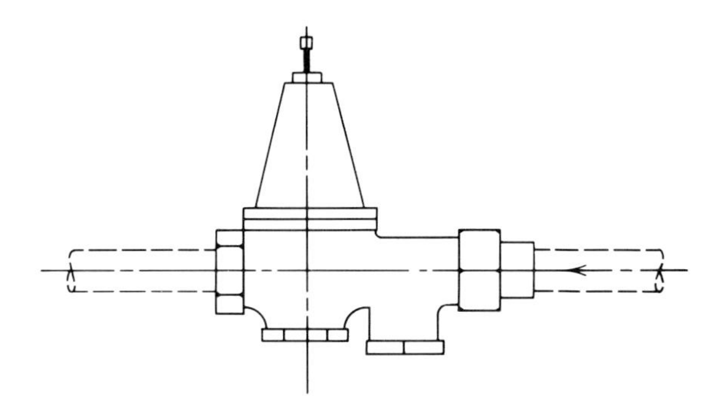

**Figure 5-35.** Pressure reducing valve.

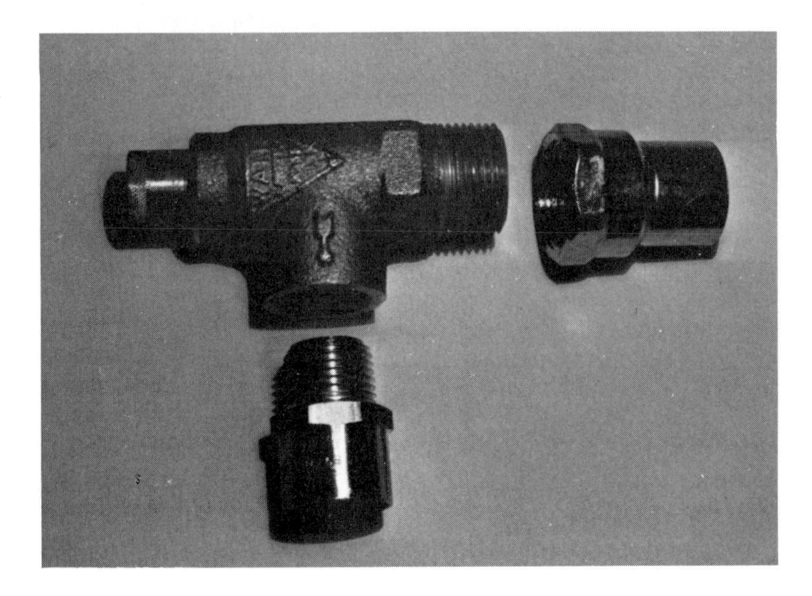

**Figure 5-36.** Adjustable pressure relief valve with male and female adapters.

pressure of the collector plate being exceeded, thus avoiding damage to the collector plate. These relief valves are normally adjusted to 75 psi.

The relief valve can be installed anywhere along the closed loop; however, there should be a relief valve between all closed valves to the collector array. Collectors must not be allowed to be isolated without a relief valve in the line. If the relief valve is set greater than the normal system pressure test prior to fill, it need not be removed from the collector loop. Discharge from a pressure relief valve will be very hot and should be connected, therefore, to a waste drain or a container. The relief valve should never be installed on the roof because its discharge could damage or discolor roofing materials depending on the transfer fluid used.

**(8) Temperature and Pressure Relief Valve (Figure 5-37).** Temperature and pressure relief valves are similar to pressure relief valves, but contain a

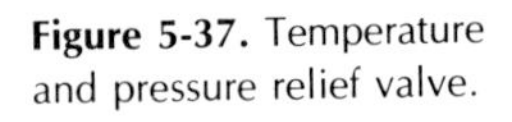

**Figure 5-37.** Temperature and pressure relief valve.

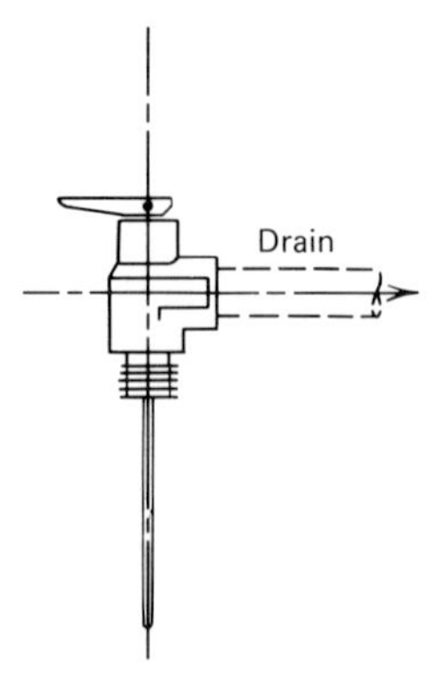

temperature-sensing element at the valve inlet that extends about 6 in. into the top of each storage tank where the hottest water is stored. Valve limits are usually set to 125 psi and 210° F. Ratings should be listed on the valve.

Temperature and pressure relief valves should be connected to within 6 in. of a waste drain or dry well to prevent unexpected discharge from scalding occupants or service personnel. Operating a hot water tank or two tanks in series without a temperature and pressure relief valve is extremely dangerous. Therefore, it is required by most local building codes.

**(9) Solenoid and Motorized Valves (Figure 5-38).** Solenoid valves and motorized valves are electrically operated valves used in drain down systems to start, stop, and divert the flow of heat transfer fluid. When a freezing condition exists or electrical power fails, the valve that allows fluid to circulate in the collector loop closes, and the valve that is used to drain the system opens. This procedure is reversed during normal operating conditions.

**(10) Vacuum Relief Valve (Figure 5-39).** Vacuum relief valves are used in drain down systems to relieve an unwanted vacuum condition. This valve permits the system to drain efficiently using gravity by admitting atmospheric pressure into the return piping. Depending on local codes, it is also

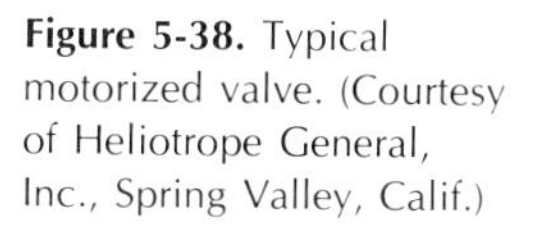

**Figure 5-38.** Typical motorized valve. (Courtesy of Heliotrope General, Inc., Spring Valley, Calif.)

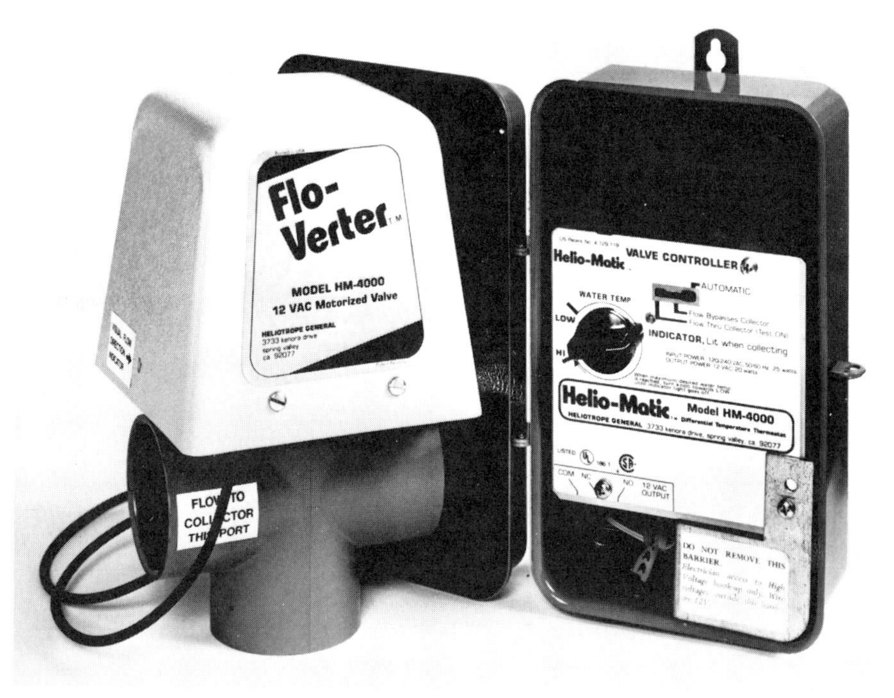

**Figure 5-39.** Vacuum relief valve.

used above the cold water inlet of storage tanks to eliminate any vacuum conditions that could collapse the tanks.

## PUMPS AND BLOWERS

### Description

There are many types and general classes of pumps and blowers, but we shall not attempt to discuss all variations available. In general, pumps and blowers serve the same basic function, which is to move a transfer fluid from one point to another. Pumps typically used in solar domestic hot water systems are the centrifugal type as shown in Figure 5-40.

The ***mechanical seal*** type of centrifugal pump shown in Figure 5-41 has been used in hydronic heating systems for years; however, they are very inefficient because most of the energy is used to turn the shaft against the shaft seal. These seals are also a common source of leaks, and the use of this type of pump should be avoided in the collector loop.

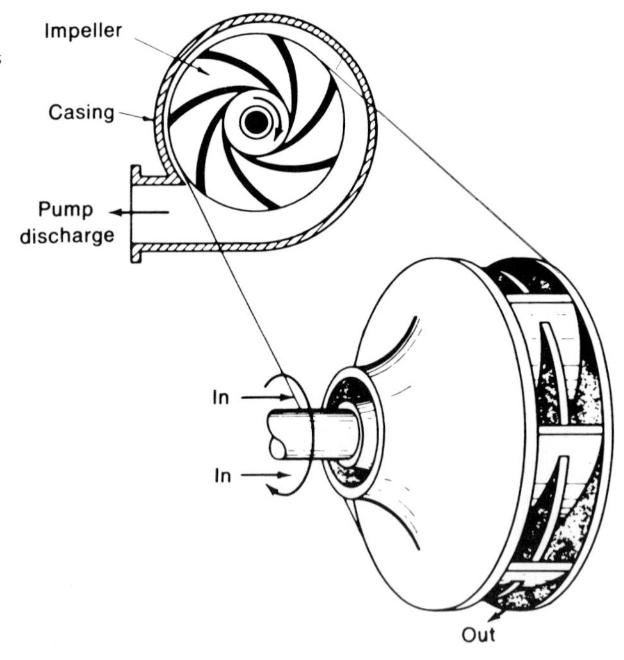


**Figure 5-40.** Centrifugal pump (*Source. Fundamentals of Solar Heating*, D.O.E. 1978)

**Figure 5-41.** Mechanical seal centrifugal pump. (Reprinted with permission from *The Solar Decision Book*, John Wiley & Sons, Inc. © 1979)

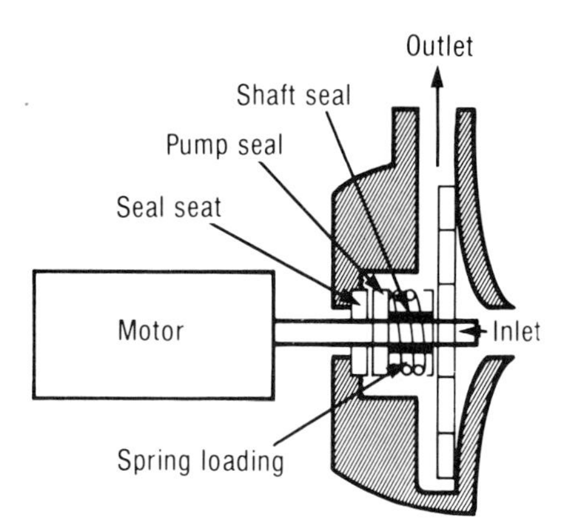


The shaft seal has now been eliminated in the newer versions by use of either a ***magnetic coupling*** or by "***canning***" the motor so its moving parts can be wetted by the circulating fluid without problems. A cross-sectional view of each of these types is illustrated in Figure 5-42.

A centrifugal pump moves a fluid by sucking liquid into the center of a rapidly rotating disc (impeller) which has a series of blades. Creating a high velocity the centrifugal force developed slings the liquid from the tips of the blades through the outlet. The frictional heat between the impeller and liquid is radiated out the pump body. These pumps are nonpriming and as such the liquid supply to the pump must be higher than the inlet of the pump. The material used for the pump housing depends on whether the collector system is open or closed loop. Closed loops should use a circulator with an iron housing whereas open loops must use a pump in which all water-touched parts are manufactured of stainless steel or bronze. Failure to use a stainless steel or bronze body pump in an open loop system with water means an unnecessarily short life for the pump. A typical canned rotor centrifugal pump suitable for solar applications is shown in Figure 5-43.

**Figure 5-42.** Canned rotor and magnetic drive centrifugal pumps. (Reprinted with permission from *The Solar Decision Book*, John Wiley & Sons, Inc., © 1979)

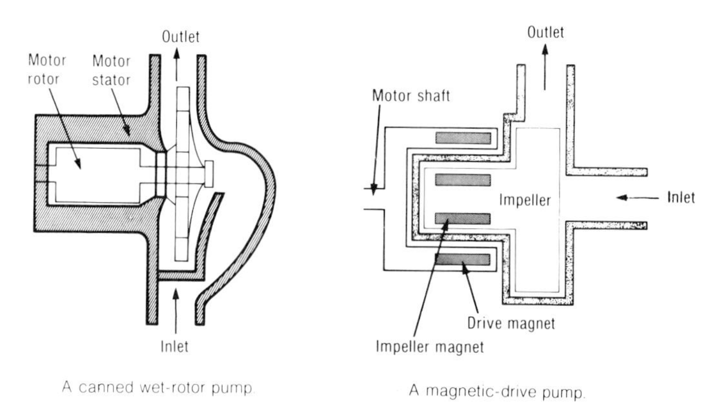

**Figure 5-43.** "Canned" rotor centrifugal pump. (Courtesy of Grundfos Pump Corp., Clovis, Calif.)

Two types of centrifugal blowers are used to circulate air in the air to liquid heat type of solar domestic hot water system. These types are the forward curved or squirrel cage design and backward-curved design. Backward-curved centrifugal blowers are quieter, more efficient, and not as sensitive to underloading or overloading as are the foward-curved fans. Physically larger, they are also more expensive because of close running clearances.

## Design Elements

The pump's capacity to do work is expressed as the amount of fluid that can be moved at a given resistance to flow in a given period of time. Manufacturer's performance curves are plotted as fluid flow (gallons per minute) on the horizontal axis (abscissa) versus head (equivalent resistance to height in feet) on the vertical axis (ordinate). Figure 5-44 depicts a typical pump's work capacity.

The sum of all flow restrictions is called the total head, $H_T$. Total head

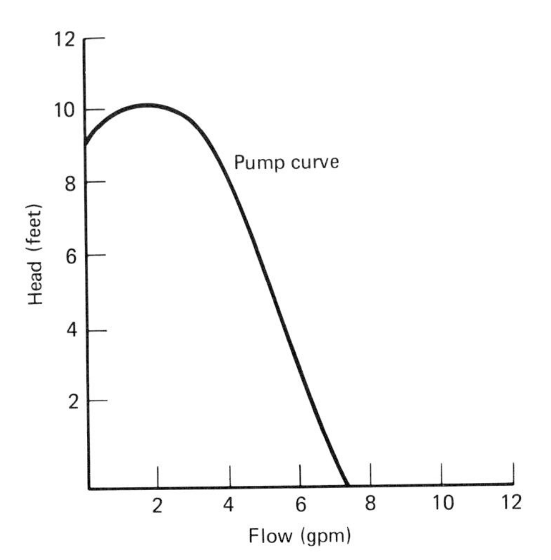


**Figure 5-44.** Performance curve for a hypothetical pump. (*Source*. U.S. Dept. of Housing & Urban Development, *Installation Guidelines for Solar DHW Systems in One and Two Family Dwellings*)

$H_T$ is comprised of (1) static head, $H_S$, (2) friction head, $H_F$, (3) pressure head, $H_P$, and (4) velocity head, $H_V$. Total head is represented as Equation 5-4.

***Equation 5-4***

$$H_T = H_S + H_F + H_P + H_V$$

Velocity head $H_V$ is analogous to velocity pressure in a duct. This value is very small in relation to the other factors that comprise the total head. The pressure head $H_P$ is a measure of the flow restriction due to pressure maintained in the pipe loop. Such a head would be present if it were required to pump water from a lower open tank to a closed tank maintained at a pressure above atmosphere by the use of compressed air. Because neither of these factors is significant to the following discussion, we shall eliminate these terms and rewrite Equation 5-4 as Equation 5-5:

***Equation 5-5***

$$H_T = H_S + H_F$$

***Static head*** $H_S$ is the height that water must be pumped in an open loop system. It occurs when the pump must lift water against the force of gravity. Note that a column of water 2.31 ft high exerts a pressure at its base of 1 lb per square inch (1 psia). Unlike friction head, static head has no relationship to the length of the heat transfer fluid circulation loop. The static head is simply equivalent to the ***difference*** between the level that the water stands in a system and the highest point to which the water must be lifted ***multiplied*** by a factor of 0.43.

For example, if a column of water to a collector array was 25 ft to the highest point, a gauge at the bottom of this vertical piping would indicate (25 ft × 0.43) 10.75 psia. Static head is a factor in an open loop system such as a drain back system because each time the pump starts it must move fluid to the top of the collector array. Static head is *not* a factor in closed loop systems. It is this factor alone which determines that a larger pump of sufficient horsepower be used, rather than a circulator of smaller capacity or a series of circulators.

***Friction head***, $H_F$, is a measure of the restriction of flow caused by the friction of a fluid as it travels through the system. Unlike static head, friction head is related directly to flow rate. Friction head can be determined by adding up the total length of pipe in a system plus the equivalent lengths of pipe for each type of fitting and valve used as summarized in Table 5-10. Once the total equivalent length of pipe has been determined, the friction head $H_F$ can be found once the flow rate is specified.

***Flow rate*** through a solar domestic hot water system is usually between 0.015 and 0.04 gal/min/ft² of effective collector area. (The average flow rate specified by manufacturers is approximately 0.02 gal/min/ft².) This figure is based on establishing a 30° F temperature difference across the collector plate and a maximum energy removal of 225 BTU/ft²-hr. To solve for this flow rate for water we have

$$\text{Flow rate in } \frac{(\text{gallon})}{\text{ft}^2\text{-min}} \times (30^\circ \text{ F}) \times \frac{(1 \text{ BTU})}{\text{lb-}^\circ\text{F}} \times \frac{(8.33 \text{ lb})}{\text{gal}} = \frac{(225 \text{ BTU})}{\text{ft}^2\text{-hr}} \times \frac{(1 \text{ hr})}{60 \text{ min}}$$

And

$$\text{Flow rate} = 0.015 \frac{\text{gal}}{\text{min-ft}^2}$$

**Table 5-10**
**Equivalent Lengths of Pipe for Fittings and Valves**

| | *Equivalent Length of Pipe in Feet* | | | | | |
|---|---|---|---|---|---|---|
| *Fitting Size (in.)* | *90° Ell* | *45° Ell* | *90° Tee Straight Run* | *90° Tee Side Branch* | *Coupling* | *Gate Valve* |
| ⅜ | 0.5 | 0.3 | 0.15 | 0.75 | 0.15 | 0.1 |
| ½ | 1.0 | 0.6 | 0.3 | 1.5 | .03 | 0.2 |
| ¾ | 1.25 | 0.75 | 0.4 | 2.0 | 0.4 | 0.25 |
| 1 | 1.5 | 1.0 | 0.45 | 2.5 | 0.45 | 0.3 |
| 1¼ | 2.0 | 1.2 | 0.6 | 3.0 | 0.6 | 0.4 |

*Source:* Courtesy of Copper Development Association, Inc., Stamford, Conn.

Knowing the specific heat and weight of the fluid, we can now find the idealized flow rate for any transfer fluid.

The friction head can now be found using the flow rate and equivalent length of pipe in conjunction with the nomograph of Figure 5-45.

Let us illustrate this with an example. Assume we have the system shown in Figure 5-46. We shall assume the system is closed loop with an idealized flow rate of 1 gal/min for the transfer fluid used. We shall also assume we have 40 ft of supply and return pipe to storage.

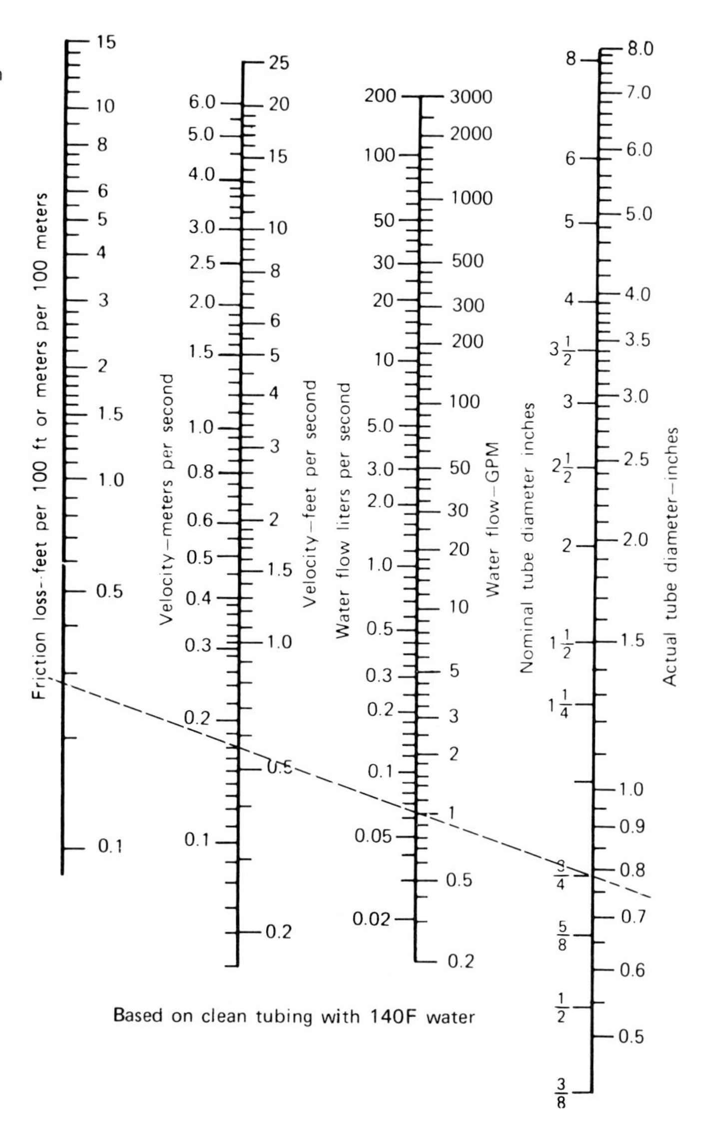


**Figure 5-45.** Friction head loss due to flow of water in type L pipe (Nomograph). (Reprinted with permission from *Solar Thermal Engineering*, John Wiley & Sons, Inc., © 1980)

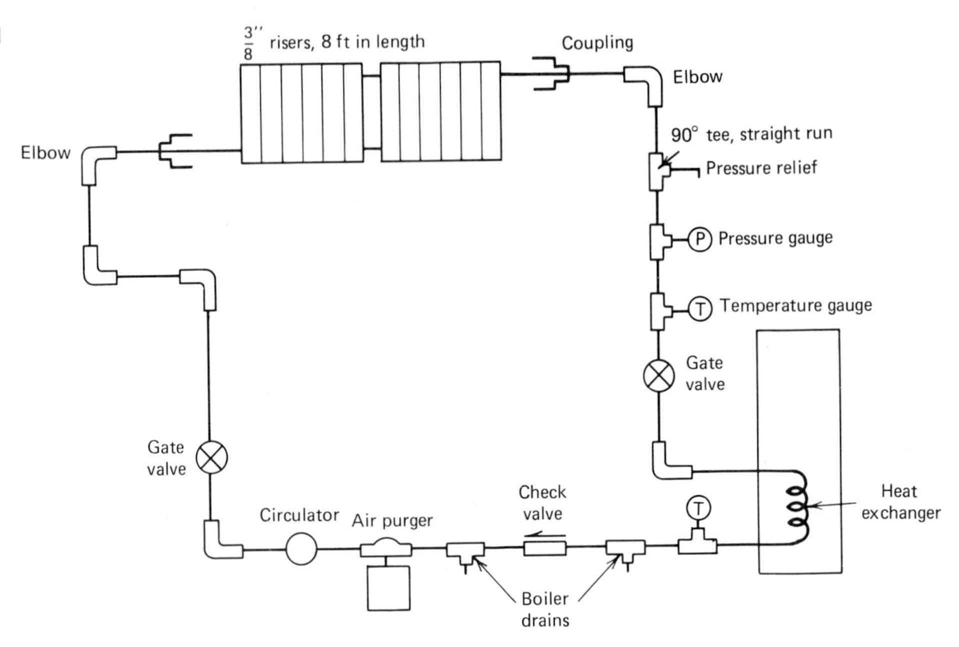


**Figure 5-46.** A typical collector loop.

With the aid of Table 5-10 we find the equivalent length of piping for a typical system illustrated in Figure 5-46.

| | | |
|---|---|---|
| (40 ft) ¾ in pipe | 40.00 ft | |
| (2) Couplings | 0.80 ft | |
| (10) ⅜ in risers | 0.46 ft | (per manufacturer's data) |
| (6) 90° elbows | 7.50 ft | |
| (6) 90° tee, straight run | 2.40 ft | |
| (2) Gate valves | 0.50 ft | |
| (1) Air purger | 7.50 ft | |
| (1) Check valve | 7.50 ft | |
| (1) Heat exchanger | 10.00 ft | (per manufacturer's data) |
| Total equivalent feet = | 76.66 ft | |

From the nomograph of Figure 5-45 we have 0.27 feet of friction head for every 100 ft of piping. Because our total equivalent length of piping is 76.66 ft we have (76.66 ft/100 ft $\times$ 0.27 ft) a friction head of 0.21 ft. Because the static head is zero we have a total head loss of 0.21 ft.

Now that we know how to find the static head and the friction head, we can determine the total head loss at various flow rates and plot a system curve as illustrated in Figure 5-47. The intersection of these two curves will determine the capacity at which the pump will operate.

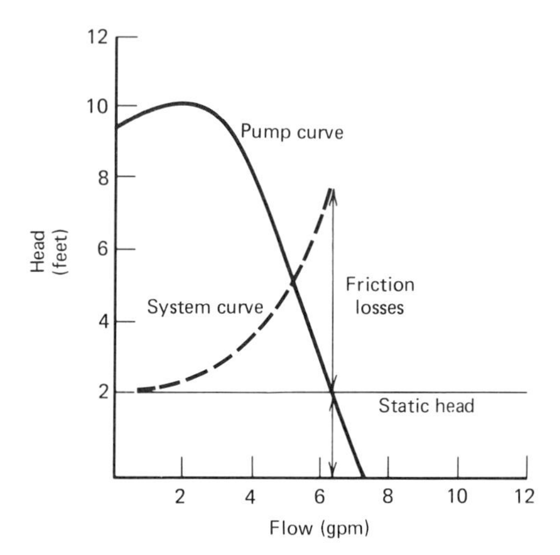


**Figure 5-47.** Pump and system curves.

## STORAGE TANKS

To maximize the use of the available insolation, it is necessary to have a method of energy storage that can retain and release solar—derived energy on demand. Storage tanks can be constructed of glass-lined steel, stone-lined steel, copper, high temperature fiberglass, or polyethylene, and normally are available in standard sizes of 65, 80, 100, and 120 gal. The determination of the size of storage tank needed is, of course, dependent on the amount of hot water consumed, which is discussed in Chapter 6, "Collector Array Sizing and Siting."

Storage tanks that have internal finned copper heat exchangers (illustrated in Figure 5-48) for closed loop system use should be stone-lined. The reason is that when an internal heat exchanger is used with glass-lined steel tanks, the presence of such a large quantity of dissimilar metal accelerates electrolytic corrosion. The magnesium anode normally installed in glass-lined tanks to protect against corrosion will then have a relatively short life, and the tank could fail prematurely. Let us explain this situation.

A metal is more positive (anodic) or more negative (cathodic) than another metal because of its chemical structure. The various metals can be ranked from positive to negative in what is termed a ***galvanic series***. The more positive the material, the more it will corrode. A galvanic series of common metals and alloys is shown in Table 5-11.

To prevent corrosion, a metal that is more anodic than carbon steel is placed in a glass-lined hot water tank when manufactured. Magnesium is the most anodic and is usually chosen. The magnesium ions flow into solution before the carbon steel becomes the anode and starts to corrode, and the life of the hot water tank is thus prolonged.

Stone-lined tanks, on the other hand, depend on an entirely different method of protecting the steel tank walls from the corrosive effect of the hot water. The cement lining, which is approximately ½ in. thick, absorbs water

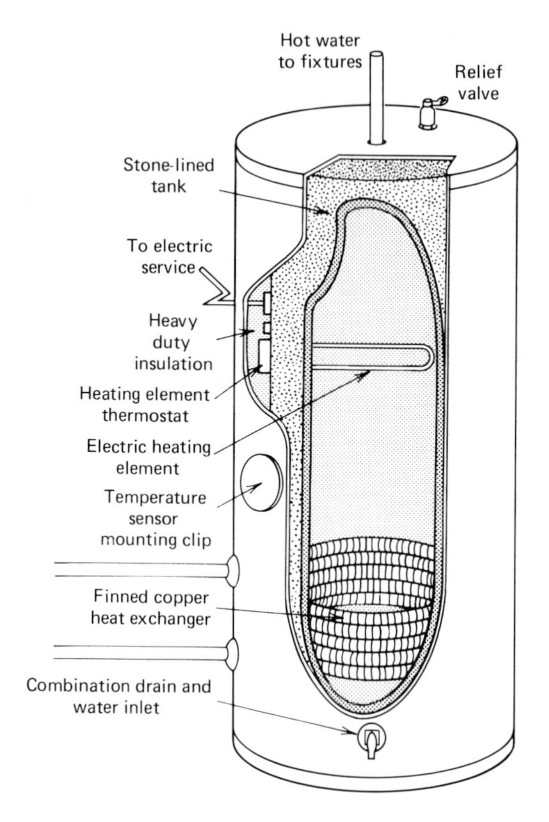


**Figure 5-48.** Stone-lined tank with heat exchanger. (Courtesy of Ford Products Corp., Valley Cottage, N.Y.)

like a sponge during the first filling of the tank. This same layer of water is kept in contact with the tank walls by capillary attraction for the entire life of the tank, and loses its oxygen to become dead, noncorrosive water. Calcium and carbonates are also picked up from the cement, further reducing corrosive effects. Because the tappings of the tank are brass and the lining is neutral, the copper heat exchanger has little effect on tank life. In addition,

**Table 5-11**

| *Corroded End (Anodic)* |
|---|
| Magnesium |
| Zinc |
| Aluminum |
| Carbon Steel |
| Brass |
| Tin |
| Copper |
| Bronze |
| Stainless Steel |
| *Protected End (Cathodic)* |

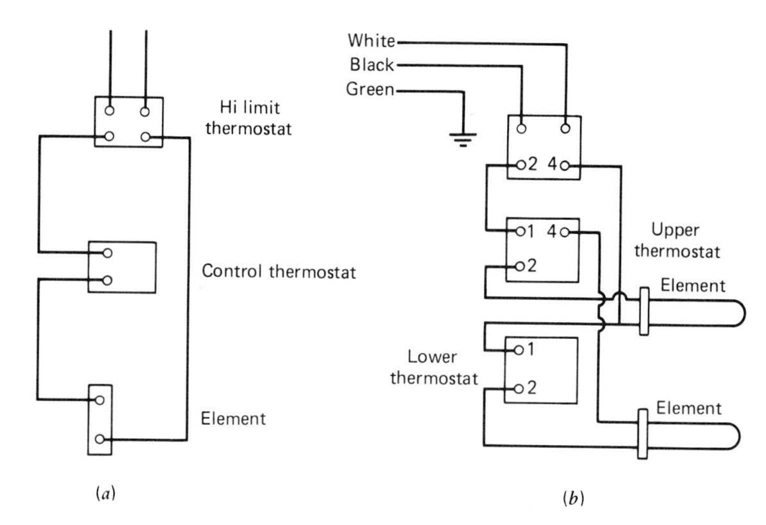


**Figure 5-49.** Typical resistance element wiring diagram. (*a*) Single element. (*b*) Dual element.

the stone lining also serves as an insulator and as a heat sink for heat generated by the collector heat transfer fluid.

The solar storage tank for domestic hot water should ideally be the last tank in line for user consumption. This situation is, of course, dependent on the type of generic system used. Specifically, for closed loop freeze-resistant systems, a single tank system can perform 10 to 20 percent better than systems in which the solar tank is merely a preheater. Such a tank typically has one 4500-W electric element at the top of the tank that is used during periods of noninsolation. Wiring diagrams for the electric backup element(s) are usually provided by the manufacturers as depicted in Figure 5-49.

It is interesting to note the amount of time it takes an electric hot water heater to completely heat a tank from 40 to 140° F with one electric element of 4500 W. One can appreciate the amount of electricity used by observing Table 5-12.

Table 5-12 is derived with the following assumptions:

Temperature difference (140°-40° F) = 100° F
Specific heat of water = 1 BTU/lb-°F
1 kW-hr = 3414 BTU

**Table 5-12**
**Time Required to Electrically Heat Water**

| *Tank Size (gal)* | *Single Heating Element (Watts)* | *BTU Consumed* | *Time to Heat (hr)* |
|---|---|---|---|
| 65 | 4500 | 54,145 | 3.5 |
| 80 | 4500 | 66,640 | 4.3 |
| 100 | 4500 | 83,300 | 5.4 |
| 120 | 4500 | 99,960 | 6.5 |

For example, the total energy consumption needed to heat 65 gal of water from 40° to 140° F is obtained from Equation 5-6.

***Equation 5-6***

$$\text{BTU consumed} = (65\text{ gal})(100°\text{ F})(1\text{ BTU/lb-F°})(8.33\text{ lb/gal}) = 54{,}145\text{ BTU}$$

Because we can use 4.5 kW in the heating element, we can determine the number of BTU that this is equivalent to per hour in Equation 5-7.

***Equation 5-7***

$$\text{BTU/hr} = (4.5\text{ kW})\frac{(3414\text{ BTU})}{\text{kW-hr}} = 15{,}363\text{ BTU/hr}$$

To determine the total time necessary to heat the water to 140° F we need only to divide Equation 5-6 by Equation 5-7.

$$\frac{\text{Equation 5-6}}{\text{Equation 5-7}} = \frac{54{,}145\text{ BTU}}{15{,}363\text{ BTU/hr}}$$

and

$$\text{total hours to heat} = 3.5\text{ hr}$$

Finally, adequate tank insulation should be a minimum of R-11. If the storage tank does not have an R-11 insulation value, it can be achieved by wrapping the tank with an additional insulative jacket either from a commercially available kit or from a Kraft backed or aluminum foil backed fiberglass roll.

Table 5-13 illustrates storage tank heat loss in degrees Fahrenheit per hour and degrees Fahrenheit per day for the four common cylindrical tank sizes. We will assume equal heat loss distribution on all surface areas, a tank temperature of 140° F, an ambient air temperature of 50° F, and a wall

**Table 5-13**
**Storage Tank Heat Loss**

| *Tank Size* | *Weight of Water (lb)* | *Typical Diameter (in.)* | *Typical Height (in.)* | *Total Surface Area (ft²)* | *Total Volume (ft³)* | *Heat Loss (BTU/hr)* | *Temp Loss (°F/hr)* | *Temp Loss (°F/day)* |
|---|---|---|---|---|---|---|---|---|
| 65 | 541.5 | 24 | 60 | 37.7 | 15.7 | 305.4 | .56 | 13.5 |
| 80 | 666.4 | 26 | 60 | 41.4 | 18.4 | 335.3 | .50 | 12.0 |
| 100 | 833.0 | 26 | 69 | 46.5 | 21.2 | 376.7 | .45 | 10.8 |
| 120 | 999.6 | 28 | 69 | 50.7 | 24.6 | 410.7 | .41 | 9.8 |

insulation of R-11. Heat loss $Q$ of a storage tank can be expressed as Equation 5-8 as discussed in Chapter 2.

***Equation 5-8***

$$Q = UA\ \Delta T$$

*where* $R = 11$ or $U = 0.09 = 1/\mathrm{R}$
$A$ = The total surface area of a cylinder as shown in Appendix B
$\Delta T$ = Temperature of water at 140° F minus the ambient air at 50° F

Solving for $Q$ (heat loss) we have

$$Q = (0.09\ \mathrm{BTU/ft^2\text{-}hr\text{-}^\circ F})(37.7\ \mathrm{ft^2})(90^\circ\ \mathrm{F})$$
$$Q = 305.4\ \mathrm{BTU/hr} \quad \text{or} \quad 7328.9,\ \mathrm{BTU/day}$$

Because one gallon of water weighs 8.33 lb, a 65-gal tank would therefore contain 541.5 lb of water. Because it takes one BTU/lb to raise the water temperature 1° F, a loss of 541.5 BTU will lower the tank 1° F. Therefore, a 65-gal tank will lose (7328.9 BTU ÷ 541.5 BTU/°F) 13.5° F per day. It is interesting to note in Table 5-13 that as the volume to surface area ratio is increased, the total temperature drop is decreased.

## REVIEW QUESTIONS

1. Which of the following generic solar hot water systems do not require the use of a differential controller?
   a. Drain down
   b. Phase change
   c. Drain back
   d. Thermosiphon
   e. Closed loop freeze-resistant
2. Which of the following generic solar hot water systems do not require the use of a heat exchanger?
   a. Drain down
   b. Phase change
   c. Drain back
   d. Thermosiphon
   e. Closed loop freeze-resistant
3. Which of the following heat exchanger categories is the most effective type of heat exchanger?
   a. Counterflow
   b. Mixed flow

c. Coil in tank

d. Air-to-liquid

4. The addition of a heat exchanger will compromise an ideal situation whereby the collector inlet temperature should be equal to the lowest temperature in the storage tank.

   True or False

5. What is the maximum amount of heat transferred to storage through a heat exchanger if the transfer fluid is water, the temperature difference from inlet to outlet of the heat exchanger is 40° F, and the flow rate through the heat exchanger is 2 gal/min?

   a. 4,800 BTU/hr

   b. 22,391 BTU/hr

   c. 39,984 BTU/hr

   d. 19,757 BTU/hr

   e. None of the above

6. If the effectiveness of the heat exchanger in Review Question 5 is 0.75, what is the actual amount of heat transferred to storage?

   a. 11,995 BTU/hr

   b. 29,988 BTU/hr

   c. 16,793 BTU/hr

   d. 34,575 BTU/hr

   e. 30,000 BTU/hr

7. If the specific heat $C_p$ of fluid is 0.67 and its weight per unit volume is 12.5 lb/gal, what is the relative amount of heat supplied by one gallon of the fluid at a temperature change of one degree Fahrenheit?

   a. 7.2 BTU

   b. 4.1 BTU

   c. 16.6 BTU

   d. 18.7 BTU

   e. 8.4 BTU

8. Only CPVC pipe is recommended for use in temperatures greater than 160° F but not exceeding 200° F.

   True or False

9. Which one of the following types of rigid copper pipe is the most economical to use in solar hot water systems?

   a. Type J    b. Type K

   c. Type L    d. Type M

   e. None of the above

10. Consider a closed loop system with a 120-gal tank containing a 40 $ft^2$ single wall heat exchanger. If the total supply and return length of pipe is 80 ft, and the flow rate in the closed loop is 2.5 gal/min, what is the recommended pipe size?
    a. ½ in. b. ¾ in.
    c. 1 in. d. 1¼ in.

11. Which component in a closed loop system is used to separate air from the transfer fluid?
    a. Air vent
    b. Air purger
    c. Air flow preventer
    d. Vacuum vent
    e. None of the above

12. Pressurized closed loop systems require the use of an expansion tank to compensate for variation in volume. Nonpressurized open loop systems do not.
    True or False

13. Which is the best reason for not using an elastomeric pipe insulation outdoors between collectors?
    a. Breakdown of pipe insulation from high temperatures
    b. Insufficient insulation value
    c. Ultraviolet degradation
    d. Uneconomical

14. Collector header connections should be soldered with 95/5 tin-lead solder.
    True or False

15. What type of fitting can be used to connect one collector to another without soldering connections?
    a. Unions
    b. Straight couplings
    c. Flared fittings
    d. Straight adapters
    e. Quick disconnects

16. Which type of valve can be used to vary fluid flow without causing unnecessary restrictions?
    a. Gate valve
    b. Check valve
    c. Mixing valve

   d. Ball valve
   e. Globe valve

17. What type of valve permits the flow of fluid in one direction only?
   a. Boiler drain valve
   b. Check valve
   c. Gate valve
   d. Mixing valve
   e. Pressure reducing valve

18. In which generic system is a vacuum relief valve used in the collector to storage loop?
   a. Drain down
   b. Drain back
   c. Phase change
   d. Closed loop freeze-resistant
   e. Thermosiphon

19. Closed loop systems should use a circulator with a bronze or stainless steel housing, and open loop systems should use a circulator with an iron housing.

   True or False

20. A pump's work capacity is graphically depicted by which of the following?
   a. Force versus distance
   b. Head versus pressure
   c. Fluid flow versus pressure
   d. Pressure versus distance
   e. Fluid flow versus head

21. In an open loop system, which components of the total head most likely determine that a larger pump be used instead of a smaller circulator as used in a closed loop system?
   a. Pressure head
   b. Static head
   c. Friction head
   d. Velocity head

22. An open column of water 2.31 ft high exerts a pressure at its base of 1 psia. If an open loop collector array is 30 ft vertical for both the supply and return runs, what is the pressure at the bottom of the vertical piping?
   a. 25.8 psia

b. 11.1 psia
c. 27.6 psia
d. 12.9 psia
e. None of the above

23. If a transfer fluid has a specific heat ($C_p$) of 0.6 and weighs 12 lb/gal, what is the flow rate needed to achieve a minimum energy removal of 250 BTU/ft$^2$-hr with 35° F across the collector plate?
    a. 0.017 gal/min-ft$^2$
    b. 0.015 gal/min-ft$^2$
    c. 0.012 gal/min-ft$^2$
    d. 0.019 gal/min-ft$^2$
    e. none of the above

24. If the total energy consumption to heat an 80 gal hot water tank is 60,000 BTU and the tank has two 4500 W electric elements, what is the total time necessary to heat the water?
    a. 1 hr  b. 2 hr
    c. 3 hr  d. 4 hr
    e. None of the above

25. Assuming an ambient air temperature of 50° F and an equal heat loss distribution on all surface areas, what is the heat loss of an 80 gal storage tank which has a total surface area of 41.4 ft$^2$, a tank temperature of 150° F, and a wall insulation of R-22?
    a. 335.3 BTU/hr
    b. 376.4 BTU/hr
    c. 94.1 BTU/hr
    d. 150.5 BTU/hr
    e. None of the above

# Collector Array Sizing and Siting

Sizing and siting the collector array are actually the first steps in establishing the solar system. We must determine how many collectors are necessary to provide us with an adequate supply of hot water. This is called ***sizing***. We must also determine the correct azimuth and tilt of the collector array, as well as account for possible shading which might result from surrounding obstructions so that we can collect as much solar energy as possible for our region. This is called ***siting***.

## SIZING

First we should establish how much hot water we shall need each day. This depends on the number of people in the family. On the average we can assume 20 gal of water use per person per day for the first two members of a family and an additional 15 gal for every person thereafter. These usage figures will vary from family to family and are dependent on the amount of laundry, dishwashing, and personal hygiene, as depicted in Table 6-1. Using the "20-20-15-15—" supposition, the average family of four would require 70 gal of hot water per day.

**Table 6-1**
**Hot Water Consumption Table**

| | *Gallons per Use Hot Water Required* | |
|---|---|---|
| *Clothes Washing Machine* | *14 lb Machine* | *18 lb Machine* |
| Hot wash/hot rinse | 38 gal | 48 gal |
| Hot wash/warm rinse | 28 gal | 36 gal |
| Hot wash/cold rinse | 19 gal | 24 gal |
| Warm wash/cold rinse | 10 gal | 12 gal |
| *Dishwashing* | *Small* | *Large* |
| Dishwashing machine | 10 gal | 15 gal |
| Sink washing | 4 to 8 gal | |
| *Personal Hygiene* | | |
| Tub bathing | 12 to 30 gal | |
| Wet Shaving/hair washing | 2 to 4 gal | |
| Showering | 2 to 6 gal per minute | |

The daily BTU requirement for domestic hot water heating can be found by multiplying the daily hot water consumption $Wc$ in gallons by 8.33 lb of water per gallon, by the water's specific heat $C_p$ (1 BTU per pound per degree increase in temperature), and by the average temperature increase required in the water. If a dishwasher is used without a preheater, then the storage temperature $T_s$ should be maintained at approximately 135° F. Let us assume that the inlet water temperature $T_i$ from a well or city water, enters storage at 40° F in the winter and 50° in the summer. The average temperature increase, $T_s - T_i$, in a winter condition would be 95° F (135° −40°). If a dishwasher is used with a preheater, then the storage temperature $T_s$ need be maintained at only 120° F. The average temperature, $T_s - T_i$, increase in this circumstance for a winter condition would be 80° F (120° −40°). Equation 6-1 illustrates the calculation of the daily BTU requirement for domestic hot water.

***Equation 6-1***

$$\text{Daily hot water BTU requirement} = (Wc)(T_s - T_i)(C_p)(8.33)$$

*where*

$Wc$ = daily hot water consumed (gallons)
$T_s - T_i$ = temperature increase required (°F)
$C_p$ = specific heat of water (1 BTU/°F-lb)
8.33 = weight of water per gallon (lb/gal)

For example, a typical family of four would require 70 gal of hot water. Assuming this family has a dishwasher with no preheater and the season is winter, the following parameters would exist:

$$C_p = 1 \text{ BTU/lb-°F}$$
$$T_s - T_i = 135° - 40° \text{ F} = 95° \text{ F}$$
$$Wc = 70 \text{ gal}$$

Using Equation 6-1, our daily hot water BTU requirement would therefore be

$$(Wc)(T_s - T_i)(C_p)\ 8.33 = (70 \text{ gals/day})(95° \text{ F})(1 \text{ BTU/lb-°F})(8.33 \text{ lb/gal}) = 55{,}395 \text{ BTU/day}$$

Now that we know how much energy is needed, we should determine how many solar collectors we shall need to maintain our daily hot water BTU requirement. We can determine collector sizing by a simple rule-of-thumb calculation method assuming an average collector system efficiency of 50 percent and using an average horizontal insolation obtained from Figure 6-1 for each particular region. The amount of energy delivered by the collector array can be obtained by Equation 6-2 for this method.

***Equation 6-2***

$$\text{BTU's delivered from collector array} = (n_s)(Ac)(I)$$

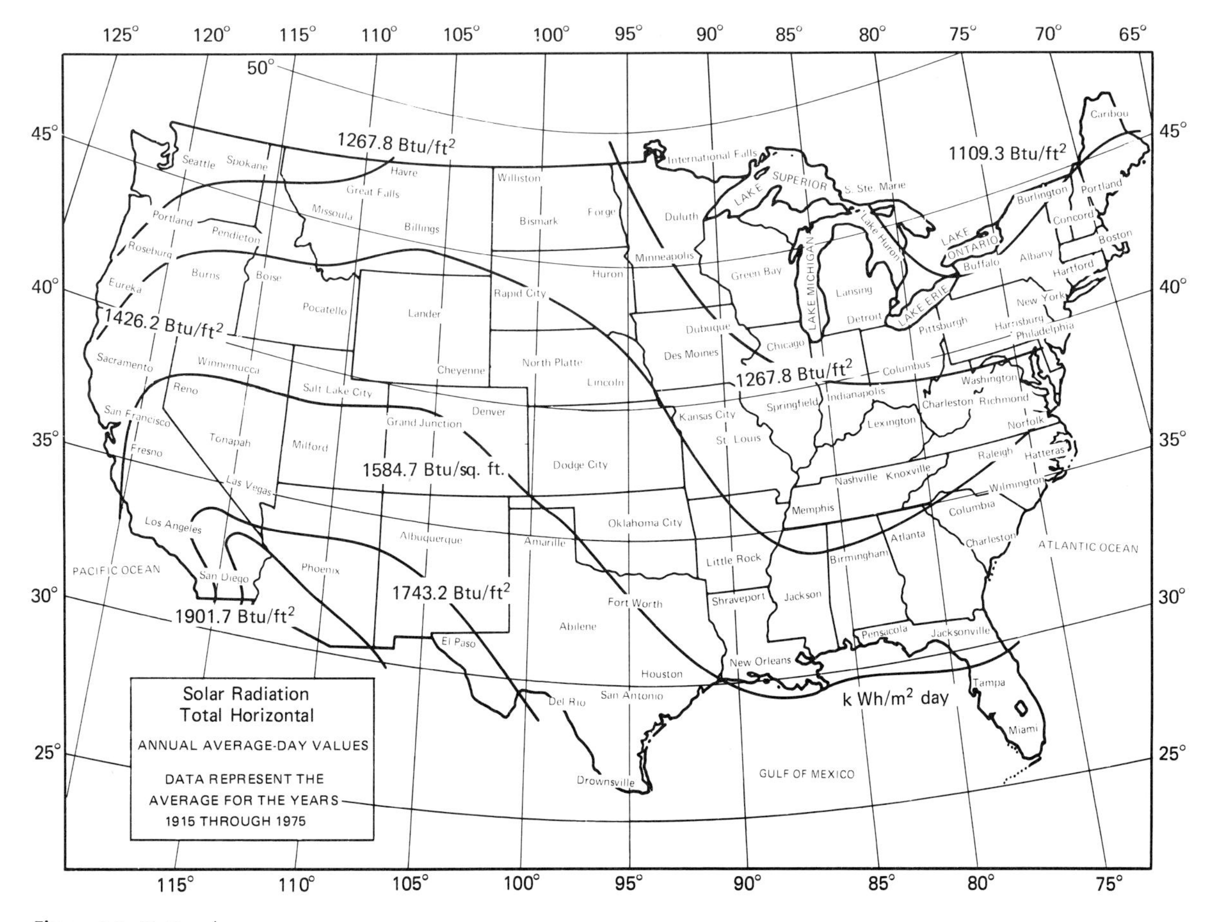


**Figure 6-1.** National insolation data. (Courtesy of Northern Energy Corporation)

*where* $n_s$ = the collector system efficiency (unitless)
$Ac$ = the effective collector area ($ft^2$)
$I$ = the available solar radiation ($BTU/ft^2$-hr)

This method will not give an exact BTU delivered figure because it does not take into account ground reflection, clearness factors, and system losses. These factors will vary in each situation and are not actually necessary in the determination of sizing the normal residential solar domestic hot water collector array.

## Rule of Thumb Sizing Using Total Horizontal Average Insolation

For this approximation method we shall assume an average collector system efficiency $n_s$ of 50 percent. We shall obtain an average solar radiation factor $I$ from Figure 6-1. Now to determine the area of collector needed to meet our daily BTU hot water consumption, we set Equation 6-1 to equal Equation 6-2 and algebraically solve for the area $Ac$ needed as shown:

***Equation 6-1 = Equation 6-2***

$$(Wc)(T_s - T_i)(C_p)(8.33) = (n_s)(Ac)(I)$$

Let us illustrate this with an example. Using our typical family of four, we found the average BTU hot water consumption was 55,395 BTU/day. Assuming they live in Billings, Montana, we find from Figure 6-1 an average insolation value ($I$) of 1426.2 BTU/ft²-day. Therefore,

$$(Wc)(T_s - T_i)(C_p)\ 8.33 = 55{,}395 = (n_s)(Ac)(I)$$

*where* $I = 1426.2$ BTU/ft²
$n_s = 0.5$

Solving for the collector area ($Ac$) we have

$$Ac = \frac{55{,}395 \text{ BTU}}{(0.5)(1426.2 \text{ BTU/ft}^2)}$$

and

$$Ac = 77.7 \text{ ft}^2$$

The collector area should be 77.7 ft². If a manufacturer's solar aperture or effective collector area is 22.2 ft² (i.e., 3 ft × 8 ft panel) per collector, then we would need 3.5 collectors. This, of course, would round up to 4 collectors. We should take note that the solar radiation ($I$) number obtained from Figure 6-1 is for ***average total horizontal***. In no case should the collectors be in a horizontal or 0 deg tilt. This value of insolation is low in comparison with the actual value at the correct tilt angle; however, we have used this value to obtain our first approximation of collector sizing and to illustrate the importance of the proper tilt angle which we shall discuss later in this chapter.

## Rule of Thumb Sizing Using Tilt Angle at Latitude

We can refine our collector sizing by averaging the surface daily totals in Table 6-2 with the collector tilt angle equal to latitude to obtain our average insolation received.

For Billings, Montana, we would use the closest latitude to 46 deg obtained from Table 6-2 which is 48 deg. The average of the surface daily totals in this case is 1955 BTU/ft². Refining our results, we have

$$(Wc)(T_s - T_i)(C_p)\ 8.33 = 55{,}395 = (n_s)(Ac)(I)$$

*where* $I = 1955$ BTU/ft²
$n_s = 0.5$

Solving for the collector area $Ac$, we have

Table 6-2
Solar Positions and Insolation Values for Various Latitudes

## Solar Positions and Insolation Values for Various Latitudes[a]

### 24 Degrees North Latitude

| DATE | SOLAR TIME AM | SOLAR TIME PM | SOLAR POSITION ALT | SOLAR POSITION AZM | BTUH/SQ. FT. TOTAL INSOLATION ON SURFACES NORMAL | HORIZ. | SOUTH FACING SURFACE ANGLE WITH HORIZ. 14 | 24 | 34 | 54 | 90 |
|---|---|---|---|---|---|---|---|---|---|---|---|
| JAN 21 | 7 | 5 | 4.8 | 65.6 | 71 | 10 | 17 | 21 | 25 | 28 | 31 |
| | 8 | 4 | 16.9 | 58.3 | 239 | 83 | 110 | 126 | 137 | 145 | 127 |
| | 9 | 3 | 27.9 | 48.8 | 288 | 151 | 188 | 207 | 221 | 228 | 176 |
| | 10 | 2 | 37.2 | 36.1 | 308 | 204 | 246 | 268 | 282 | 287 | 207 |
| | 11 | 1 | 43.6 | 19.6 | 317 | 237 | 283 | 306 | 319 | 324 | 226 |
| | 12 | | 46.0 | 0.0 | 320 | 249 | 296 | 319 | 332 | 336 | 232 |
| | SURFACE DAILY TOTALS | | | | 2766 | 1622 | 1984 | 2174 | 2300 | 2360 | 1766 |
| FEB 21 | 7 | 5 | 9.3 | 74.6 | 158 | 35 | 44 | 49 | 53 | 56 | 46 |
| | 8 | 4 | 22.3 | 67.2 | 263 | 116 | 135 | 145 | 150 | 151 | 102 |
| | 9 | 3 | 34.4 | 57.6 | 298 | 187 | 213 | 225 | 230 | 228 | 141 |
| | 10 | 2 | 45.1 | 44.2 | 314 | 241 | 273 | 286 | 291 | 287 | 168 |
| | 11 | 1 | 53.0 | 25.0 | 321 | 276 | 310 | 324 | 328 | 323 | 185 |
| | 12 | | 56.0 | 0.0 | 324 | 288 | 323 | 337 | 341 | 335 | 191 |
| | SURFACE DAILY TOTALS | | | | 3036 | 1998 | 2276 | 2396 | 2446 | 2424 | 1476 |
| MAR 21 | 7 | 5 | 13.7 | 83.8 | 194 | 60 | 63 | 64 | 62 | 59 | 27 |
| | 8 | 4 | 27.2 | 76.8 | 267 | 141 | 150 | 152 | 149 | 142 | 64 |
| | 9 | 3 | 40.2 | 67.9 | 295 | 212 | 226 | 229 | 225 | 214 | 95 |
| | 10 | 2 | 52.3 | 54.8 | 309 | 266 | 285 | 288 | 283 | 270 | 120 |
| | 11 | 1 | 61.9 | 33.4 | 315 | 300 | 322 | 326 | 320 | 305 | 135 |
| | 12 | | 66.0 | 0.0 | 317 | 312 | 334 | 339 | 333 | 317 | 140 |
| | SURFACE DAILY TOTALS | | | | 3078 | 2270 | 2428 | 2456 | 2412 | 2298 | 1022 |
| APR 21 | 6 | 6 | 4.7 | 100.6 | 40 | 7 | 5 | 4 | 4 | 3 | 2 |
| | 7 | 5 | 18.3 | 94.9 | 203 | 83 | 77 | 70 | 62 | 51 | 10 |
| | 8 | 4 | 32.0 | 89.0 | 256 | 160 | 157 | 149 | 137 | 122 | 16 |
| | 9 | 3 | 45.6 | 81.9 | 280 | 227 | 227 | 220 | 206 | 186 | 41 |
| | 10 | 2 | 59.0 | 71.8 | 292 | 278 | 282 | 275 | 259 | 237 | 61 |
| | 11 | 1 | 71.1 | 51.6 | 298 | 310 | 316 | 309 | 293 | 269 | 74 |
| | 12 | | 77.6 | 0.0 | 299 | 321 | 328 | 321 | 305 | 280 | 79 |
| | SURFACE DAILY TOTALS | | | | 3036 | 2454 | 2458 | 2374 | 2228 | 2016 | 488 |
| MAY 21 | 6 | 6 | 8.0 | 108.4 | 86 | 22 | 15 | 10 | 9 | 9 | 5 |
| | 7 | 5 | 21.2 | 103.2 | 203 | 98 | 85 | 73 | 59 | 44 | 12 |
| | 8 | 4 | 34.6 | 98.5 | 248 | 171 | 159 | 145 | 127 | 106 | 15 |
| | 9 | 3 | 48.3 | 93.6 | 269 | 233 | 224 | 210 | 190 | 165 | 16 |
| | 10 | 2 | 62.0 | 87.7 | 280 | 281 | 275 | 261 | 239 | 211 | 22 |
| | 11 | 1 | 75.5 | 76.9 | 286 | 311 | 307 | 293 | 270 | 240 | 34 |
| | 12 | | 86.0 | 0.0 | 288 | 322 | 317 | 304 | 281 | 250 | 37 |
| | SURFACE DAILY TOTALS | | | | 3032 | 2556 | 2447 | 2286 | 2072 | 1800 | 246 |
| JUN 21 | 6 | 6 | 9.3 | 111.6 | 97 | 29 | 20 | 12 | 12 | 11 | 7 |
| | 7 | 5 | 22.3 | 106.8 | 201 | 103 | 87 | 73 | 58 | 41 | 13 |
| | 8 | 4 | 35.5 | 102.6 | 242 | 173 | 158 | 142 | 122 | 99 | 16 |
| | 9 | 3 | 49.0 | 98.7 | 263 | 234 | 221 | 204 | 182 | 155 | 18 |
| | 10 | 2 | 62.6 | 95.0 | 274 | 280 | 269 | 253 | 229 | 199 | 18 |
| | 11 | 1 | 76.3 | 90.8 | 279 | 309 | 300 | 283 | 259 | 227 | 19 |
| | 12 | | 89.4 | 0.0 | 281 | 319 | 310 | 294 | 269 | 236 | 22 |
| | SURFACE DAILY TOTALS | | | | 2994 | 2574 | 2422 | 2230 | 1992 | 1700 | 204 |
| JUL 21 | 6 | 6 | 8.2 | 109.0 | 81 | 23 | 16 | 11 | 10 | 9 | 6 |
| | 7 | 5 | 21.4 | 103.8 | 195 | 98 | 85 | 73 | 59 | 44 | 13 |
| | 8 | 4 | 34.8 | 99.2 | 239 | 169 | 157 | 143 | 125 | 104 | 16 |
| | 9 | 3 | 48.4 | 94.5 | 261 | 231 | 221 | 207 | 187 | 161 | 18 |
| | 10 | 2 | 62.1 | 89.0 | 272 | 278 | 270 | 256 | 235 | 206 | 21 |
| | 11 | 1 | 75.7 | 79.2 | 278 | 307 | 302 | 287 | 265 | 235 | 32 |
| | 12 | | 86.6 | 0.0 | 280 | 317 | 312 | 298 | 275 | 245 | 36 |
| | SURFACE DAILY TOTALS | | | | 2932 | 2526 | 2412 | 2250 | 2036 | 1766 | 246 |
| AUG 21 | 6 | 6 | 5.0 | 101.3 | 35 | 7 | 5 | 4 | 4 | 4 | 2 |
| | 7 | 5 | 18.5 | 95.6 | 186 | 82 | 76 | 69 | 60 | 50 | 11 |
| | 8 | 4 | 32.2 | 89.7 | 241 | 158 | 154 | 146 | 134 | 118 | 16 |
| | 9 | 3 | 45.9 | 82.9 | 265 | 223 | 222 | 214 | 200 | 181 | 39 |
| | 10 | 2 | 59.3 | 73.0 | 278 | 273 | 275 | 268 | 252 | 230 | 58 |
| | 11 | 1 | 71.6 | 53.2 | 284 | 304 | 309 | 301 | 285 | 261 | 71 |
| | 12 | | 78.3 | 0.0 | 286 | 315 | 320 | 313 | 296 | 272 | 75 |
| | SURFACE DAILY TOTALS | | | | 2864 | 2408 | 2402 | 2316 | 2168 | 1958 | 470 |
| SEP 21 | 7 | 5 | 13.7 | 83.8 | 173 | 57 | 60 | 60 | 59 | 56 | 26 |
| | 8 | 4 | 27.2 | 76.8 | 248 | 136 | 144 | 146 | 143 | 136 | 62 |
| | 9 | 3 | 40.2 | 67.9 | 278 | 205 | 218 | 221 | 217 | 206 | 93 |
| | 10 | 2 | 52.3 | 54.8 | 292 | 258 | 275 | 278 | 273 | 261 | 116 |
| | 11 | 1 | 61.9 | 33.4 | 299 | 291 | 311 | 315 | 309 | 295 | 131 |
| | 12 | | 66.0 | 0.0 | 301 | 302 | 323 | 327 | 321 | 306 | 136 |
| | SURFACE DAILY TOTALS | | | | 2878 | 2194 | 2342 | 2366 | 2322 | 2212 | 992 |
| OCT 21 | 7 | 5 | 9.1 | 74.1 | 138 | 32 | 40 | 45 | 48 | 50 | 42 |
| | 8 | 4 | 22.0 | 66.7 | 247 | 111 | 129 | 139 | 144 | 145 | 99 |
| | 9 | 3 | 34.1 | 57.1 | 284 | 180 | 206 | 217 | 223 | 221 | 138 |
| | 10 | 2 | 44.7 | 43.8 | 301 | 234 | 265 | 277 | 282 | 279 | 165 |
| | 11 | 1 | 52.5 | 24.7 | 309 | 268 | 301 | 315 | 319 | 314 | 182 |
| | 12 | | 55.5 | 0.0 | 311 | 279 | 314 | 328 | 332 | 327 | 188 |
| | SURFACE DAILY TOTALS | | | | 2868 | 1928 | 2198 | 2314 | 2364 | 2346 | 1442 |
| NOV 21 | 7 | 5 | 4.9 | 65.8 | 67 | 10 | 16 | 20 | 24 | 27 | 29 |
| | 8 | 4 | 17.0 | 58.4 | 232 | 82 | 108 | 123 | 135 | 142 | 124 |
| | 9 | 3 | 28.0 | 48.9 | 282 | 150 | 186 | 205 | 217 | 224 | 172 |
| | 10 | 2 | 37.3 | 36.3 | 303 | 203 | 244 | 265 | 278 | 283 | 204 |
| | 11 | 1 | 43.8 | 19.7 | 312 | 236 | 280 | 302 | 316 | 320 | 222 |
| | 12 | | 46.2 | 0.0 | 315 | 247 | 293 | 315 | 328 | 332 | 228 |
| | SURFACE DAILY TOTALS | | | | 2706 | 1610 | 1962 | 2146 | 2268 | 2324 | 1730 |
| DEC 21 | 7 | 5 | 3.2 | 62.6 | 30 | 3 | 7 | 9 | 11 | 12 | 14 |
| | 8 | 4 | 14.9 | 55.3 | 225 | 71 | 99 | 116 | 129 | 139 | 130 |
| | 9 | 3 | 25.5 | 46.0 | 281 | 137 | 176 | 198 | 214 | 223 | 184 |
| | 10 | 2 | 34.3 | 33.7 | 304 | 189 | 234 | 258 | 275 | 283 | 217 |
| | 11 | 1 | 40.4 | 18.2 | 314 | 221 | 270 | 295 | 312 | 320 | 236 |
| | 12 | | 42.6 | 0.0 | 317 | 232 | 282 | 308 | 325 | 332 | 243 |
| | SURFACE DAILY TOTALS | | | | 2624 | 1474 | 1852 | 2058 | 2204 | 2286 | 1808 |

**Notes:** a. From ASHRAE *Transactions*; ground reflection not included.

Sun
90°
0°
Tilt Angle 0°

90°
45°
0°
Tilt Angle 45°

90°
Lat. ± 10°
0°
Optimum Tilt Angle

(Table 6-2, continued)



## 32 Degrees North Latitude

| DATE | SOLAR TIME AM | SOLAR TIME PM | SOLAR POSITION ALT | SOLAR POSITION AZM | BTUH/SQ. FT. TOTAL INSOLATION ON SURFACES NORMAL | HORIZ. | SOUTH FACING SURFACE ANGLE WITH HORIZ. 22 | 32 | 42 | 52 | 90 |
|---|---|---|---|---|---|---|---|---|---|---|---|
| JAN 21 | 7 | 5 | 1.4 | 65.2 | 1 | 0 | 0 | 0 | 0 | 1 | 1 |
| | 8 | 4 | 12.5 | 56.5 | 203 | 56 | 93 | 106 | 116 | 123 | 115 |
| | 9 | 3 | 22.5 | 46.0 | 269 | 118 | 175 | 193 | 206 | 212 | 181 |
| | 10 | 2 | 30.6 | 33.1 | 295 | 167 | 235 | 256 | 269 | 274 | 221 |
| | 11 | 1 | 36.1 | 17.5 | 306 | 198 | 273 | 295 | 308 | 312 | 245 |
| | 12 | | 38.0 | 0.0 | 310 | 209 | 285 | 308 | 321 | 324 | 253 |
| | SURFACE DAILY TOTALS | | | | 2458 | 1288 | 1839 | 2008 | 2118 | 2166 | 1779 |
| FEB 21 | 7 | 5 | 7.1 | 73.5 | 121 | 22 | 34 | 37 | 40 | 42 | 38 |
| | 8 | 4 | 19.0 | 64.4 | 247 | 95 | 127 | 136 | 140 | 141 | 108 |
| | 9 | 3 | 29.9 | 53.4 | 288 | 161 | 206 | 217 | 222 | 220 | 158 |
| | 10 | 2 | 39.1 | 39.4 | 306 | 212 | 266 | 278 | 283 | 279 | 193 |
| | 11 | 1 | 45.6 | 21.4 | 315 | 244 | 304 | 317 | 321 | 315 | 214 |
| | 12 | | 48.0 | 0.0 | 317 | 255 | 316 | 330 | 334 | 328 | 222 |
| | SURFACE DAILY TOTALS | | | | 2872 | 1724 | 2188 | 2300 | 2345 | 2322 | 1644 |
| MAR 21 | 7 | 5 | 12.7 | 81.9 | 185 | 54 | 60 | 60 | 59 | 56 | 32 |
| | 8 | 4 | 25.1 | 73.0 | 260 | 129 | 146 | 147 | 144 | 137 | 78 |
| | 9 | 3 | 36.8 | 62.1 | 290 | 194 | 222 | 224 | 220 | 209 | 119 |
| | 10 | 2 | 47.3 | 47.5 | 304 | 245 | 280 | 283 | 278 | 265 | 150 |
| | 11 | 1 | 55.0 | 26.8 | 311 | 277 | 317 | 321 | 315 | 300 | 170 |
| | 12 | | 58.0 | 0.0 | 313 | 287 | 329 | 333 | 327 | 312 | 177 |
| | SURFACE DAILY TOTALS | | | | 3012 | 2084 | 2378 | 2403 | 2358 | 2246 | 1276 |
| APR 21 | 6 | 6 | 6.1 | 99.9 | 66 | 14 | 9 | 6 | 6 | 5 | 3 |
| | 7 | 5 | 18.8 | 92.2 | 206 | 86 | 78 | 71 | 62 | 51 | 10 |
| | 8 | 4 | 31.5 | 84.0 | 255 | 158 | 156 | 148 | 136 | 120 | 35 |
| | 9 | 3 | 43.9 | 74.2 | 278 | 220 | 225 | 217 | 203 | 183 | 68 |
| | 10 | 2 | 55.7 | 60.3 | 290 | 267 | 279 | 272 | 256 | 234 | 95 |
| | 11 | 1 | 65.4 | 37.5 | 295 | 297 | 313 | 306 | 290 | 265 | 112 |
| | 12 | | 69.6 | 0.0 | 297 | 307 | 325 | 318 | 301 | 276 | 118 |
| | SURFACE DAILY TOTALS | | | | 3076 | 2390 | 2444 | 2356 | 2206 | 1994 | 764 |
| MAY 21 | 6 | 6 | 10.4 | 107.2 | 119 | 36 | 21 | 13 | 13 | 12 | 7 |
| | 7 | 5 | 22.8 | 100.1 | 211 | 107 | 88 | 75 | 60 | 44 | 13 |
| | 8 | 4 | 35.4 | 92.9 | 250 | 175 | 159 | 145 | 127 | 105 | 15 |
| | 9 | 3 | 48.1 | 84.7 | 269 | 233 | 223 | 209 | 188 | 163 | 33 |
| | 10 | 2 | 60.6 | 73.3 | 280 | 277 | 273 | 259 | 237 | 208 | 56 |
| | 11 | 1 | 72.0 | 51.9 | 285 | 305 | 305 | 290 | 268 | 237 | 72 |
| | 12 | | 78.0 | 0.0 | 286 | 315 | 315 | 301 | 278 | 247 | 77 |
| | SURFACE DAILY TOTALS | | | | 3112 | 2582 | 2454 | 2284 | 2064 | 1788 | 469 |
| JUN 21 | 6 | 6 | 12.2 | 110.2 | 131 | 45 | 26 | 16 | 15 | 14 | 9 |
| | 7 | 5 | 24.3 | 103.4 | 210 | 115 | 91 | 76 | 59 | 41 | 14 |
| | 8 | 4 | 36.9 | 96.8 | 245 | 180 | 159 | 143 | 122 | 99 | 16 |
| | 9 | 3 | 49.6 | 89.4 | 264 | 236 | 221 | 204 | 181 | 153 | 19 |
| | 10 | 2 | 62.2 | 79.7 | 274 | 279 | 268 | 251 | 227 | 197 | 41 |
| | 11 | 1 | 74.2 | 60.9 | 279 | 306 | 299 | 282 | 257 | 224 | 56 |
| | 12 | | 81.5 | 0.0 | 280 | 315 | 309 | 292 | 267 | 234 | 60 |
| | SURFACE DAILY TOTALS | | | | 3084 | 2634 | 2436 | 2234 | 1990 | 1690 | 370 |
| JUL 21 | 6 | 6 | 10.7 | 107.7 | 113 | 37 | 22 | 14 | 13 | 12 | 8 |
| | 7 | 5 | 23.1 | 100.6 | 203 | 107 | 87 | 75 | 60 | 44 | 14 |
| | 8 | 4 | 35.7 | 93.6 | 241 | 174 | 158 | 143 | 125 | 104 | 16 |
| | 9 | 3 | 48.4 | 85.5 | 261 | 231 | 220 | 205 | 185 | 159 | 31 |
| | 10 | 2 | 60.9 | 74.3 | 271 | 274 | 269 | 254 | 232 | 204 | 54 |
| | 11 | 1 | 72.4 | 53.3 | 277 | 302 | 300 | 285 | 262 | 232 | 69 |
| | 12 | | 78.6 | 0.0 | 279 | 311 | 310 | 296 | 273 | 242 | 74 |
| | SURFACE DAILY TOTALS | | | | 3012 | 2558 | 2422 | 2250 | 2030 | 1754 | 458 |

| DATE | SOLAR TIME AM | SOLAR TIME PM | SOLAR POSITION ALT | SOLAR POSITION AZM | BTUH/SQ. FT. TOTAL INSOLATION ON SURFACES NORMAL | HORIZ. | SOUTH FACING SURFACE ANGLE WITH HORIZ. 22 | 32 | 42 | 52 | 90 |
|---|---|---|---|---|---|---|---|---|---|---|---|
| AUG 21 | 6 | 6 | 6.5 | 100.5 | 59 | 14 | 9 | 7 | 6 | 6 | 4 |
| | 7 | 5 | 19.1 | 92.8 | 190 | 85 | 77 | 69 | 60 | 50 | 12 |
| | 8 | 4 | 31.8 | 84.7 | 240 | 156 | 152 | 144 | 132 | 116 | 33 |
| | 9 | 3 | 44.3 | 75.0 | 263 | 216 | 220 | 212 | 197 | 178 | 65 |
| | 10 | 2 | 56.1 | 61.3 | 276 | 262 | 272 | 264 | 249 | 226 | 91 |
| | 11 | 1 | 66.0 | 38.4 | 282 | 292 | 305 | 298 | 281 | 257 | 107 |
| | 12 | | 70.3 | 0.0 | 284 | 302 | 317 | 309 | 292 | 268 | 113 |
| | SURFACE DAILY TOTALS | | | | 2902 | 2352 | 2388 | 2296 | 2144 | 1934 | 736 |
| SEP 21 | 7 | 5 | 12.7 | 81.9 | 163 | 51 | 56 | 56 | 55 | 52 | 30 |
| | 8 | 4 | 25.1 | 73.0 | 240 | 124 | 140 | 141 | 138 | 131 | 75 |
| | 9 | 3 | 36.8 | 62.1 | 272 | 188 | 213 | 215 | 211 | 201 | 114 |
| | 10 | 2 | 47.3 | 47.5 | 287 | 237 | 270 | 273 | 268 | 255 | 145 |
| | 11 | 1 | 55.0 | 26.8 | 294 | 268 | 306 | 309 | 303 | 289 | 164 |
| | 12 | | 58.0 | 0.0 | 296 | 278 | 318 | 321 | 315 | 300 | 171 |
| | SURFACE DAILY TOTALS | | | | 2808 | 2014 | 2288 | 2308 | 2264 | 2154 | 1226 |
| OCT 21 | 7 | 5 | 6.8 | 73.1 | 99 | 19 | 29 | 32 | 34 | 36 | 32 |
| | 8 | 4 | 18.7 | 64.0 | 229 | 90 | 120 | 128 | 133 | 134 | 104 |
| | 9 | 3 | 29.5 | 53.0 | 273 | 155 | 198 | 208 | 213 | 212 | 153 |
| | 10 | 2 | 38.7 | 39.1 | 293 | 204 | 257 | 269 | 273 | 270 | 188 |
| | 11 | 1 | 45.1 | 21.1 | 302 | 236 | 294 | 307 | 311 | 306 | 209 |
| | 12 | | 47.5 | 0.0 | 304 | 247 | 306 | 320 | 324 | 318 | 217 |
| | SURFACE DAILY TOTALS | | | | 2696 | 1654 | 2100 | 2208 | 2252 | 2232 | 1588 |
| NOV 21 | 7 | 5 | 1.5 | 65.4 | 2 | 0 | 0 | 0 | 1 | 1 | 1 |
| | 8 | 4 | 12.7 | 56.6 | 196 | 55 | 91 | 104 | 113 | 119 | 111 |
| | 9 | 3 | 22.6 | 46.1 | 263 | 118 | 173 | 190 | 202 | 208 | 176 |
| | 10 | 2 | 30.8 | 33.2 | 289 | 166 | 233 | 252 | 265 | 270 | 217 |
| | 11 | 1 | 36.2 | 17.6 | 301 | 197 | 270 | 291 | 303 | 307 | 241 |
| | 12 | | 38.2 | 0.0 | 304 | 207 | 282 | 304 | 316 | 320 | 249 |
| | SURFACE DAILY TOTALS | | | | 2406 | 1280 | 1816 | 1980 | 2084 | 2130 | 1742 |
| DEC 21 | 8 | 4 | 10.3 | 53.8 | 176 | 41 | 77 | 90 | 101 | 108 | 107 |
| | 9 | 3 | 19.8 | 43.6 | 257 | 102 | 161 | 180 | 195 | 204 | 183 |
| | 10 | 2 | 27.6 | 31.2 | 288 | 150 | 221 | 244 | 259 | 267 | 226 |
| | 11 | 1 | 32.7 | 16.4 | 301 | 180 | 258 | 282 | 298 | 305 | 251 |
| | 12 | | 34.6 | 0.0 | 304 | 190 | 271 | 295 | 311 | 318 | 259 |
| | SURFACE DAILY TOTALS | | | | 2348 | 1136 | 1704 | 1888 | 2016 | 2086 | 1794 |

## 40 Degrees North Latitude

| DATE | SOLAR TIME AM | SOLAR TIME PM | SOLAR POSITION ALT | SOLAR POSITION AZM | BTUH/SQ. FT. TOTAL INSOLATION ON SURFACES NORMAL | HORIZ. | SOUTH FACING SURFACE ANGLE WITH HORIZ. 30 | 40 | 50 | 60 | 90 |
|---|---|---|---|---|---|---|---|---|---|---|---|
| JAN 21 | 8 | 4 | 8.1 | 55.3 | 142 | 28 | 65 | 74 | 81 | 85 | 84 |
| | 9 | 3 | 16.8 | 44.0 | 239 | 83 | 155 | 171 | 182 | 187 | 171 |
| | 10 | 2 | 23.8 | 30.9 | 274 | 127 | 218 | 237 | 249 | 254 | 223 |
| | 11 | 1 | 28.4 | 16.0 | 289 | 154 | 257 | 277 | 290 | 293 | 253 |
| | 12 | | 30.0 | 0.0 | 294 | 164 | 270 | 291 | 303 | 306 | 263 |
| | SURFACE DAILY TOTALS | | | | 2182 | 948 | 1660 | 1810 | 1906 | 1944 | 1726 |
| FEB 21 | 7 | 5 | 4.8 | 72.7 | 69 | 10 | 19 | 21 | 23 | 24 | 22 |
| | 8 | 4 | 15.4 | 62.2 | 224 | 73 | 114 | 122 | 126 | 127 | 107 |
| | 9 | 3 | 25.0 | 50.2 | 274 | 132 | 195 | 205 | 209 | 208 | 167 |
| | 10 | 2 | 32.8 | 35.9 | 295 | 178 | 256 | 267 | 271 | 267 | 210 |
| | 11 | 1 | 38.1 | 18.9 | 305 | 206 | 293 | 306 | 310 | 304 | 236 |
| | 12 | | 40.0 | 0.0 | 308 | 216 | 306 | 319 | 323 | 317 | 245 |
| | SURFACE DAILY TOTALS | | | | 2640 | 1414 | 2060 | 2162 | 2202 | 2176 | 1730 |
| MAR 21 | 7 | 5 | 11.4 | 80.2 | 171 | 46 | 55 | 55 | 54 | 51 | 35 |
| | 8 | 4 | 22.5 | 69.6 | 250 | 114 | 140 | 141 | 138 | 131 | 89 |
| | 9 | 3 | 32.8 | 57.3 | 282 | 173 | 215 | 217 | 213 | 202 | 138 |
| | 10 | 2 | 41.6 | 41.9 | 297 | 218 | 273 | 276 | 271 | 258 | 176 |
| | 11 | 1 | 47.7 | 22.6 | 305 | 247 | 310 | 313 | 307 | 293 | 200 |
| | 12 | | 50.0 | 0.0 | 307 | 257 | 322 | 326 | 320 | 305 | 208 |
| | SURFACE DAILY TOTALS | | | | 2916 | 1852 | 2308 | 2330 | 2284 | 2174 | 1484 |
| APR 21 | 6 | 6 | 7.4 | 98.9 | 89 | 20 | 11 | 8 | 7 | 7 | 4 |
| | 7 | 5 | 18.9 | 89.5 | 206 | 87 | 77 | 70 | 61 | 50 | 12 |
| | 8 | 4 | 30.3 | 79.3 | 252 | 152 | 153 | 145 | 133 | 117 | 53 |
| | 9 | 3 | 41.3 | 67.2 | 274 | 207 | 221 | 213 | 199 | 179 | 93 |
| | 10 | 2 | 51.2 | 51.4 | 286 | 250 | 275 | 267 | 252 | 229 | 126 |
| | 11 | 1 | 58.7 | 29.2 | 292 | 277 | 308 | 301 | 285 | 260 | 147 |
| | 12 | | 61.6 | 0.0 | 293 | 287 | 320 | 313 | 296 | 271 | 154 |
| | SURFACE DAILY TOTALS | | | | 3092 | 2274 | 2412 | 2320 | 2168 | 1956 | 1022 |
| MAY 21 | 5 | 7 | 1.9 | 114.7 | 1 | 0 | 0 | 0 | 0 | 0 | 0 |
| | 6 | 6 | 12.7 | 105.6 | 144 | 49 | 25 | 15 | 14 | 13 | 9 |
| | 7 | 5 | 24.0 | 96.6 | 216 | 214 | 89 | 76 | 60 | 44 | 13 |
| | 8 | 4 | 35.4 | 87.2 | 250 | 175 | 158 | 144 | 125 | 104 | 25 |
| | 9 | 3 | 46.8 | 76.0 | 267 | 227 | 221 | 206 | 186 | 160 | 60 |
| | 10 | 2 | 57.5 | 60.9 | 277 | 267 | 270 | 255 | 233 | 205 | 89 |
| | 11 | 1 | 66.2 | 37.1 | 283 | 293 | 301 | 287 | 264 | 234 | 108 |
| | 12 | | 70.0 | 0.0 | 284 | 301 | 312 | 297 | 274 | 243 | 114 |
| | SURFACE DAILY TOTQLS | | | | 3160 | 2552 | 2442 | 2264 | 2040 | 1760 | 724 |
| JUN 21 | 5 | 7 | 4.2 | 117.3 | 22 | 4 | 3 | 3 | 2 | 2 | 1 |
| | 6 | 6 | 14.8 | 108.4 | 155 | 60 | 30 | 18 | 17 | 16 | 10 |
| | 7 | 5 | 26.0 | 99.7 | 216 | 123 | 92 | 77 | 59 | 41 | 14 |
| | 8 | 4 | 37.4 | 90.7 | 246 | 182 | 159 | 142 | 121 | 97 | 16 |
| | 9 | 3 | 48.8 | 80.2 | 263 | 233 | 219 | 202 | 179 | 151 | 47 |
| | 10 | 2 | 59.8 | 65.8 | 272 | 272 | 266 | 248 | 224 | 194 | 74 |
| | 11 | 1 | 69.2 | 41.9 | 277 | 296 | 296 | 278 | 253 | 221 | 92 |
| | 12 | | 73.5 | 0.0 | 279 | 304 | 306 | 289 | 263 | 230 | 98 |
| | SURFACE DAILY TOTALS | | | | 3180 | 2648 | 2434 | 2224 | 1974 | 1670 | 610 |
| JUL 21 | 5 | 7 | 2.3 | 115.2 | 2 | 0 | 0 | 0 | 0 | 0 | 0 |
| | 6 | 6 | 13.1 | 106.1 | 138 | 50 | 26 | 17 | 15 | 14 | 9 |
| | 7 | 5 | 24.3 | 97.2 | 208 | 114 | 89 | 75 | 60 | 44 | 14 |
| | 8 | 4 | 35.8 | 87.8 | 241 | 174 | 157 | 142 | 124 | 102 | 24 |
| | 9 | 3 | 47.2 | 76.7 | 259 | 225 | 218 | 203 | 182 | 157 | 58 |
| | 10 | 2 | 57.9 | 61.7 | 269 | 265 | 266 | 251 | 229 | 200 | 86 |
| | 11 | 1 | 66.7 | 37.9 | 275 | 290 | 296 | 281 | 258 | 228 | 104 |
| | 12 | | 70.6 | 0.0 | 276 | 298 | 307 | 292 | 269 | 238 | 111 |
| | SURFACE DAILY TOTALS | | | | 3062 | 2534 | 2409 | 2230 | 2006 | 1728 | 702 |
| AUG 21 | 6 | 6 | 7.9 | 99.5 | 81 | 21 | 12 | 9 | 8 | 7 | 5 |
| | 7 | 5 | 19.3 | 90.0 | 191 | 87 | 76 | 69 | 60 | 49 | 12 |
| | 8 | 4 | 30.7 | 79.9 | 237 | 150 | 150 | 141 | 129 | 113 | 50 |
| | 9 | 3 | 41.8 | 67.9 | 260 | 205 | 216 | 207 | 193 | 173 | 89 |
| | 10 | 2 | 51.7 | 52.1 | 272 | 246 | 267 | 259 | 244 | 221 | 120 |
| | 11 | 1 | 59.3 | 29.7 | 278 | 273 | 300 | 292 | 276 | 252 | 140 |
| | 12 | | 62.3 | 0.0 | 280 | 282 | 311 | 303 | 287 | 262 | 147 |
| | SURFACE DAILY TOTALS | | | | 2916 | 2244 | 2354 | 2258 | 2104 | 1894 | 978 |
| SEP 21 | 7 | 5 | 11.4 | 80.2 | 149 | 43 | 51 | 51 | 49 | 47 | 32 |
| | 8 | 4 | 22.5 | 69.6 | 230 | 109 | 133 | 134 | 131 | 124 | 84 |
| | 9 | 3 | 32.8 | 57.3 | 263 | 167 | 206 | 208 | 203 | 193 | 132 |
| | 10 | 2 | 41.6 | 41.9 | 280 | 211 | 262 | 265 | 260 | 247 | 168 |
| | 11 | 1 | 47.7 | 22.6 | 287 | 239 | 298 | 301 | 295 | 281 | 192 |
| | 12 | | 50.0 | 0.0 | 290 | 249 | 310 | 313 | 307 | 292 | 200 |
| | SURFACE DAILY TOTALS | | | | 2708 | 1788 | 2210 | 2228 | 2182 | 2074 | 1416 |
| OCT 21 | 7 | 5 | 4.5 | 72.3 | 48 | 7 | 14 | 15 | 17 | 17 | 16 |
| | 8 | 4 | 15.0 | 61.9 | 204 | 68 | 106 | 113 | 117 | 118 | 100 |
| | 9 | 3 | 24.5 | 49.8 | 257 | 126 | 185 | 195 | 200 | 198 | 160 |
| | 10 | 2 | 32.4 | 35.6 | 280 | 170 | 245 | 257 | 261 | 257 | 203 |
| | 11 | 1 | 37.6 | 18.7 | 291 | 199 | 283 | 295 | 299 | 294 | 229 |
| | 12 | | 39.5 | 0.0 | 294 | 208 | 295 | 308 | 312 | 306 | 238 |
| | SURFACE DAILY TOTALS | | | | 2454 | 1348 | 1962 | 2060 | 2098 | 2074 | 1654 |
| NOV 21 | 8 | 4 | 8.2 | 55.4 | 136 | 28 | 63 | 72 | 78 | 82 | 81 |
| | 9 | 3 | 17.0 | 44.1 | 232 | 82 | 152 | 167 | 178 | 183 | 167 |
| | 10 | 2 | 24.0 | 31.0 | 268 | 126 | 215 | 233 | 245 | 249 | 219 |
| | 11 | 1 | 28.6 | 16.1 | 283 | 153 | 254 | 273 | 285 | 288 | 248 |
| | 12 | | 30.2 | 0.0 | 288 | 163 | 267 | 287 | 298 | 301 | 258 |
| | SURFACE DAILY TOTALS | | | | 2128 | 942 | 1636 | 1778 | 1870 | 1908 | 1686 |
| DEC | 8 | 4 | 5.5 | 53.0 | 89 | 14 | 39 | 45 | 50 | 54 | 56 |
| | 9 | 3 | 14.0 | 41.9 | 217 | 65 | 135 | 152 | 164 | 171 | 163 |
| | 10 | 2 | 20.7 | 29.4 | 261 | 107 | 200 | 221 | 235 | 242 | 221 |
| | 11 | 1 | 25.0 | 15.2 | 280 | 134 | 239 | 262 | 276 | 283 | 252 |
| | 12 | | 26.6 | 0.0 | 285 | 143 | 253 | 275 | 290 | 296 | 263 |
| | SURFACE DAILY TOTALS | | | | 1978 | 782 | 1480 | 1634 | 1740 | 1796 | 1646 |

(Table 6-2, continued)

## 48 Degrees North Latitude

| DATE | SOLAR TIME AM | SOLAR TIME PM | SOLAR POSITION ALT | SOLAR POSITION AZM | BTUH/SQ. FT. TOTAL INSOLATION ON SURFACES NORMAL | HORIZ. | SOUTH FACING SURFACE ANGLE WITH HORIZ. 38 | 48 | 58 | 68 | 90 |
|---|---|---|---|---|---|---|---|---|---|---|---|
| JAN 21 | 8 | 4 | 3.5 | 54.6 | 37 | 4 | 17 | 19 | 21 | 22 | 22 |
| | 9 | 3 | 11.0 | 42.6 | 185 | 46 | 120 | 132 | 140 | 145 | 139 |
| | 10 | 2 | 16.9 | 29.4 | 239 | 83 | 190 | 206 | 216 | 220 | 206 |
| | 11 | 1 | 20.7 | 15.1 | 261 | 107 | 231 | 249 | 260 | 263 | 243 |
| | 12 | | 22.0 | 0.0 | 267 | 115 | 245 | 264 | 275 | 278 | 255 |
| | SURFACE DAILY TOTALS | | | | 1710 | 596 | 1360 | 1478 | 1550 | 1578 | 1478 |
| FEB 21 | 7 | 5 | 2.4 | 72.2 | 12 | 1 | 3 | 4 | 4 | 4 | 4 |
| | 8 | 4 | 11.6 | 60.5 | 188 | 49 | 95 | 102 | 105 | 106 | 96 |
| | 9 | 3 | 19.7 | 47.7 | 251 | 100 | 178 | 187 | 191 | 190 | 167 |
| | 10 | 2 | 26.2 | 33.3 | 278 | 139 | 240 | 251 | 255 | 251 | 217 |
| | 11 | 1 | 30.5 | 17.2 | 290 | 165 | 278 | 290 | 294 | 288 | 247 |
| | 12 | | 32.0 | 0.0 | 293 | 173 | 291 | 304 | 307 | 301 | 258 |
| | SURFACE DAILY TOTALS | | | | 2330 | 1080 | 1880 | 1972 | 2024 | 1978 | 1720 |
| MAR 21 | 7 | 5 | 10.0 | 78.7 | 153 | 37 | 49 | 49 | 47 | 45 | 35 |
| | 8 | 4 | 19.5 | 66.8 | 236 | 96 | 131 | 132 | 129 | 122 | 96 |
| | 9 | 3 | 28.2 | 53.4 | 270 | 147 | 205 | 207 | 203 | 193 | 152 |
| | 10 | 2 | 35.4 | 37.8 | 287 | 187 | 263 | 266 | 261 | 248 | 195 |
| | 11 | 1 | 40.3 | 19.8 | 295 | 212 | 300 | 303 | 297 | 283 | 223 |
| | 12 | | 42.0 | 0.0 | 298 | 220 | 312 | 315 | 309 | 294 | 232 |
| | SURFACE DAILY TOTALS | | | | 2780 | 1578 | 2208 | 2228 | 2182 | 2074 | 1632 |
| APR 21 | 6 | 6 | 8.6 | 97.8 | 108 | 27 | 13 | 9 | 8 | 7 | 5 |
| | 7 | 5 | 18.6 | 86.7 | 205 | 85 | 76 | 69 | 59 | 48 | 21 |
| | 8 | 4 | 28.5 | 74.9 | 247 | 142 | 149 | 141 | 129 | 113 | 69 |
| | 9 | 3 | 37.8 | 61.2 | 268 | 191 | 216 | 208 | 194 | 174 | 115 |
| | 10 | 2 | 45.8 | 44.6 | 280 | 228 | 268 | 260 | 245 | 223 | 152 |
| | 11 | 1 | 51.5 | 24.0 | 286 | 252 | 301 | 294 | 278 | 254 | 177 |
| | 12 | | 53.6 | 0.0 | 288 | 260 | 313 | 305 | 289 | 264 | 185 |
| | SURFACE DAILY TOTALS | | | | 3076 | 2106 | 2358 | 2266 | 2114 | 1902 | 1262 |
| MAY 21 | 5 | 7 | 5.2 | 114.3 | 41 | 9 | 4 | 4 | 4 | 3 | 2 |
| | 6 | 6 | 14.7 | 103.7 | 162 | 61 | 27 | 16 | 15 | 13 | 10 |
| | 7 | 5 | 24.6 | 93.0 | 219 | 118 | 89 | 75 | 60 | 43 | 13 |
| | 8 | 4 | 34.7 | 81.6 | 248 | 171 | 156 | 142 | 123 | 101 | 45 |
| | 9 | 3 | 44.3 | 68.3 | 264 | 217 | 217 | 202 | 182 | 156 | 86 |
| | 10 | 2 | 53.0 | 51.3 | 274 | 252 | 265 | 251 | 229 | 200 | 120 |
| | 11 | 1 | 59.5 | 28.6 | 279 | 274 | 296 | 281 | 258 | 228 | 141 |
| | 12 | | 62.0 | 0.0 | 280 | 281 | 306 | 292 | 269 | 238 | 149 |
| | SURFACE DAILY TOTALS | | | | 3254 | 2482 | 2418 | 2234 | 2010 | 1728 | 982 |
| JUN 21 | 5 | 7 | 7.9 | 116.5 | 77 | 21 | 9 | 9 | 8 | 7 | 5 |
| | 6 | 6 | 17.2 | 106.2 | 172 | 74 | 33 | 19 | 18 | 16 | 12 |
| | 7 | 5 | 27.0 | 95.8 | 220 | 129 | 93 | 77 | 59 | 39 | 15 |
| | 8 | 4 | 37.1 | 84.6 | 246 | 181 | 157 | 140 | 119 | 95 | 35 |
| | 9 | 3 | 46.9 | 71.6 | 261 | 225 | 216 | 198 | 175 | 147 | 74 |
| | 10 | 2 | 55.8 | 54.8 | 269 | 259 | 262 | 244 | 220 | 189 | 105 |
| | 11 | 1 | 62.7 | 31.2 | 274 | 280 | 291 | 273 | 248 | 216 | 126 |
| | 12 | | 65.5 | 0.0 | 275 | 287 | 301 | 283 | 258 | 225 | 133 |
| | SURFACE DAILY TOTALS | | | | 3312 | 2626 | 2420 | 2204 | 1950 | 1644 | 874 |
| JUL 21 | 5 | 7 | 5.7 | 114.7 | 43 | 10 | 5 | 5 | 4 | 4 | 3 |
| | 6 | 6 | 15.2 | 104.1 | 156 | 62 | 28 | 18 | 16 | 15 | 11 |
| | 7 | 5 | 25.1 | 93.5 | 211 | 118 | 89 | 75 | 59 | 42 | 14 |
| | 8 | 4 | 35.1 | 82.1 | 240 | 171 | 154 | 140 | 121 | 99 | 43 |
| | 9 | 3 | 44.8 | 68.8 | 256 | 215 | 214 | 199 | 178 | 153 | 83 |
| | 10 | 2 | 53.5 | 51.9 | 266 | 250 | 261 | 246 | 224 | 195 | 116 |
| | 11 | 1 | 60.1 | 29.0 | 271 | 272 | 291 | 276 | 253 | 223 | 137 |
| | 12 | | 62.6 | 0.0 | 272 | 279 | 301 | 286 | 263 | 232 | 144 |
| | SURFACE DAILY TOTALS | | | | 3158 | 2474 | 2386 | 2200 | 1974 | 1694 | 956 |

| DATE | SOLAR TIME AM | SOLAR TIME PM | SOLAR POSITION ALT | SOLAR POSITION AZM | BTUH/SQ. FT. TOTAL INSOLATION ON SURFACES NORMAL | HORIZ. | SOUTH FACING SURFACE ANGLE WITH HORIZ. 38 | 48 | 58 | 68 | 90 |
|---|---|---|---|---|---|---|---|---|---|---|---|
| AUG 21 | 6 | 6 | 9.1 | 98.3 | 99 | 28 | 14 | 10 | 9 | 8 | 6 |
| | 7 | 5 | 19.1 | 87.2 | 190 | 85 | 75 | 67 | 58 | 47 | 20 |
| | 8 | 4 | 29.0 | 75.4 | 232 | 141 | 145 | 137 | 125 | 109 | 65 |
| | 9 | 3 | 38.4 | 61.8 | 254 | 189 | 210 | 201 | 187 | 168 | 110 |
| | 10 | 2 | 46.4 | 45.1 | 266 | 225 | 260 | 252 | 237 | 214 | 146 |
| | 11 | 1 | 52.2 | 24.3 | 272 | 248 | 293 | 285 | 268 | 244 | 169 |
| | 12 | | 54.3 | 0.0 | 274 | 256 | 304 | 296 | 279 | 255 | 177 |
| | SURFACE DAILY TOTALS | | | | 2898 | 2086 | 2300 | 2200 | 2046 | 1836 | 1208 |
| SEP 21 | 7 | 5 | 10.0 | 78.7 | 131 | 35 | 44 | 44 | 43 | 40 | 31 |
| | 8 | 4 | 19.5 | 66.8 | 215 | 92 | 124 | 124 | 121 | 115 | 90 |
| | 9 | 3 | 28.2 | 53.4 | 251 | 142 | 196 | 197 | 193 | 183 | 143 |
| | 10 | 2 | 35.4 | 37.8 | 269 | 181 | 251 | 254 | 248 | 236 | 185 |
| | 11 | 1 | 40.3 | 19.8 | 278 | 205 | 287 | 289 | 284 | 269 | 212 |
| | 12 | | 42.0 | 0.0 | 280 | 213 | 299 | 302 | 296 | 281 | 221 |
| | SURFACE DAILY TOTALS | | | | 2568 | 1522 | 2102 | 2118 | 2070 | 1966 | 1546 |
| OCT 21 | 7 | 5 | 2.0 | 71.9 | 4 | 0 | 1 | 1 | 1 | 1 | 1 |
| | 8 | 4 | 11.2 | 60.2 | 165 | 44 | 86 | 91 | 95 | 95 | 87 |
| | 9 | 3 | 19.3 | 47.4 | 233 | 94 | 167 | 176 | 180 | 178 | 157 |
| | 10 | 2 | 25.7 | 33.1 | 262 | 133 | 228 | 239 | 242 | 239 | 207 |
| | 11 | 1 | 30.0 | 17.1 | 274 | 157 | 266 | 277 | 281 | 276 | 237 |
| | 12 | | 31.5 | 0.0 | 278 | 166 | 279 | 291 | 294 | 288 | 247 |
| | SURFACE DAILY TOTALS | | | | 2154 | 1022 | 1774 | 1860 | 1890 | 1866 | 1626 |
| NOV 21 | 8 | 4 | 3.6 | 54.7 | 36 | 5 | 17 | 19 | 21 | 22 | 22 |
| | 9 | 3 | 11.2 | 42.7 | 179 | 46 | 117 | 129 | 137 | 141 | 135 |
| | 10 | 2 | 17.1 | 29.5 | 233 | 83 | 186 | 202 | 212 | 215 | 201 |
| | 11 | 1 | 20.9 | 15.1 | 255 | 107 | 227 | 245 | 255 | 258 | 238 |
| | 12 | | 22.2 | 0.0 | 261 | 115 | 241 | 259 | 270 | 272 | 250 |
| | SURFACE DAILY TOTALS | | | | 1668 | 596 | 1336 | 1448 | 1518 | 1544 | 1442 |
| DEC 21 | 9 | 3 | 8.0 | 40.9 | 140 | 27 | 87 | 98 | 105 | 110 | 109 |
| | 10 | 2 | 13.6 | 28.2 | 214 | 63 | 164 | 180 | 192 | 197 | 190 |
| | 11 | 1 | 17.3 | 14.4 | 242 | 86 | 207 | 226 | 239 | 244 | 231 |
| | 12 | | 18.6 | 0.0 | 250 | 94 | 222 | 241 | 254 | 260 | 244 |
| | SURFACE DAILY TOTALS | | | | 1444 | 446 | 1136 | 1250 | 1326 | 1364 | 1304 |

Example: Billings, Montana

Average surface daily totals at 48° tilt:

(1478 + 1972 + 2228 + 2266 + 2234 + 2204 + 2200
+ 2200 + 2118 + 1860 + 1448 + 1250) ÷ 12 = 1955 BTU/ft$^2$ –day

# 56 Degrees North Latitude

| DATE | SOLAR TIME AM | SOLAR TIME PM | SOLAR POSITION ALT | SOLAR POSITION AZM | BTUH/SQ. FT. TOTAL INSOLATION ON SURFACES NORMAL | HORIZ. | SOUTH FACING SURFACE ANGLE WITH HORIZ. 46 | 56 | 66 | 76 | 90 |
|---|---|---|---|---|---|---|---|---|---|---|---|
| JAN 21 | 9 | 3 | 5.0 | 41.8 | 78 | 11 | 50 | 55 | 59 | 60 | 60 |
| | 10 | 2 | 9.9 | 28.5 | 170 | 39 | 135 | 146 | 154 | 156 | 153 |
| | 11 | 1 | 12.9 | 14.5 | 207 | 58 | 183 | 197 | 206 | 208 | 201 |
| | 12 | | 14.0 | 0.0 | 217 | 65 | 198 | 214 | 222 | 225 | 217 |
| | SURFACE DAILY TOTALS | | | | 1126 | 282 | 934 | 1010 | 1058 | 1074 | 1044 |
| FEB 21 | 8 | 4 | 7.6 | 59.4 | 129 | 25 | 65 | 69 | 72 | 72 | 69 |
| | 9 | 3 | 14.2 | 45.9 | 214 | 65 | 151 | 159 | 162 | 161 | 151 |
| | 10 | 2 | 19.4 | 31.5 | 250 | 98 | 215 | 225 | 228 | 224 | 208 |
| | 11 | 1 | 22.8 | 16.1 | 266 | 119 | 254 | 265 | 268 | 263 | 243 |
| | 12 | | 24.0 | 0.0 | 270 | 126 | 268 | 279 | 282 | 276 | 255 |
| | SURFACE DAILY TOTALS | | | | 1986 | 740 | 1640 | 1716 | 1742 | 1716 | 1598 |
| MAR 21 | 7 | 5 | 8.3 | 77.5 | 128 | 28 | 40 | 40 | 39 | 37 | 32 |
| | 8 | 4 | 16.2 | 64.4 | 215 | 75 | 119 | 120 | 117 | 111 | 97 |
| | 9 | 3 | 23.3 | 50.3 | 253 | 118 | 192 | 193 | 189 | 180 | 154 |
| | 10 | 2 | 29.0 | 34.9 | 272 | 151 | 249 | 251 | 246 | 234 | 205 |
| | 11 | 1 | 32.7 | 17.9 | 282 | 172 | 285 | 288 | 282 | 268 | 236 |
| | 12 | | 34.0 | 0.0 | 284 | 179 | 297 | 300 | 294 | 280 | 246 |
| | SURFACE DAILY TOTALS | | | | 2586 | 1268 | 2066 | 2084 | 2040 | 1938 | 1700 |
| APR 21 | 5 | 7 | 1.4 | 108.8 | 0 | 0 | 0 | 0 | 0 | 0 | 0 |
| | 6 | 6 | 9.6 | 96.5 | 122 | 32 | 14 | 9 | 8 | 7 | 6 |
| | 7 | 5 | 18.0 | 84.1 | 201 | 81 | 74 | 66 | 57 | 46 | 29 |
| | 8 | 4 | 26.1 | 70.9 | 239 | 129 | 143 | 135 | 123 | 108 | 82 |
| | 9 | 3 | 33.6 | 56.3 | 260 | 169 | 208 | 200 | 186 | 167 | 133 |
| | 10 | 2 | 39.9 | 39.7 | 272 | 201 | 259 | 251 | 236 | 214 | 174 |
| | 11 | 1 | 44.1 | 20.7 | 278 | 220 | 292 | 284 | 268 | 245 | 200 |
| | 12 | | 45.6 | 0.0 | 280 | 227 | 303 | 295 | 279 | 255 | 209 |
| | SURFACE DAILY TOTALS | | | | 3024 | 1892 | 2282 | 2186 | 2038 | 1830 | 1458 |
| MAY 21 | 4 | 8 | 1.2 | 125.5 | 0 | 0 | 0 | 0 | 0 | 0 | 0 |
| | 5 | 7 | 8.5 | 113.4 | 93 | 25 | 10 | 9 | 8 | 7 | 6 |
| | 6 | 6 | 16.5 | 101.5 | 175 | 71 | 28 | 17 | 15 | 13 | 11 |
| | 7 | 5 | 24.8 | 89.3 | 219 | 119 | 88 | 74 | 58 | 41 | 16 |
| | 8 | 4 | 33.1 | 76.3 | 244 | 163 | 153 | 138 | 119 | 98 | 63 |
| | 9 | 3 | 40.9 | 61.6 | 259 | 201 | 212 | 197 | 176 | 151 | 109 |
| | 10 | 2 | 47.6 | 44.2 | 268 | 231 | 259 | 244 | 222 | 194 | 146 |
| | 11 | 1 | 52.3 | 23.4 | 273 | 249 | 288 | 274 | 251 | 222 | 170 |
| | 12 | | 54.0 | 0.0 | 275 | 255 | 299 | 284 | 261 | 231 | 178 |
| | SURFACE DAILY TOTALS | | | | 3340 | 2374 | 2374 | 2188 | 1962 | 1682 | 1218 |
| JUN 21 | 4 | 8 | 4.2 | 127.2 | 21 | 4 | 2 | 2 | 2 | 2 | 1 |
| | 5 | 7 | 11.4 | 115.3 | 122 | 40 | 14 | 13 | 11 | 10 | 8 |
| | 6 | 6 | 19.3 | 103.6 | 185 | 86 | 34 | 19 | 17 | 15 | 12 |
| | 7 | 5 | 27.6 | 91.7 | 222 | 132 | 92 | 76 | 57 | 38 | 15 |
| | 8 | 4 | 35.9 | 78.8 | 243 | 175 | 154 | 137 | 116 | 92 | 55 |
| | 9 | 3 | 43.8 | 64.1 | 257 | 212 | 211 | 193 | 170 | 143 | 98 |
| | 10 | 2 | 50.7 | 46.4 | 265 | 240 | 255 | 238 | 214 | 184 | 133 |
| | 11 | 1 | 55.6 | 24.9 | 269 | 258 | 284 | 267 | 242 | 210 | 156 |
| | 12 | | 57.5 | 0.0 | 271 | 264 | 294 | 276 | 251 | 219 | 164 |
| | SURFACE DAILY TOTALS | | | | 3438 | 2562 | 2388 | 2166 | 1910 | 1606 | 1120 |
| JUL 21 | 4 | 8 | 1.7 | 125.8 | 0 | 0 | 0 | 0 | 0 | 0 | 0 |
| | 5 | 7 | 9.0 | 113.7 | 91 | 27 | 11 | 10 | 9 | 8 | 6 |
| | 6 | 6 | 17.0 | 101.9 | 169 | 72 | 30 | 18 | 16 | 14 | 12 |
| | 7 | 5 | 25.3 | 89.7 | 212 | 119 | 88 | 74 | 58 | 41 | 15 |
| | 8 | 4 | 33.6 | 76.7 | 237 | 163 | 151 | 136 | 117 | 96 | 61 |
| | 9 | 3 | 41.4 | 62.0 | 252 | 201 | 208 | 193 | 173 | 147 | 106 |
| | 10 | 2 | 48.2 | 44.6 | 261 | 230 | 254 | 239 | 217 | 189 | 142 |
| | 11 | 1 | 52.9 | 23.7 | 265 | 248 | 283 | 268 | 245 | 216 | 165 |
| | 12 | | 54.6 | 0.0 | 267 | 254 | 293 | 278 | 255 | 225 | 173 |
| | SURFACE DAILY TOTALS | | | | 3240 | 2372 | 2342 | 2152 | 1926 | 1646 | 1186 |
| AUG 21 | 5 | 7 | 2.0 | 109.2 | 1 | 0 | 0 | 0 | 0 | 0 | 0 |
| | 6 | 6 | 10.2 | 97.0 | 112 | 34 | 16 | 11 | 10 | 9 | 7 |
| | 7 | 5 | 18.5 | 84.5 | 187 | 82 | 73 | 65 | 56 | 45 | 28 |
| | 8 | 4 | 26.7 | 71.3 | 225 | 128 | 140 | 131 | 119 | 104 | 78 |
| | 9 | 3 | 34.3 | 56.7 | 246 | 168 | 202 | 193 | 179 | 160 | 126 |
| | 10 | 2 | 40.5 | 40.0 | 258 | 199 | 251 | 242 | 227 | 206 | 166 |
| | 11 | 1 | 44.8 | 20.9 | 264 | 218 | 282 | 274 | 258 | 235 | 191 |
| | 12 | | 46.3 | 0.0 | 266 | 225 | 293 | 285 | 269 | 245 | 200 |
| | SURFACE DAILY TOTALS | | | | 2850 | 1884 | 2218 | 2118 | 1966 | 1760 | 1392 |
| SEP 21 | 7 | 5 | 8.3 | 77.5 | 107 | 25 | 36 | 36 | 34 | 32 | 28 |
| | 8 | 4 | 16.2 | 64.4 | 194 | 72 | 111 | 111 | 108 | 102 | 89 |
| | 9 | 3 | 23.3 | 50.3 | 233 | 114 | 181 | 182 | 178 | 168 | 147 |
| | 10 | 2 | 29.0 | 34.9 | 253 | 146 | 236 | 237 | 232 | 221 | 193 |
| | 11 | 1 | 32.7 | 17.9 | 263 | 166 | 271 | 273 | 267 | 254 | 223 |
| | 12 | | 34.0 | 0.0 | 266 | 173 | 283 | 285 | 279 | 265 | 233 |
| | SURFACE DAILY TOTALS | | | | 2368 | 1220 | 1950 | 1962 | 1918 | 1820 | 1594 |
| OCT 21 | 8 | 4 | 7.1 | 59.1 | 104 | 20 | 53 | 57 | 59 | 59 | 57 |
| | 9 | 3 | 13.8 | 45.7 | 193 | 60 | 138 | 145 | 148 | 147 | 138 |
| | 10 | 2 | 19.0 | 31.3 | 231 | 92 | 201 | 210 | 213 | 210 | 195 |
| | 11 | 1 | 22.3 | 16.0 | 248 | 112 | 240 | 250 | 253 | 248 | 230 |
| | 12 | | 23.5 | 0.0 | 253 | 119 | 253 | 263 | 266 | 261 | 241 |
| | SURFACE DAILY TOTALS | | | | 1804 | 688 | 1516 | 1586 | 1612 | 1588 | 1480 |
| NOV 21 | 9 | 3 | 5.2 | 41.9 | 76 | 12 | 49 | 54 | 57 | 59 | 58 |
| | 10 | 2 | 10.0 | 28.5 | 165 | 39 | 132 | 143 | 149 | 152 | 148 |
| | 11 | 1 | 13.1 | 14.5 | 201 | 58 | 179 | 193 | 201 | 203 | 196 |
| | 12 | | 14.2 | 0.0 | 211 | 65 | 194 | 209 | 217 | 219 | 211 |
| | SURFACE DAILY TOTALS | | | | 1094 | 284 | 914 | 986 | 1032 | 1046 | 1016 |
| DEC 21 | 9 | 3 | 1.9 | 40.5 | 5 | 0 | 3 | 4 | 4 | 4 | 4 |
| | 10 | 2 | 6.6 | 27.5 | 113 | 19 | 86 | 95 | 101 | 104 | 103 |
| | 11 | 1 | 9.5 | 13.9 | 166 | 37 | 141 | 154 | 163 | 167 | 164 |
| | 12 | | 10.6 | 0.0 | 180 | 43 | 159 | 173 | 182 | 186 | 182 |
| | SURFACE DAILY TOTALS | | | | 748 | 156 | 620 | 678 | 716 | 734 | 722 |

(Table 6-2, continued)

## 64 Degrees North Latitude

| DATE | SOLAR TIME AM | SOLAR TIME PM | SOLAR POSITION ALT | SOLAR POSITION AZM | BTUH/SQ. FT. TOTAL INSOLATION ON SURFACES NORMAL | HORIZ. | SOUTH FACING SURFACE ANGLE WITH HORIZ. 54 | 64 | 74 | 84 | 90 |
|---|---|---|---|---|---|---|---|---|---|---|---|
| JAN 21 | 10 | 2 | 2.8 | 28.1 | 22 | 2 | 17 | 19 | 20 | 20 | 20 |
| | 11 | 1 | 5.2 | 14.1 | 81 | 12 | 72 | 77 | 80 | 81 | 81 |
| | 12 | | 6.0 | 0.0 | 100 | 16 | 91 | 98 | 102 | 103 | 103 |
| | SURFACE DAILY TOTALS | | | | 306 | 45 | 268 | 290 | 302 | 306 | 304 |
| FEB 21 | 8 | 4 | 3.4 | 58.7 | 35 | 4 | 17 | 19 | 19 | 19 | 19 |
| | 9 | 3 | 8.6 | 44.8 | 147 | 31 | 103 | 108 | 111 | 110 | 107 |
| | 10 | 2 | 12.6 | 30.3 | 199 | 55 | 170 | 178 | 181 | 178 | 173 |
| | 11 | 1 | 15.1 | 15.3 | 222 | 71 | 212 | 220 | 223 | 219 | 213 |
| | 12 | | 16.0 | 0.0 | 228 | 77 | 225 | 235 | 237 | 232 | 226 |
| | SURFACE DAILY TOTALS | | | | 1432 | 400 | 1230 | 1286 | 1302 | 1282 | 1252 |
| MAR 21 | 7 | 5 | 6.5 | 76.5 | 95 | 18 | 30 | 29 | 29 | 27 | 25 |
| | 8 | 4 | 20.7 | 62.6 | 185 | 54 | 101 | 102 | 99 | 94 | 89 |
| | 9 | 3 | 18.1 | 48.1 | 227 | 87 | 171 | 172 | 169 | 160 | 153 |
| | 10 | 2 | 22.3 | 32.7 | 249 | 112 | 227 | 229 | 224 | 213 | 203 |
| | 11 | 1 | 25.1 | 16.6 | 260 | 129 | 262 | 265 | 259 | 246 | 235 |
| | 12 | | 26.0 | 0.0 | 263 | 134 | 274 | 277 | 271 | 258 | 246 |
| | SURFACE DAILY TOTALS | | | | 2296 | 932 | 1856 | 1870 | 1830 | 1736 | 1656 |
| APR 21 | 5 | 7 | 4.0 | 108.5 | 27 | 5 | 2 | 2 | 2 | 1 | 1 |
| | 6 | 6 | 10.4 | 95.1 | 133 | 37 | 15 | 9 | 8 | 7 | 6 |
| | 7 | 5 | 17.0 | 81.6 | 194 | 76 | 70 | 63 | 54 | 43 | 37 |
| | 8 | 4 | 23.3 | 67.5 | 228 | 112 | 136 | 128 | 116 | 102 | 91 |
| | 9 | 3 | 29.0 | 52.3 | 248 | 144 | 197 | 189 | 176 | 158 | 145 |
| | 10 | 2 | 33.5 | 36.0 | 260 | 169 | 246 | 239 | 224 | 203 | 188 |
| | 11 | 1 | 36.5 | 18.4 | 266 | 184 | 278 | 270 | 255 | 233 | 216 |
| | 12 | | 97.6 | 0.0 | 268 | 190 | 289 | 281 | 266 | 243 | 225 |
| | SURFACE DAILY TOTALS | | | | 2982 | 1644 | 2176 | 2082 | 1936 | 1736 | 1594 |
| MAY 21 | 4 | 8 | 5.8 | 125.1 | 51 | 11 | 5 | 4 | 4 | 3 | 3 |
| | 5 | 7 | 11.6 | 112.1 | 132 | 42 | 13 | 11 | 10 | 9 | 8 |
| | 6 | 6 | 17.9 | 99.1 | 185 | 79 | 29 | 16 | 14 | 12 | 11 |
| | 7 | 5 | 24.5 | 85.7 | 218 | 117 | 86 | 72 | 56 | 39 | 28 |
| | 8 | 4 | 30.9 | 71.5 | 239 | 152 | 148 | 133 | 115 | 94 | 80 |
| | 9 | 3 | 36.8 | 56.1 | 252 | 182 | 204 | 190 | 170 | 145 | 128 |
| | 10 | 2 | 41.6 | 38.9 | 261 | 205 | 249 | 235 | 213 | 186 | 167 |
| | 11 | 1 | 44.9 | 20.1 | 265 | 219 | 278 | 264 | 242 | 213 | 193 |
| | 12 | | 46.0 | 0.0 | 267 | 224 | 288 | 274 | 251 | 222 | 201 |
| | SURFACE DAILY TOTALS | | | | 3470 | 2236 | 2312 | 2124 | 1898 | 1624 | 1436 |
| JUN 21 | 3 | 9 | 4.2 | 139.4 | 21 | 4 | 2 | 2 | 2 | 2 | 1 |
| | 4 | 8 | 9.0 | 126.4 | 93 | 27 | 10 | 9 | 8 | 7 | 6 |
| | 5 | 7 | 14.7 | 113.6 | 154 | 60 | 16 | 15 | 13 | 11 | 10 |
| | 6 | 6 | 21.0 | 100.8 | 194 | 96 | 34 | 19 | 17 | 14 | 13 |
| | 7 | 5 | 27.5 | 87.5 | 221 | 132 | 91 | 74 | 55 | 36 | 23 |
| | 8 | 4 | 34.0 | 73.3 | 239 | 166 | 150 | 133 | 112 | 88 | 73 |
| | 9 | 3 | 39.9 | 57.8 | 251 | 195 | 204 | 187 | 164 | 137 | 119 |
| | 10 | 2 | 44.9 | 40.4 | 258 | 217 | 247 | 230 | 206 | 177 | 157 |
| | 11 | 1 | 48.3 | 20.9 | 262 | 231 | 275 | 258 | 233 | 202 | 181 |
| | 12 | | 49.5 | 0.0 | 263 | 235 | 284 | 267 | 242 | 211 | 189 |
| | SURFACE DAILY TOTALS | | | | 3650 | 2488 | 2342 | 2118 | 1862 | 1558 | 1356 |
| JUL 21 | 4 | 8 | 6.4 | 125.3 | 53 | 13 | 6 | 5 | 5 | 4 | 4 |
| | 5 | 7 | 12.1 | 112.4 | 128 | 44 | 14 | 13 | 11 | 10 | 9 |
| | 6 | 6 | 18.4 | 99.4 | 179 | 81 | 30 | 17 | 16 | 13 | 12 |
| | 7 | 5 | 25.0 | 86.0 | 211 | 118 | 86 | 72 | 56 | 38 | 28 |
| | 8 | 4 | 31.4 | 71.8 | 231 | 152 | 146 | 131 | 113 | 91 | 77 |
| | 9 | 3 | 37.3 | 56.3 | 245 | 182 | 201 | 186 | 166 | 141 | 124 |
| | 10 | 2 | 42.2 | 39.2 | 253 | 204 | 245 | 230 | 208 | 181 | 162 |
| | 11 | 1 | 45.4 | 20.2 | 257 | 218 | 273 | 258 | 236 | 207 | 187 |
| | 12 | | 46.6 | 0.0 | 259 | 223 | 282 | 267 | 245 | 216 | 195 |
| | SURFACE DAILY TOTALS | | | | 3372 | 2248 | 2280 | 2090 | 1864 | 1588 | 1400 |
| AUG 21 | 5 | 7 | 4.6 | 108.8 | 29 | 6 | 3 | 3 | 2 | 2 | 2 |
| | 6 | 6 | 11.0 | 95.5 | 123 | 39 | 16 | 11 | 10 | 8 | 7 |
| | 7 | 5 | 17.6 | 81.9 | 181 | 77 | 69 | 61 | 52 | 42 | 35 |
| | 8 | 4 | 23.9 | 67.8 | 214 | 113 | 132 | 123 | 112 | 97 | 87 |
| | 9 | 3 | 29.6 | 52.6 | 234 | 144 | 190 | 182 | 169 | 150 | 138 |
| | 10 | 2 | 34.2 | 36.2 | 246 | 168 | 237 | 229 | 215 | 194 | 179 |
| | 11 | 1 | 37.2 | 18.5 | 252 | 183 | 268 | 260 | 244 | 222 | 205 |
| | 12 | | 38.3 | 0.0 | 254 | 188 | 278 | 270 | 255 | 232 | 215 |
| | SURFACE DAILY TOTALS | | | | 2808 | 1646 | 2108 | 1008 | 1860 | 1662 | 1522 |
| SEP 21 | 7 | 5 | 6.5 | 76.5 | 77 | 16 | 25 | 25 | 24 | 23 | 21 |
| | 8 | 4 | 12.7 | 72.6 | 163 | 51 | 92 | 92 | 90 | 85 | 81 |
| | 9 | 3 | 18.1 | 48.1 | 206 | 83 | 159 | 159 | 156 | 147 | 141 |
| | 10 | 2 | 22.3 | 32.7 | 229 | 108 | 212 | 213 | 209 | 198 | 189 |
| | 11 | 1 | 25.1 | 16.6 | 240 | 124 | 246 | 248 | 243 | 230 | 220 |
| | 12 | | 26.0 | 0.0 | 244 | 129 | 258 | 260 | 254 | 241 | 230 |
| | SURFACE DAILY TOTALS | | | | 2074 | 892 | 1726 | 1736 | 1696 | 1608 | 1532 |
| OCT 21 | 8 | 4 | 3.0 | 58.5 | 17 | 2 | 9 | 9 | 10 | 10 | 10 |
| | 9 | 3 | 8.1 | 44.6 | 122 | 26 | 86 | 91 | 93 | 92 | 90 |
| | 10 | 2 | 12.1 | 30.2 | 176 | 50 | 152 | 159 | 161 | 159 | 155 |
| | 11 | 1 | 14.6 | 15.2 | 201 | 65 | 193 | 201 | 203 | 200 | 295 |
| | 12 | | 15.5 | 0.0 | 208 | 71 | 207 | 215 | 217 | 213 | 208 |
| | SURFACE DAILY TOTALS | | | | 1238 | 358 | 1088 | 1136 | 1152 | 1134 | 1106 |
| NOV 21 | 10 | 2 | 3.0 | 28.1 | 23 | 3 | 18 | 20 | 21 | 21 | 21 |
| | 11 | 1 | 5.4 | 14.2 | 79 | 12 | 70 | 76 | 79 | 80 | 79 |
| | 12 | | 6.2 | 0.0 | 97 | 17 | 89 | 96 | 100 | 101 | 100 |
| | SURFACE DAILY TOTALS | | | | 302 | 46 | 266 | 286 | 298 | 302 | 300 |
| DEC 21 | 11 | 1 | 1.8 | 13.7 | 4 | 0 | 3 | 4 | 4 | 4 | 4 |
| | 12 | | 2.6 | 0.0 | 16 | 2 | 14 | 15 | 16 | 17 | 17 |
| | SURFACE DAILY TOTALS | | | | 24 | 2 | 20 | 22 | 24 | 24 | 24 |

$$Ac = \frac{55{,}395 \text{ BTU}}{(0.5)(1955 \text{ BTU/ft}^2)}$$

and

$$Ac = 56.7 \text{ ft}^2$$

The collector area should be 56.7 $ft^2$. If a manufacturer's solar aperture or effective collector area is 22.2 $ft^2$ per collector then we would need 2.6 collectors. This, of course, would round up to 3 collectors.

### Sizing Summary

We seem to have a 1-collector difference between these methods. Using an average horizontal insolation factor, we find 4 collectors are needed. Using an insolation factor dependent of the tilt angle, we find 3 collectors are needed. Because the insolation factor from Table 6-2 is a more accurate depiction of available solar energy and a factor of 2.6 is well below the next whole number, a 3-collector system should suffice. An important thing to remember is *not to undersize* the system. For instance, if our results indicated that we would need 2.9 or 3.0 collectors, then a 4-collector system would be more adequate in providing our hot water needs.

This method of determining the necessary collector area and thus the collector system output demonstrates that the system parameters of water storage capacity, water storage temperature, water inlet temperature, collector system efficiency, effective collector area, and available insolation are critical factors in determining collector array sizing.

## SITING

Since we've determined the number of collectors needed to provide the daily BTU requirement for domestic hot water, we therefore know the surface area necessary for installation. This factor in itself could determine the number of collectors and therefore the solar contribution to domestic hot water. We must now determine if there is sufficient roof or ground area available to accommodate the collectors and at the same instance ensure proper collector orientation in azimuth and tilt as well as freedom from shading. Typical tools for siting include the compass and "abney" level as shown in Figure 6-2. Neither of these instruments is mandatory for siting a system; however, the compass provides a quick and accurate determination of collector orientation, and the "abney" level provides a quick and accurate determination of problems associated with collector shading. We shall now discuss collector orientation, collector tilt, and collector shading. Siting is comprised of these three parameters.

### Collector Orientation

The collector array should face the middle of the sun's *daily* path within $\pm 15°$ east or west (azimuth) of *true* south as illustrated in Figure 6-3.

**Figure 6-2.** Siting tools.

We shall discuss three methods by which *true* south may be determined. Method I uses a compass and a *isogonic chart*, Method II is the "Stake and Shadow" method, and Method III is the "Solar Noon" method.

**Method I.** True south should not be confused with compass or magnetic south. Because the earth's magnetic field is not aligned parallel with the earth's north-south axis, a compass will not read true. This magnetic declination will vary at each location. Points on the earth's surface which have

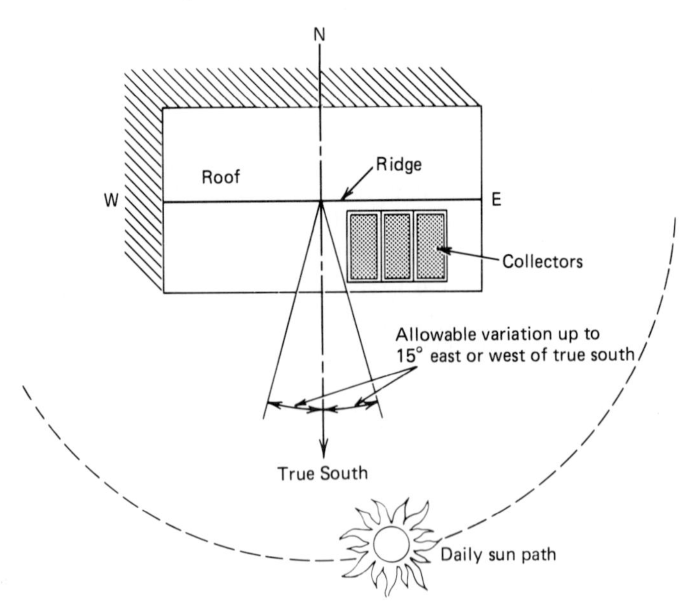


**Figure 6-3.** Collector orientation.

the same magnetic declination can be joined together by an imaginary line to form what is called an ***isogonic chart***. When the magnetic declination is zero, true and magnetic north are the same. In the United States a line of zero declination runs from the eastern end of Lake Michigan through the western edge of Florida to the Gulf of Mexico. On the west side of this zero deviation line, your compass needle will point to the east of true north. On the east side of this zero deviation line, your compass needle will point to the west of true north. This is illustrated in Figure 6-4.

The correction factors needed to adjust your compass reading for location within the United States can be obtained from the Isogonic Chart of Figure 6-5. It should be noted that there are slight annual variations to this type of chart.

As an example to correct our compass reading, we find from Figure 6-5 that Billings, Montana, has a 15 deg E deviation. This means that another 15 deg east of magnetic south should be added to the compass reading. On the other hand, a family in Boston, Massachusetts, has a 14 W deg deviation. This means that another 14 deg west of magnetic south should be added to the compass reading. When taking these compass readings, one must remember not to stand near large metallic objects or power lines so as not to distort the earth's lines of magnetic flux and therefore affect the compass reading.

**Method II.** If a compass is not available, you can find true south using the "Stake and Shadow" method, as illustrated in Figures 6-6*a* through 6-6*f*. This method uses the following materials: (1) 4-ft wooden staff, (2) two small stakes, (3) string, (4) level or plumb bob, and (5) a pointed stick. The step-by-step procedure is illustrated as follows:

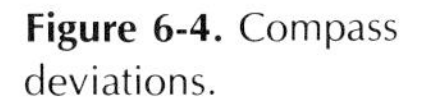
**Figure 6-4.** Compass deviations.

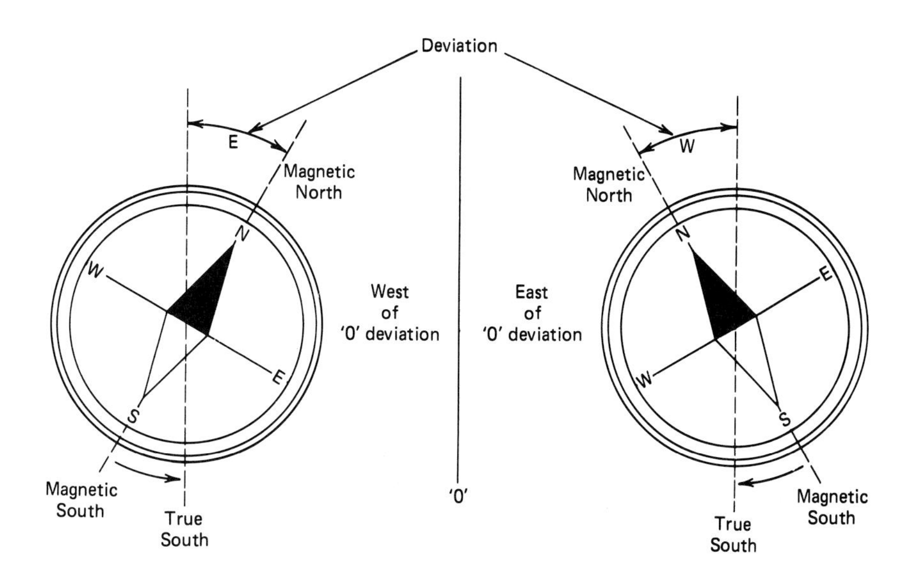

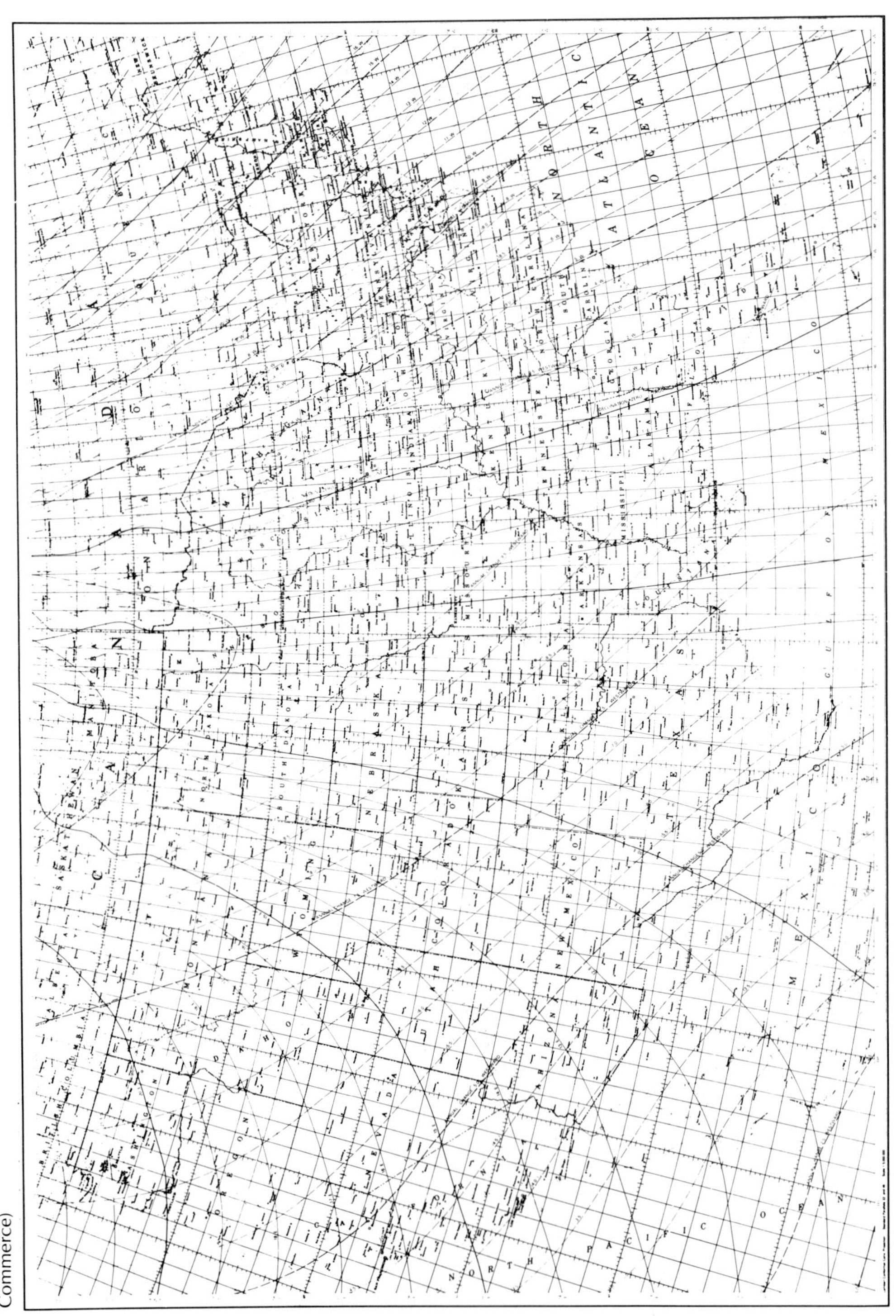


**Figure 6-5.** Isogonic chart. (*a*) Alaska and Hawaii. (*b*) Continental United States. (*Source.* Dept. of Commerce)

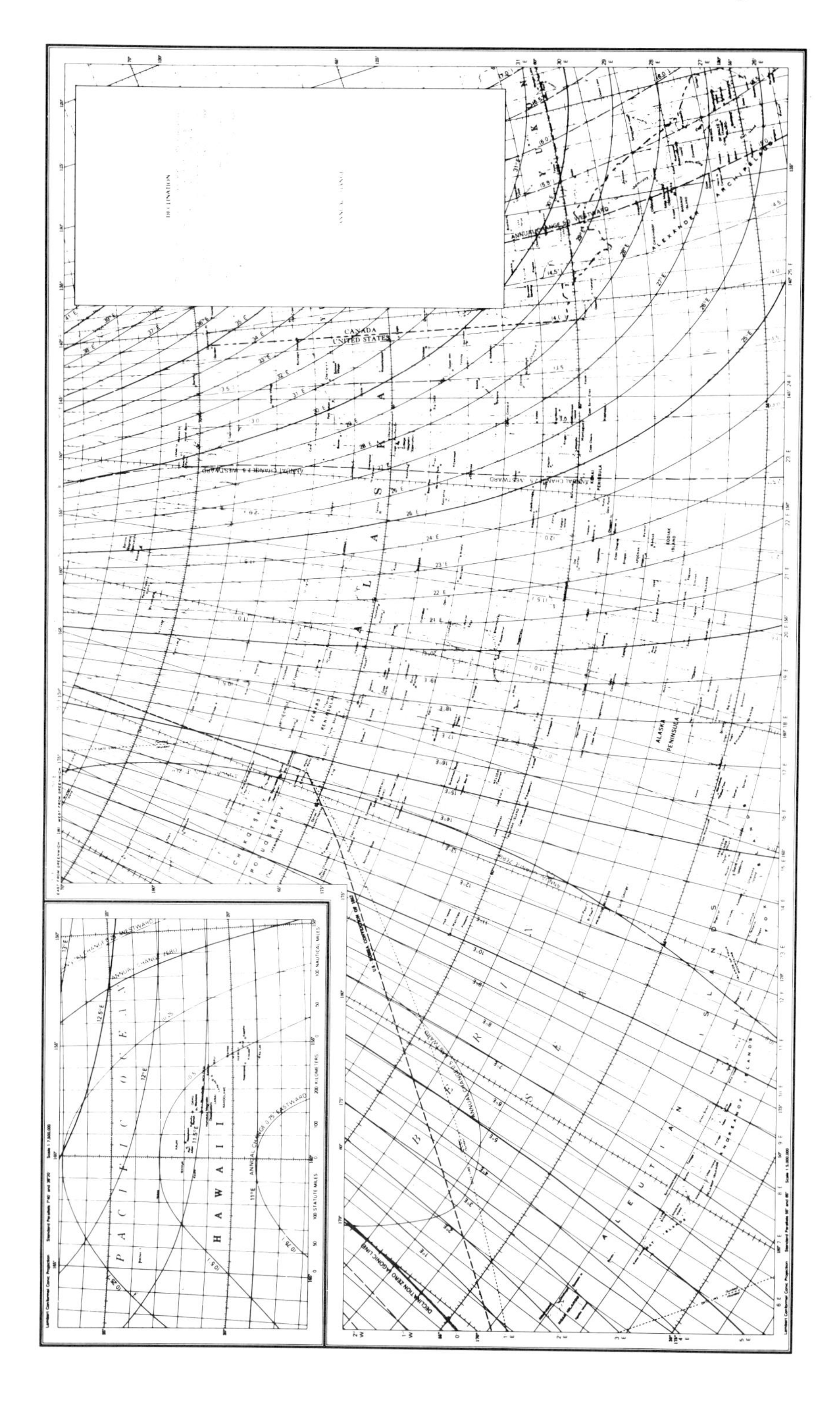
CANADA
UNITED STATES
ALASKA PENINSULA

***Step 1.*** Drive the 4-ft wooden stake into the ground in an open area. Ensure the stake is at a true vertical by using either a level or plumb bob.

**Figure 6-6*a*.** Stake and shadow method.

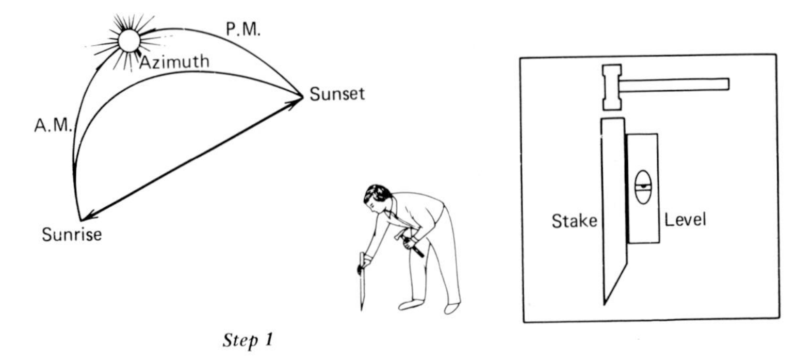


***Step 2.*** At approximately 1 to 2 hours before noon, mark the end of the stake's shadow with a small stake.

**Figure 6-6*b*.** Stake and shadow method.

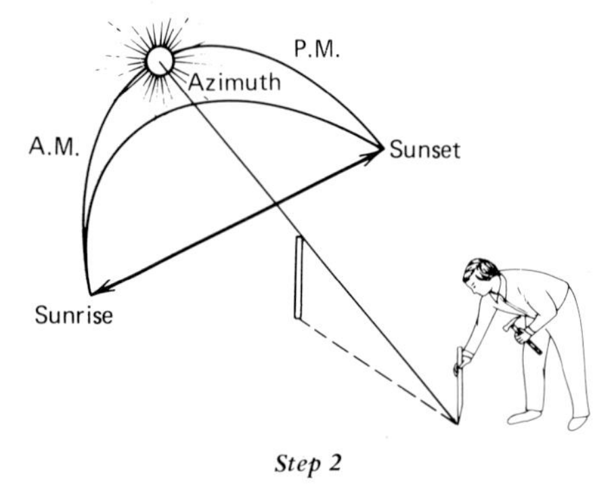


***Step 3.*** Tie a string loosely around the 4-ft stake, stretch the string to the small stake of Step 2, and tie the string to the pointed stick. Scribe an arc on the ground with the pointed stick.

**Figure 6-6*c*.** Stake and shadow method.

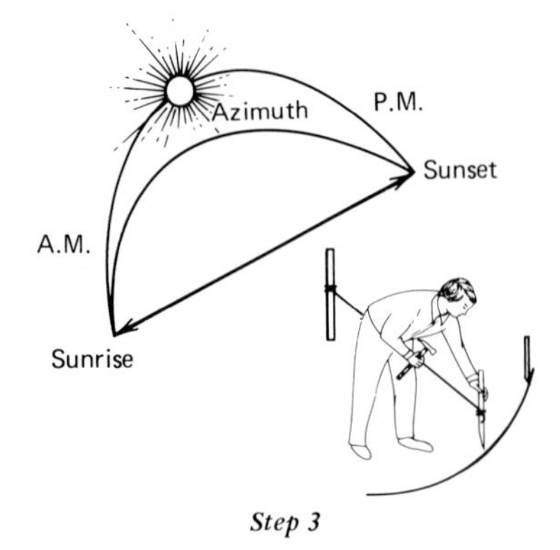

***Step 4.*** Place the other small stake where the late afternoon shadow touches the scribed arc.

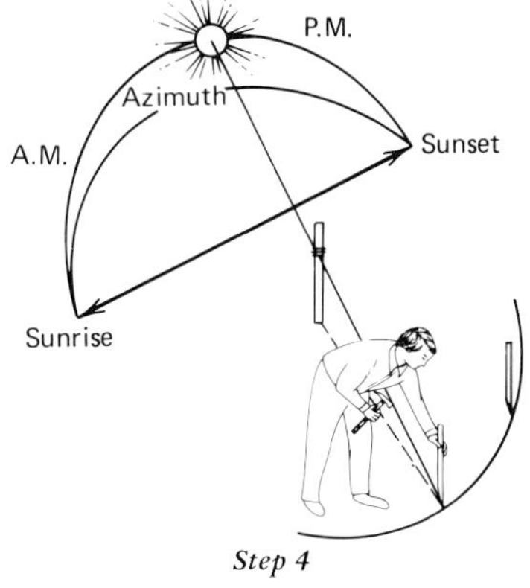


**Figure 6-6*d*.** Stake and shadow method.

***Step 5.*** Draw a straight line between the small stakes.

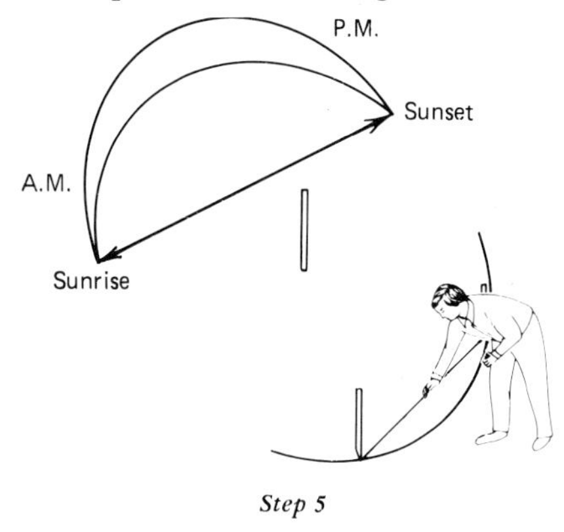


**Figure 6-6e.** Stake and shadow method.

***Step 6.*** Draw a line 90 deg (perpendicular) from the 4-ft stake through the line drawn in Step 5. This line represents the true north-south axis.

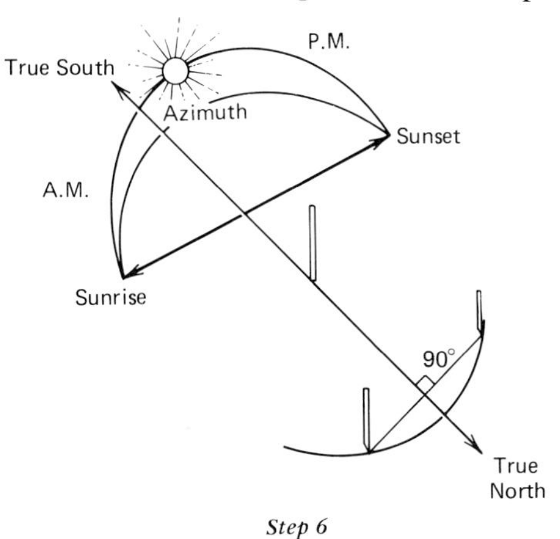


**Figure 6-6*f*.** Stake and shadow method.

**Method III.** Because Method II could essentially take you all day, there is a faster method available to determine the local north-south line. We'll call it the "solar noon" method. The shadow of a stake cast by the sun at solar noon will be on the true north-south line. Solar noon is exactly halfway between sunrise and sunset on any given day. Most local television weather reports and daily newspapers give exact times of sunrise and sunset each day. Examples of calculating solar noon are shown below and the method of determining true north-south is depicted in Figure 6-7.

**Examples of Solar Noon**

| *Sunrise* | *Sunset* | *Solar Day* |
|---|---|---|
| 1. 6:26 | 4:51 | 10 hr, 26 min |
| 2. 5:30 | 6:25 | 12 hr, 55 min |
| 3. 7:00 | 4:10 | 9 hr, 10 min |

1. One-half solar day = 5 hr, 13 min.
2. One-half solar day = 6 hr, 27½ min.
3. One-half solar day = 4 hr, 35 min.

| | *Solar Noon* |
|---|---|
| 1. 6:25 + 5 hr, 13 min | = 11:38 Standard Time |
| 2. 5:30 + 6 hr, 27½ min | = 11:57:30 Standard Time |
| 3. 7:00 + 4 hr, 35 min | = 11:35 Standard Time |

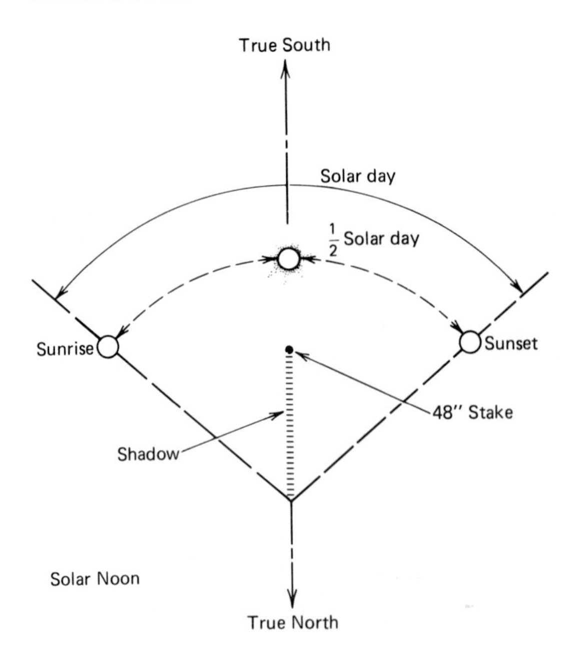


**Figure 6-7.** Solar noon method.

We have now learned three methods by which to determine true south. To reiterate, we should require an orientation of the collector to within ±15° east or west of true south. Orienting the collectors toward the east will start the system earlier in the morning, but orientation slightly to the west will increase system performance because ambient temperatures are usually higher in the afternoon, with the collectors consequently losing less heat to the surroundings. If the orientation requirement cannot be met, then either additional collectors must be considered increasing size thus impacting the economics, or the subambient phase-change type system can be used as described previously in Chapter 3.

## Collector Tilt

The collector array should face the middle of the sun's ***seasonal*** path, which is an angle from horizontal equal to the latitude. The rule generally followed for the tilt of the collector in the northern hemisphere is to position the collector at an angle of latitude plus or minus 10 deg as depicted in Figure 6-8. Variations of 10 deg in either way will not seriously affect the total ***annual collection*** of the system. However, winter ***system performance*** can be increased with the collectors at a tilt of latitude plus 10 deg, because there is less collection time available during this period, and also because there is greater heat loss due to lower ambient temperatures.

**Figure 6-8.** Collector tilt.

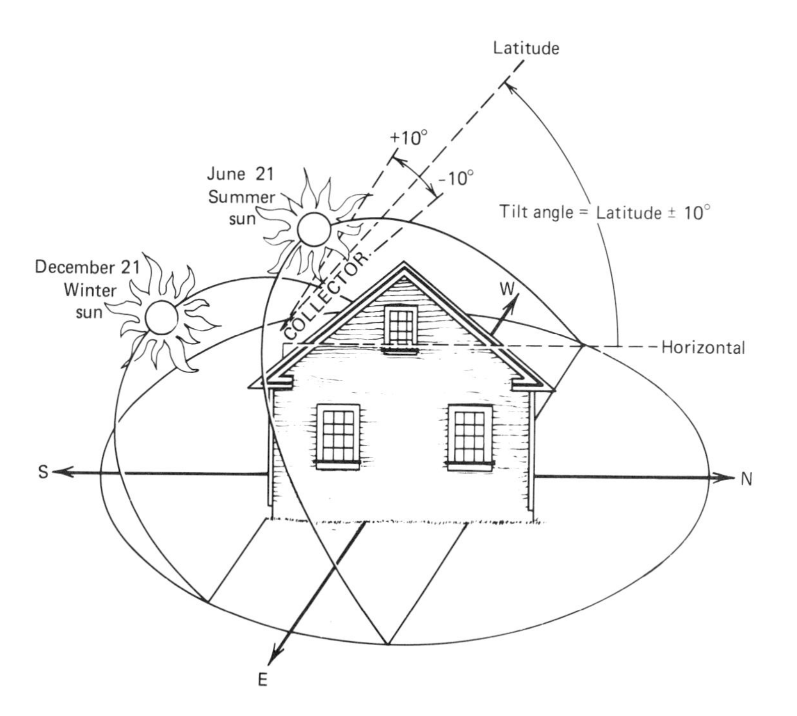

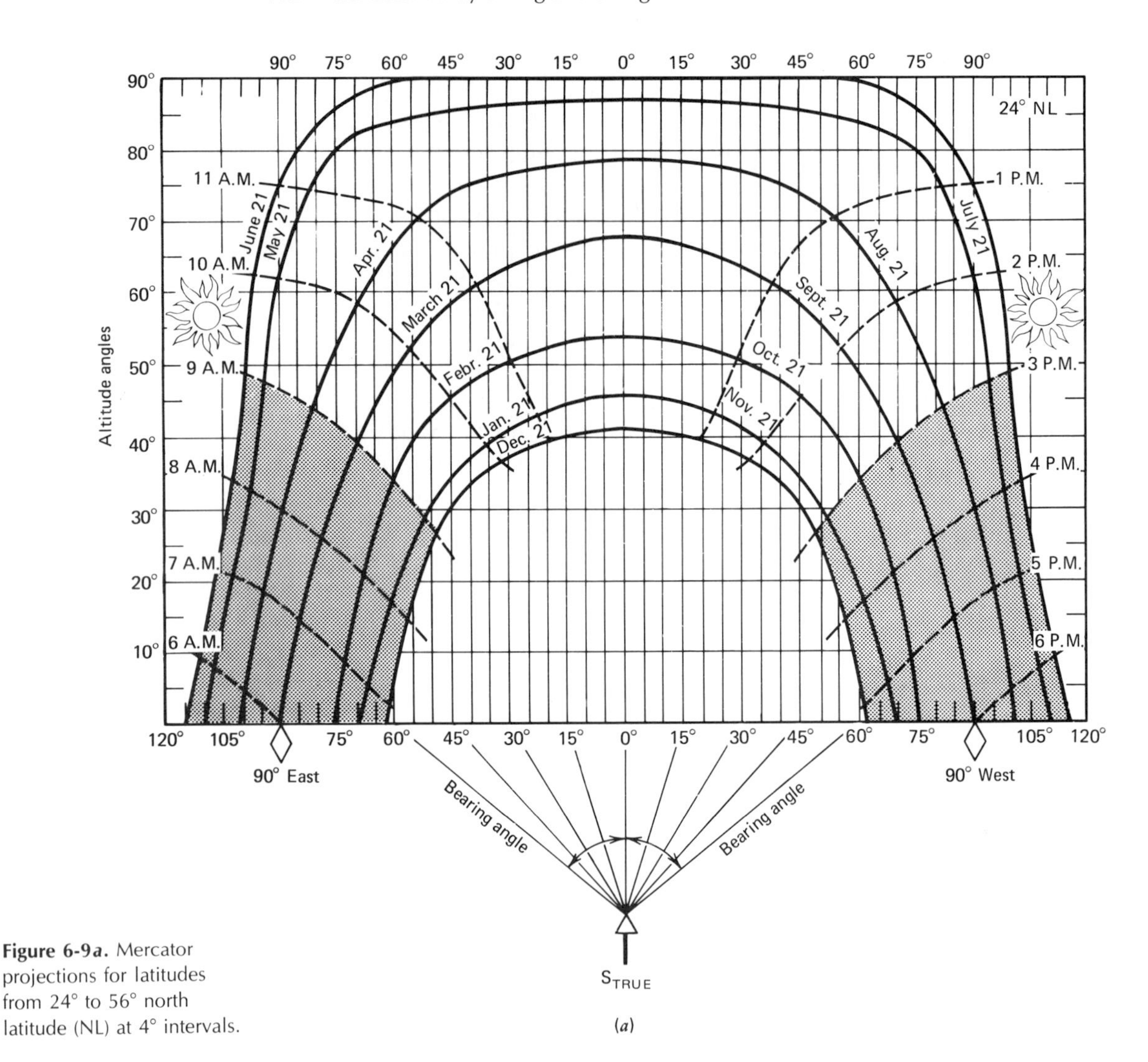


(*a*)

**Figure 6-9*a*.** Mercator projections for latitudes from 24° to 56° north latitude (NL) at 4° intervals.

## Collector Shading

Freedom from shading is a very important consideration. No more than 5 percent of the collector array should be shaded between 9 A.M. and 3 P.M. when the greatest solar potential exists. By knowing the altitude and azimuth of the sun throughout the year, you can determine if a shading problem might exist for your particular site. ***Daily*** and ***seasonal*** variation in the angle of the sun as illustrated in collector ***orientation*** and collector ***tilt***, respectively, are summarized in a concept called the ***sun path***, which we discussed

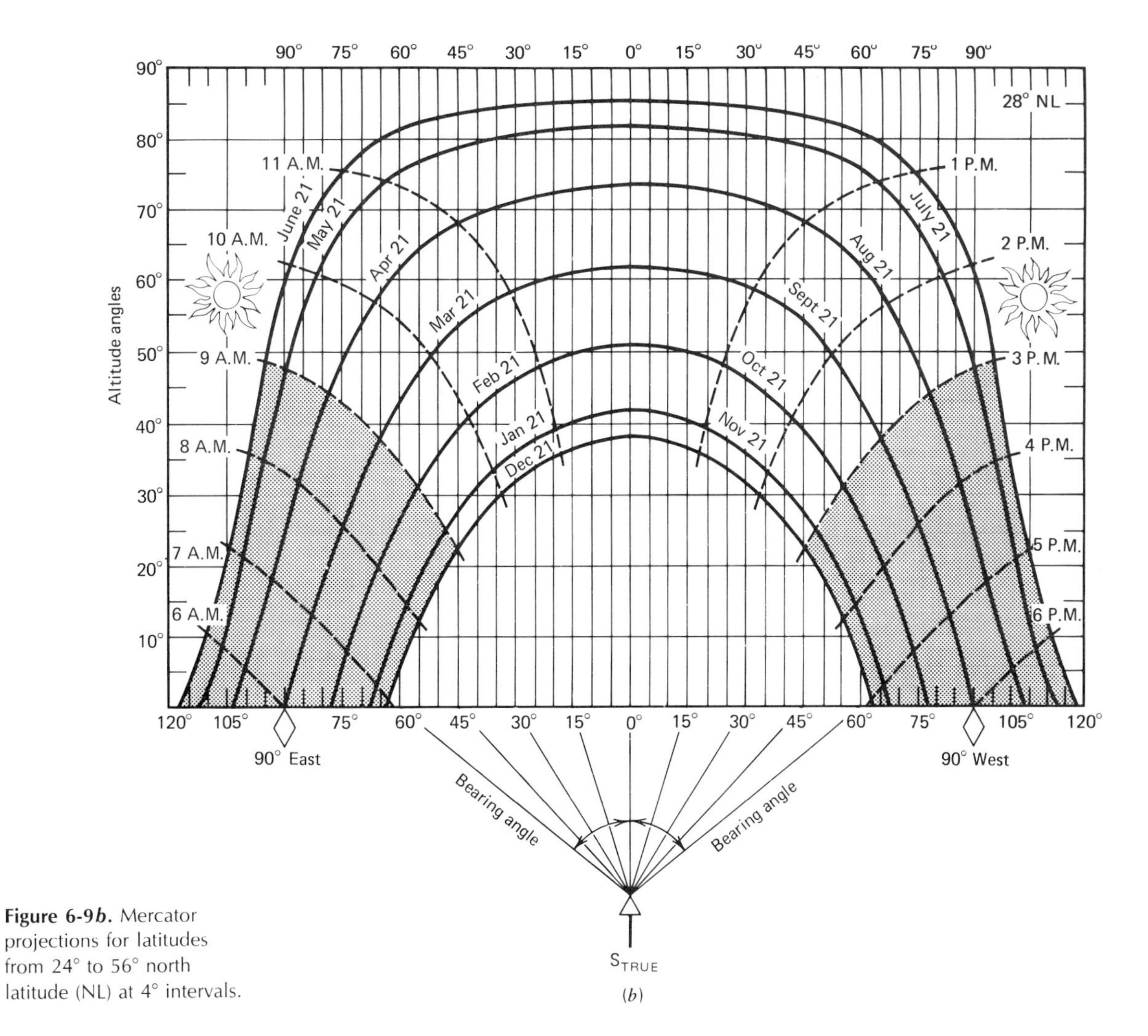


**Figure 6-9*b*.** Mercator projections for latitudes from 24° to 56° north latitude (NL) at 4° intervals.

in Chapter 1. As illustrated in Chapter 1, a "solar window" is actually a plot of the sun's path during the year. From these sun path diagrams we can derive what is called a mercator projection, which graphically depicts altitude and azimuth for each month onto a flat map for each variation of latitude. Figures 6-9*a* through 6-9*i* are mercator projections that have been replotted from the sun path diagrams of Chapter 1. Such a map is very useful for evaluating the site surroundings with respect to the "solar window."

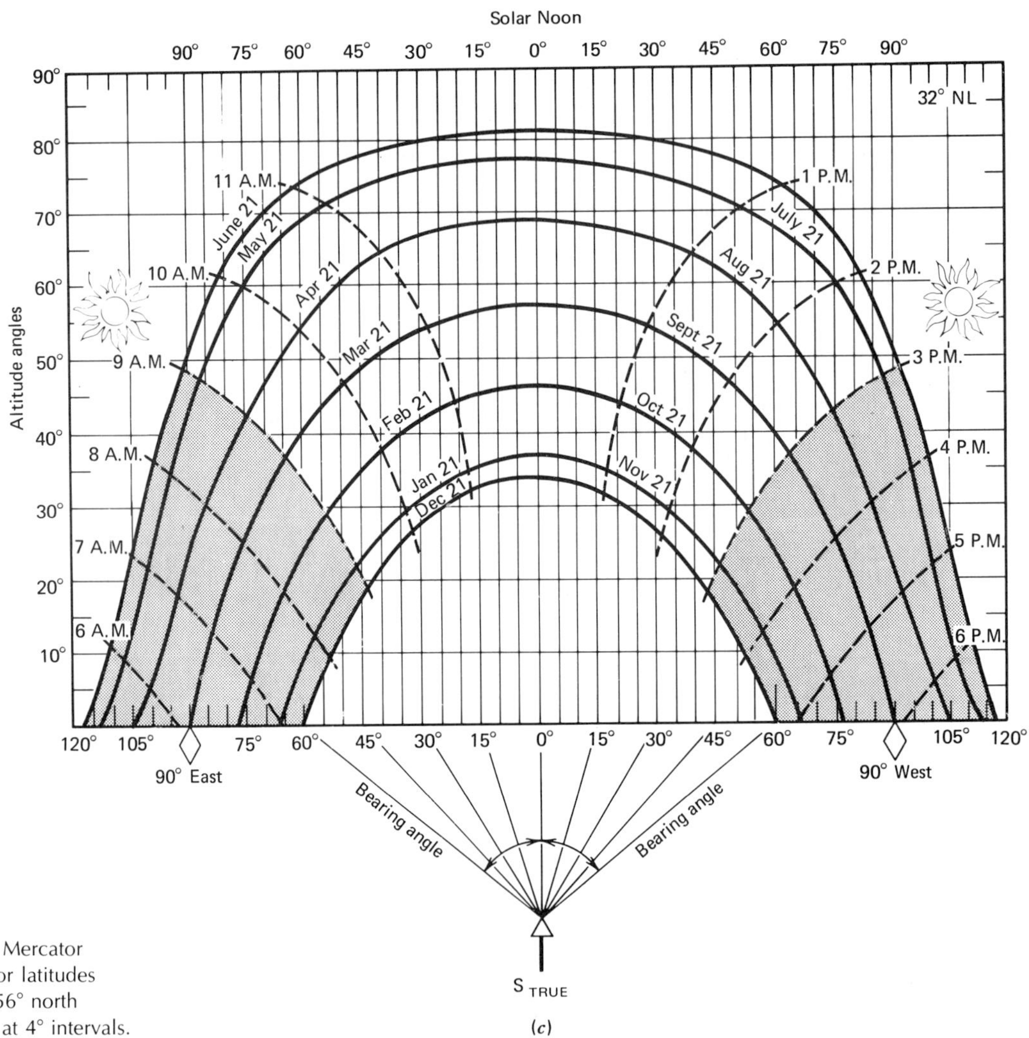


**Figure 6-9c.** Mercator projections for latitudes from 24° to 56° north latitude (NL) at 4° intervals.

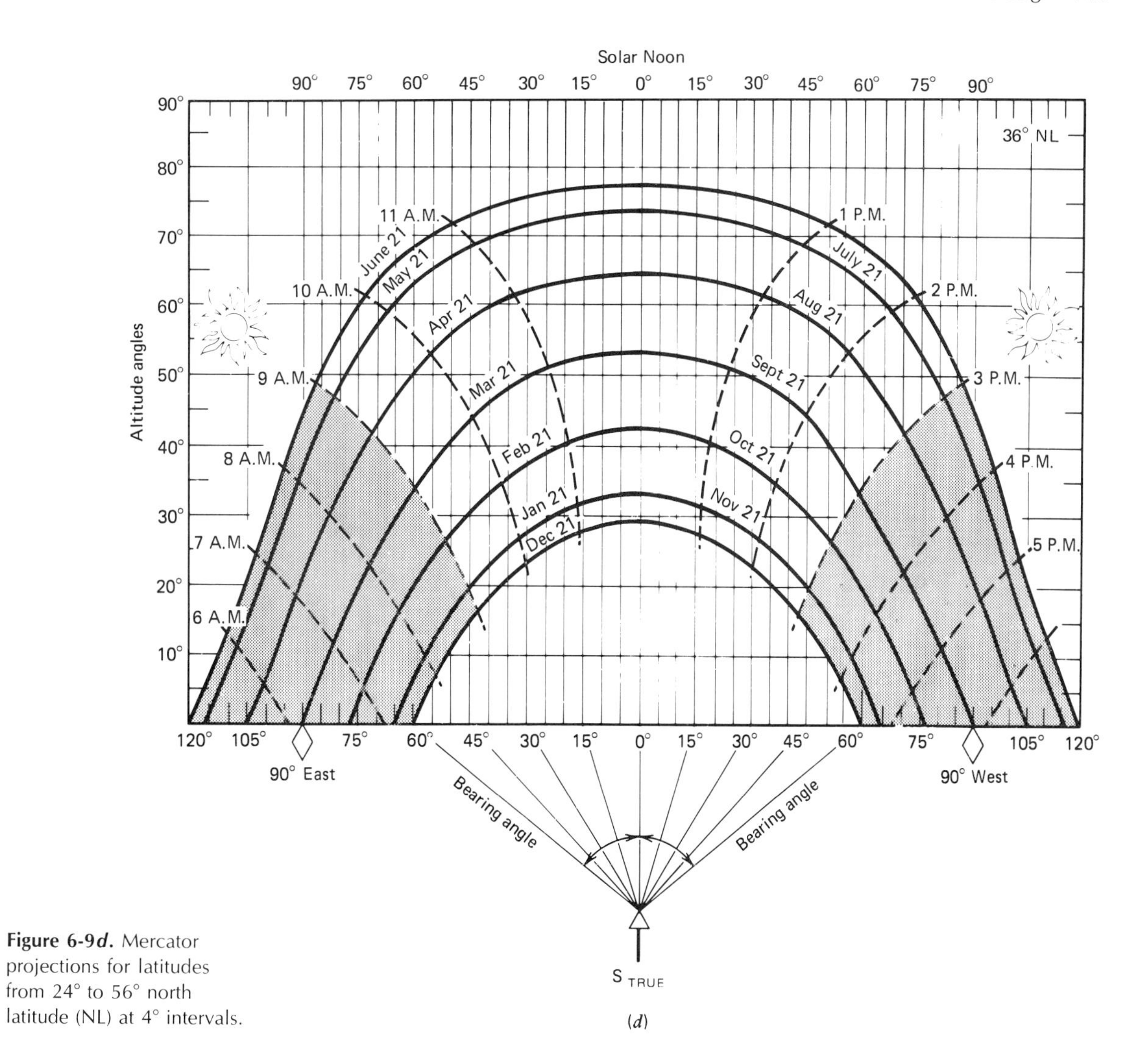


**Figure 6-9*d*.** Mercator projections for latitudes from 24° to 56° north latitude (NL) at 4° intervals.

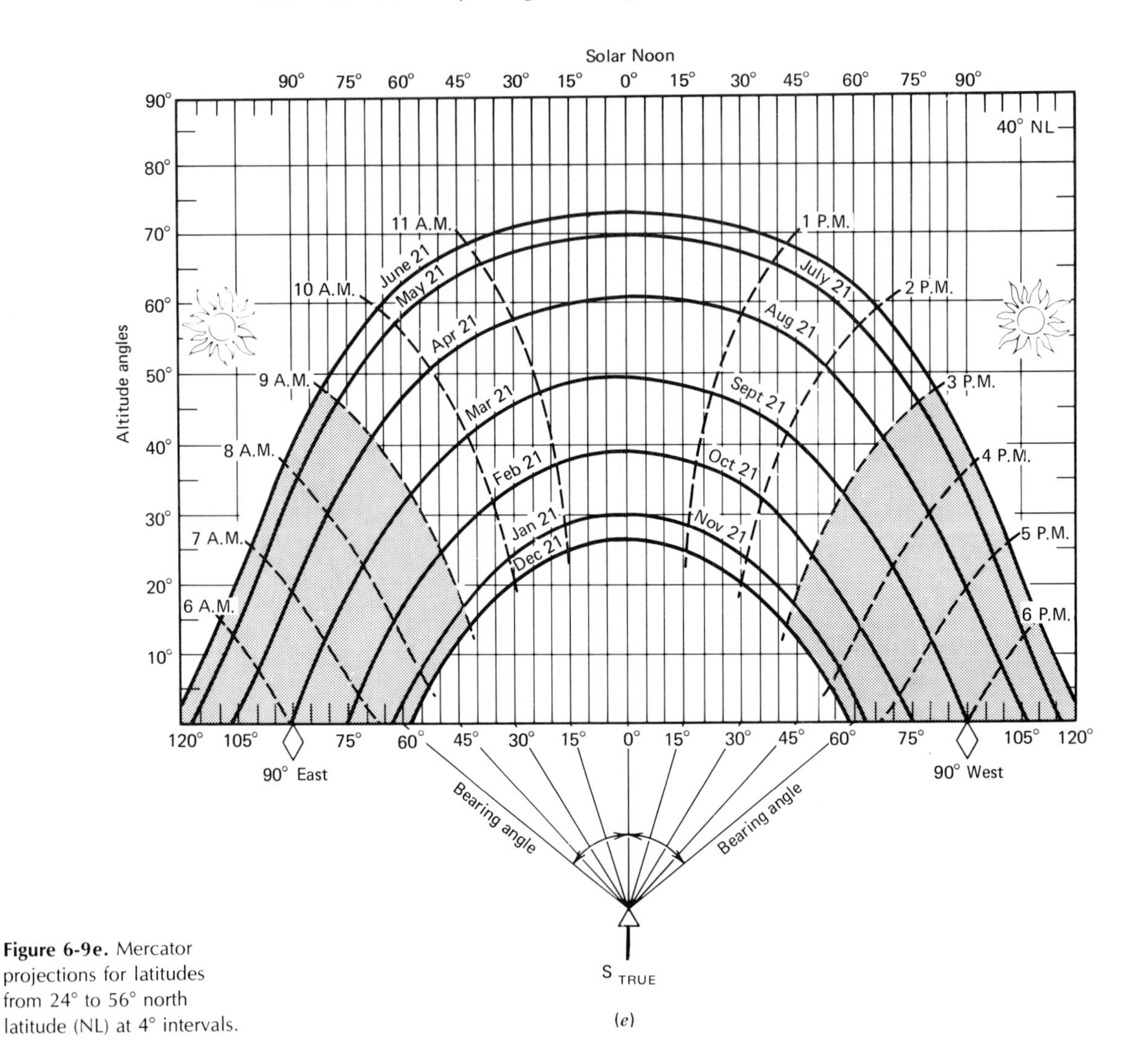


**Figure 6-9e.** Mercator projections for latitudes from 24° to 56° north latitude (NL) at 4° intervals.

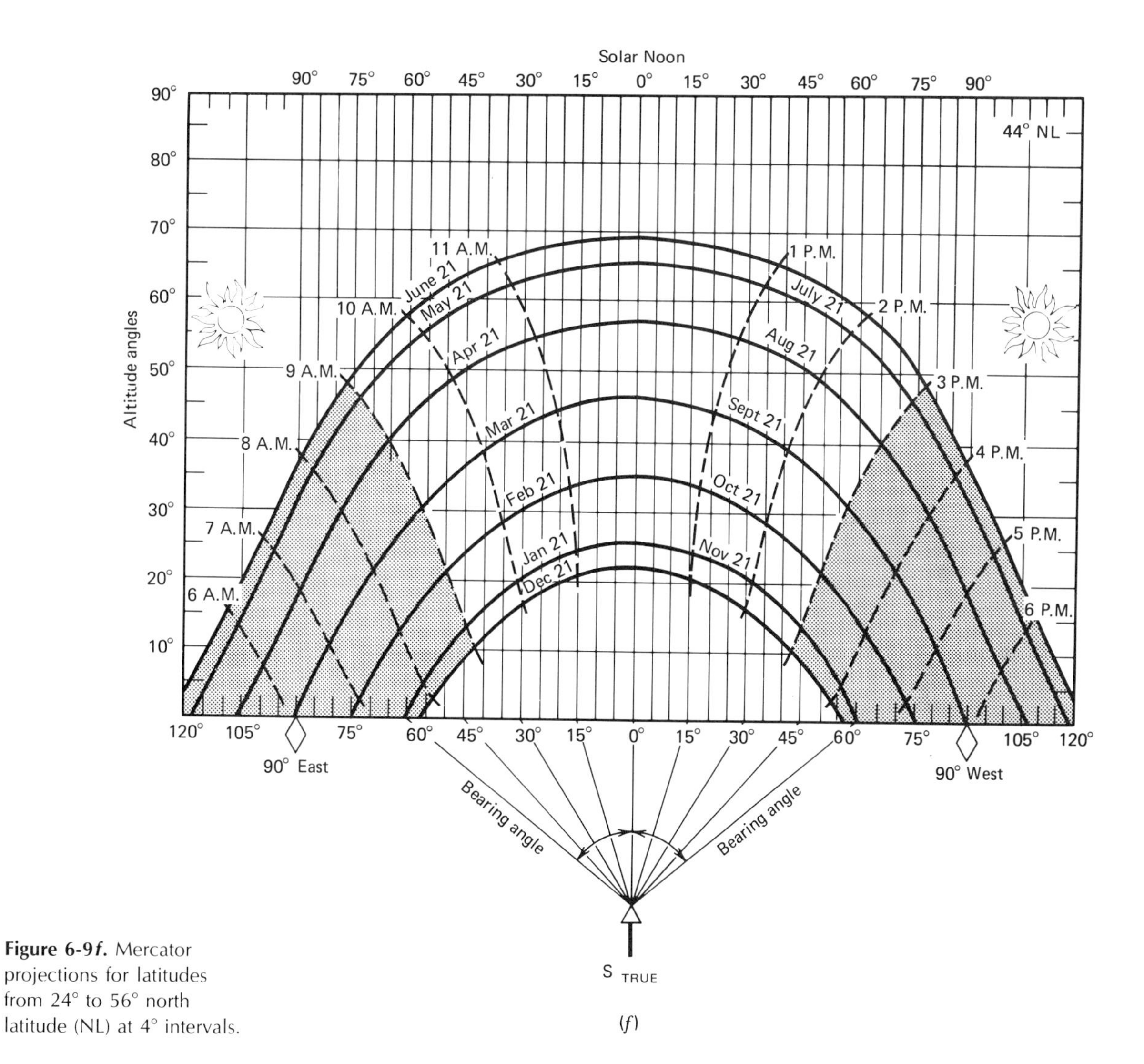


**Figure 6-9*f*.** Mercator projections for latitudes from 24° to 56° north latitude (NL) at 4° intervals.

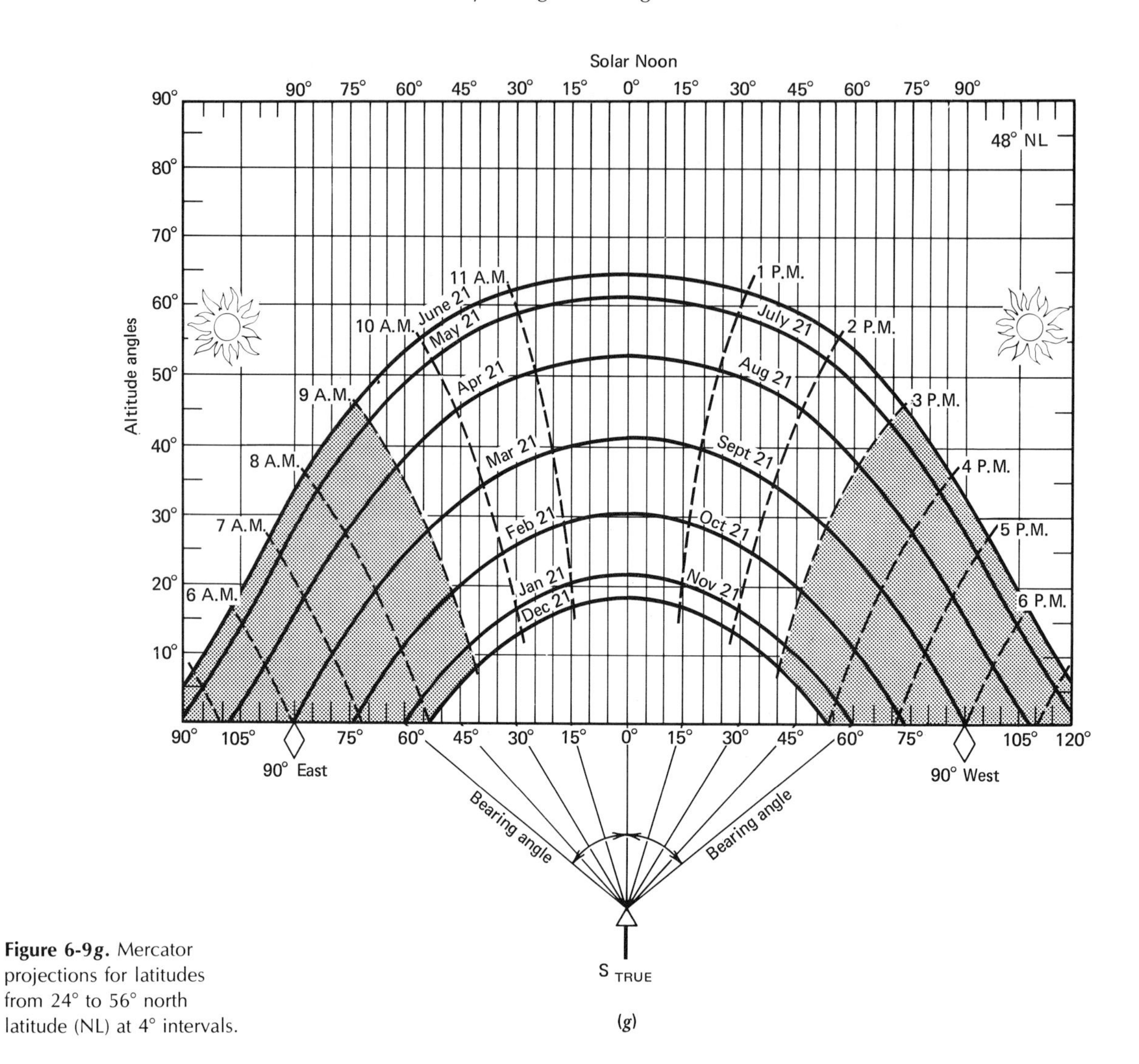


**Figure 6-9*g*.** Mercator projections for latitudes from 24° to 56° north latitude (NL) at 4° intervals.

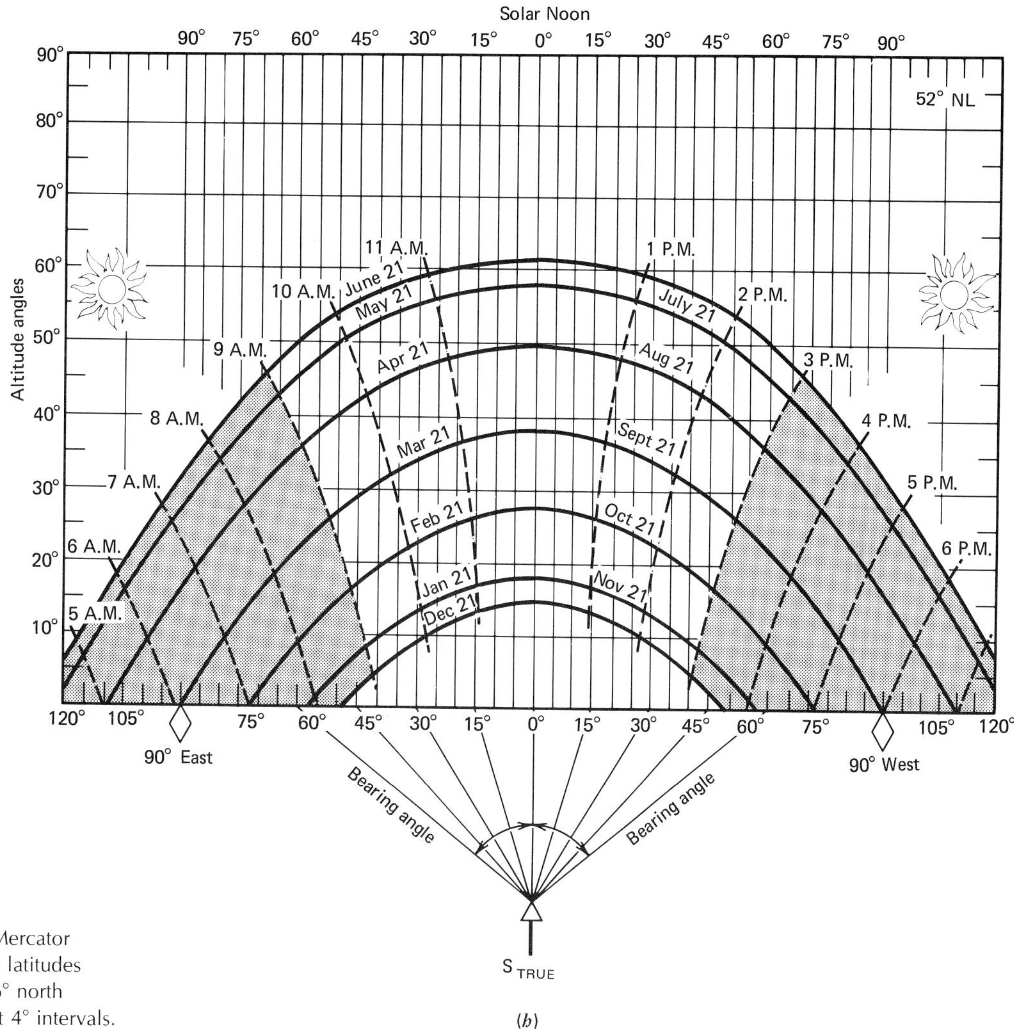


**Figure 6-9*h*.** Mercator projections for latitudes from 24° to 56° north latitude (NL) at 4° intervals.

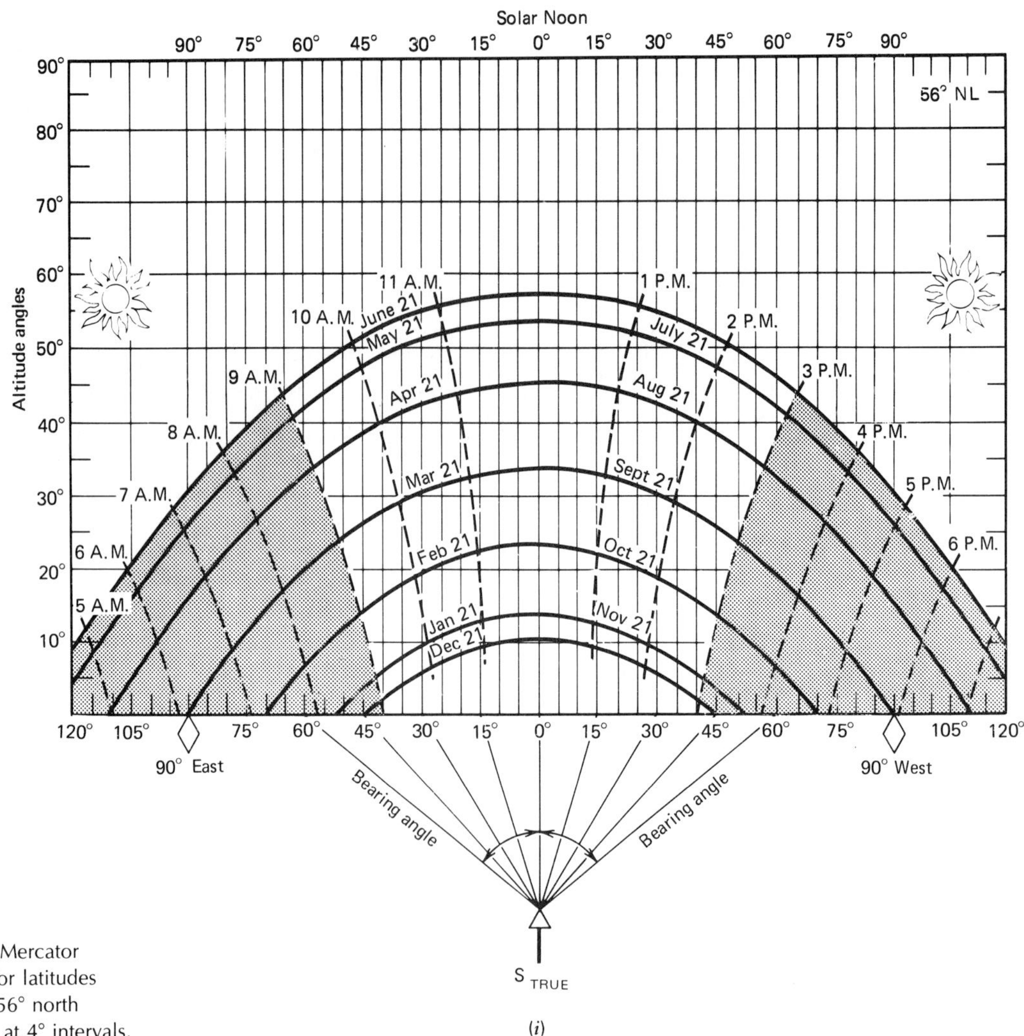


**Figure 6-9*i*.** Mercator projections for latitudes from 24° to 56° north latitude (NL) at 4° intervals.

An instrument that can be used to determine the altitude of potential obstructions is the "abney level," as shown in Figure 6-10. This is a versatile instrument because it can be used to measure slope, to determine height of objects (i.e., trees, poles), to determine elevation in relation to a point of known elevation, and to run lines of levels. By standing at your potential collector site and checking the altitude (in degrees) of all objects with an abney level as illustrated in Figure 6-11, you can plot the results directly on Figure 6-9 for your particular latitude. If this instrument is not available, a homemade plotter can be constructed as illustrated in Appendix A. A typical mercator projection plot is illustrated in Figure 6-12.

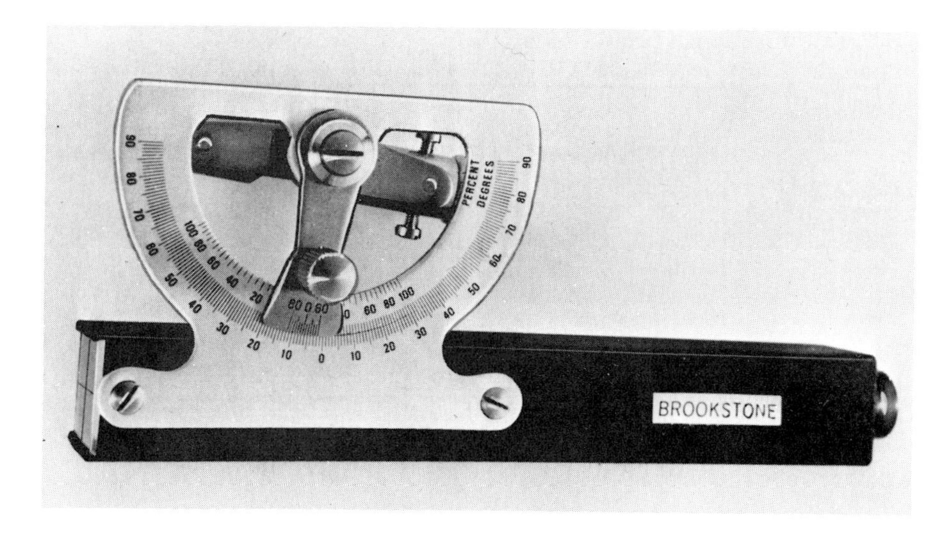


**Figure 6-10.** Abney level. (Courtesy of Brookstone Company, Peterborough, N.H.)

Note that one of the major sources of shading is trees, so the homeowner should be aware of the effect of future growth. Chimneys, dormers, adjacent roof sections, fences, topography, and other buildings may shade the collector array, especially in the winter when the sun's angles are low and shadows are long. One must therefore be careful when observing the full path of the sun to prevent shading such as illustrated in Figures 6-13*a* and 6-13*b*. By knowing the altitude and the azimuth of the sun throughout the year and with the use of Figures 6-9*a* to 6-9*i*, you can accurately determine possible shading problems.

**Figure 6-11.** Siting the collector array.

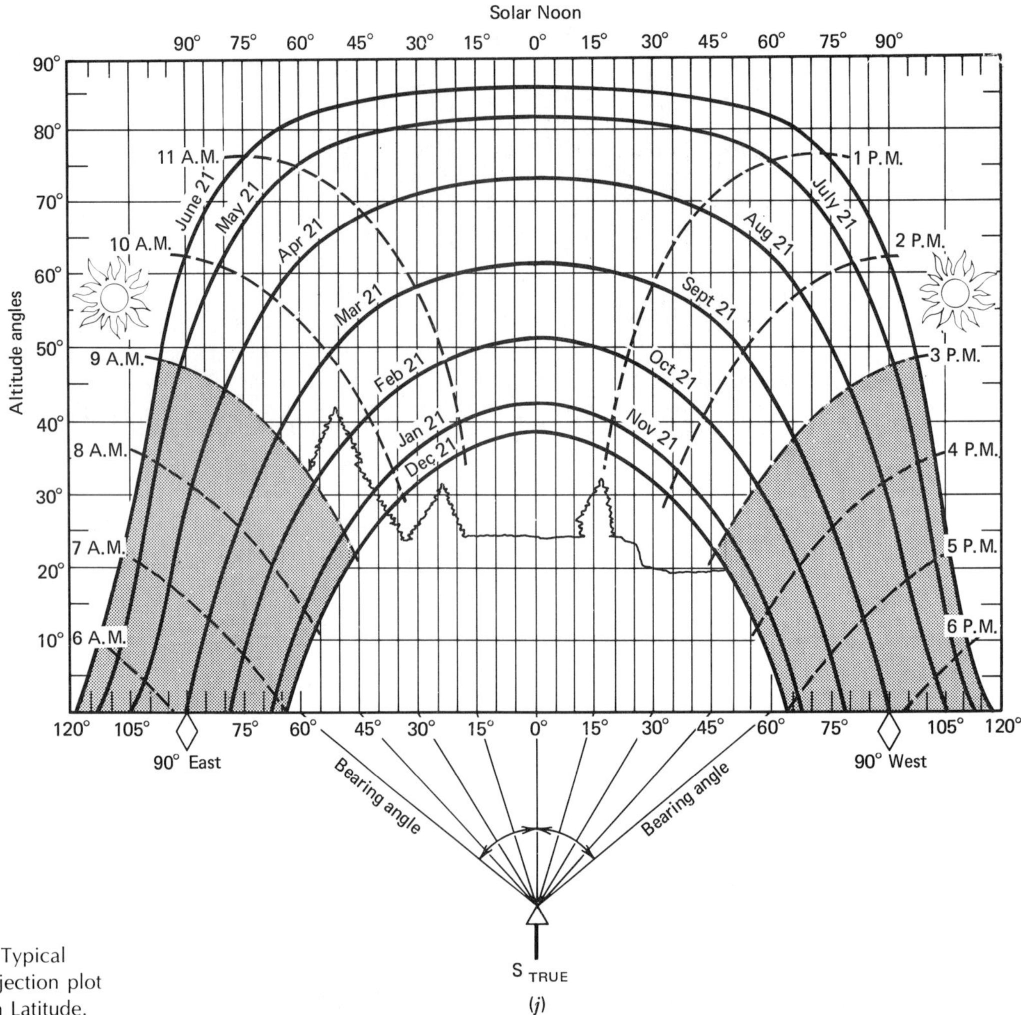


**Figure 6-12.** Typical mercator projection plot for 28° North Latitude.

One now has an explanation of all the parameters necessary to determine collector array sizing, siting, and shading relative to solar domestic hot water systems. The worksheet of Table 6-3a will allow you to summarize and record information for any particular location and to determine the proper collector sizing. Let us illustrate the use of this table by determining the hot water energy consumption, the available solar contribution, and the proper collector sizing for a typical family of four in Billings, Montana. A line-by-line description follows for the example illustrated in Table 6-3b for the month of January.

**Figure 6-13*a*.** Typical shading problems. Early Morning.

**Figure 6-13*b*.** Typical shading problems. Early Afternoon.

**Table 6-3a**
**Worksheet for Collector Sizing, Energy Consumption, and Solar Contribution**

*Latitude* ________
*Collector Tilt Angle* ________

| *Line* | *Evaluation Factors* | | *Jan.* | *Febr.* | *Mar.* | *Apr.* | *May* | *June* | *Jul.* | *Aug.* | *Sept.* | *Oct.* | *Nov.* | *Dec.* | *Total Per Year* |
|---|---|---|---|---|---|---|---|---|---|---|---|---|---|---|---|
| A | Days in month | | 31 | 28 | 31 | 30 | 31 | 30 | 31 | 31 | 30 | 31 | 30 | 31 | 365 |
| B | No. of people | | | | | | | | | | | | | | |
| [1]C | Hot water consumed (gallons; $W_c$) | Daily | | | | | | | | | | | | | |
| | | Monthly | | | | | | | | | | | | | |
| D | Storage temperature ($T_s$) | | | | | | | | | | | | | | |
| E | Inlet water temperature ($T_i$) | | | | | | | | | | | | | | |
| F | Avg. temp. increase D-E ($T_s - T_i$) | | | | | | | | | | | | | | |
| G | BTU requirement (daily) 8.33* C*F | | | | | | | | | | | | | | |
| | | | | | | | | | | | | | | | Yearly average |
| [2]H | Collector system efficiency ($n_s$) | | | | | | | | | | | | | | |
| [3]I | Available solar radiation (BTU/ft²-day) | | | | | | | | | | | | | | |
| J | Array size (ft²) G ÷ (H*I) | | | | | | | | | | | | | | [4] |
| | | | | | | | | | | | | Actual total effective collector area installed (ft²) | | | |

*Month* (spans Jan.–Dec.)

[1]Reference '20 + 20 + 15 + 15 + . . . .' supposition.
[2]Assume an average of 0.5.
[3]Reference Table 6-2.
[4]Effective collector aperture area required.

*Multiplication.
÷Division.

**Table 6-3b (Example)**
**Worksheet for Collector Sizing, Energy Consumption, and Solar Contribution**

*Latitude* 46°
*Collector Tilt Angle* 48°

| *Line* | *Evaluation Factors* | | *Jan.* | *Febr.* | *Mar.* | *Apr.* | *May* | *June* | *Jul.* | *Aug.* | *Sept.* | *Oct.* | *Nov.* | *Dec.* | *Total Per Year* |
|---|---|---|---|---|---|---|---|---|---|---|---|---|---|---|---|
| | | | *Month* | | | | | | | | | | | | |
| A | Days in month | | 31 | 28 | 31 | 30 | 31 | 30 | 31 | 31 | 30 | 31 | 30 | 31 | 365 |
| B | No. of people | | 4 | 4 | 4 | 4 | 4 | 4 | 4 | 4 | 4 | 4 | 4 | 4 | 4 |
| [1]C | Hot water consumed (gallons; $W_c$) | Daily | 70 | 70 | 70 | 70 | 70 | 70 | 70 | 70 | 70 | 70 | 70 | 70 | 25,550 |
| | | Monthly | 2170 | 1960 | 2170 | 2100 | 2170 | 2100 | 2170 | 2170 | 2100 | 2170 | 2100 | 2170 | |
| D | Storage temperature ($T_s$) | | 135 | 135 | 135 | 135 | 135 | 135 | 135 | 135 | 135 | 135 | 135 | 135 | |
| E | Inlet water temperature ($T_i$) | | 40 | 40 | 40 | 45 | 45 | 50 | 50 | 50 | 50 | 45 | 40 | 40 | |
| F | Avg. temp. increase D-E ($T_s - T_i$) | | 95 | 95 | 95 | 90 | 90 | 85 | 85 | 85 | 85 | 90 | 95 | 95 | |
| G | BTU requirement (daily) 8.33* C*F | | 55395 | 55395 | 55395 | 52479 | 52479 | 49564 | 49564 | 49564 | 49564 | 52479 | 55395 | 55395 | 19,260,000 |
| | | | | | | | | | | | | | | | Yearly average |
| [2]H | Collector system efficiency ($n_s$) | | 0.6 | 0.5 | 0.5 | 0.5 | 0.5 | 0.5 | 0.4 | 0.4 | 0.5 | 0.5 | 0.5 | 0.6 | .5 |
| [3]I | Available solar radiation (BTU/ft²-day) | | 1478 | 1972 | 2228 | 2266 | 2234 | 2204 | 2200 | 2200 | 2118 | 1860 | 1448 | 1250 | 1955 |
| J | Array size (ft²) G ÷ (H*I) | | 62.5 | 56.2 | 49.7 | 46.3 | 47.0 | 45.0 | 56.3 | 56.3 | 46.8 | 56.4 | 76.5 | 73.9 | [4] 56.1 |
| | | | | | | | | | | | | | Actual total effective collector area installed (ft²) | | 66.6 |

[1]Reference '20 + 20 + 15 + 15 + . . . .' supposition.
[2]Assume an average of 0.5.
[3]Reference Table 6-2.
[4]Effective collector aperture area required.

*Multiplication.
÷Division.

Line A = 31 days/month

Line B = number of people − (4)

Line C = hot water consumed by 4 people = 70 gal/day × 31 days/month
= 2170 gal/ month

Line D = storage temperature needed at 135° F (family has a dishwasher without a preheater)

Line E = inlet water temperature to storage is 40° F during January

Line F = Line D − Line E
= 135° F − 40° F = 95° F

Line G = 8.33 lb/gal × Line C × Line F
= 8.33 lb/gal × 70 gal/day × 95° F
= 55,395 BTU/day

Line H = collector system efficiency (conservative average)
= 0.6

Line I = available insolation at 46 deg north latitude at optimum collector tilt of 46 deg (Use Table 6-2, 48 deg north latitude at nearest optimum collector tilt of 48 deg)
= 1478 BTU/ft$^2$-day

Line J = (Line G) ÷ (Line H × Line I)
= (55,395 BTU/ day) ÷ (0.6 × 1478 BTU/ft$^2$-day)
= 62.5 ft$^2$

This procedure can be followed to determine the minimum collector array sizing required to meet the domestic hot water demand for each month. In this example, the yearly average collector area required is calculated to be 56.1 ft$^2$. If the collector effective aperature area is 22.2 ft$^2$, then three collectors should be installed, resulting an actual total effective collector area installed of 66.6 ft$^2$. The total solar contribution for heating domestic water for each month from such a collector array will be illustrated in Chapter 10.

## REVIEW QUESTIONS

1. What is the daily hot water BTU requirement for a family of four using 80 gal per day? Assume the inlet water to the storage tank to be 50° F and the storage temperature to be maintained at 125° F.

   a. 6000 BTU b. 39,620 BTU
   c. 49,980 BTU d. 83,300 BTU

2. If the efficiency ($n_s$) of a collector system is 0.6 and the average insolation is 1860 BTU/ft$^2$-day, what collector area is required to supply the heating requirement of Question 1?

   a. 26.8 ft$^2$ b. 35.5 ft$^2$
   c. 44.8 ft$^2$ d. 67.2 ft$^2$

3. The system parameters necessary to determine sizing the collector area include which of the following?
   a. Water usage
   b. Water storage
   c. Water inlet temperature
   d. Available insolation
   e. Specific heat
4. The collector array should face the middle of the sun's daily path within ±15 deg east or west of true south.
   True or False
5. Because the earth's magnetic field is aligned parallel with the earth's north-south axis, a compass will not read true.
   True or False
6. On the east side of the zero magnetic deviation line, a compass needle will point to the east of true south.
   True or False
7. A collector array is being sited in Blytheville, Arkansas. What correction to the magnetic compass reading should be made to correct for a true south reading?
   a. Add 18 deg east to the magnetic south reading
   b. Add 18 deg west to the magnetic south reading
   c. Add 3 deg east to the magnetic south reading
   d. Add 18 deg east to the magnetic north reading
   e. Add 3 deg west to the magnetic north reading
   f. Add 3 deg west to the magnetic south reading
8. If sunrise and sunset are 6:10 A.M. and 5:30 P.M. Standard Time, respectively, at what time is solar noon?
   a. 11:20 A. M.     b. 12:00 P. M.
   c. 12:20 P. M.     d. 11:50 A. M.
   e. None of the above
9. The collector array should face the middle of the sun's seasonal path which is an angle from horizontal equal to the latitude. The rule generally followed for the tilt of the collector in the northern hemisphere is to position the collector at an angle of latitude plus or minus what angle of variation?
   a. 5 deg     b. 10 deg
   c. 20 deg     d. 30 deg
10. Shading is a very important consideration when siting a collector array.

What percentage of the collector array should remain unshaded between 9 A.M. and 3 P.M.?

a. 50 percent
b. 85 percent
c. 90 percent
d. 95 percent
e. None of the above

11. What is the term that depicts altitude and azimuth for each month onto a flat map for each variation of latitude?
   a. Isogonic chart
   b. Sun path diagram
   c. "Solar window" plot
   d. Mercator projection
   e. Sky vault diagram

12. Which of the following methods would take the most time in determining true south?
   a. Using a compass and an isogonic chart
   b. Using the "season variation method"
   c. Using the "solar noon" method
   d. Using the "stake and shadow" method
   e. None of the above

13. If the orientation requirement cannot be met, then nothing can be done to use solar energy as an alternative for domestic water heating.
   True or False

14. What is the average surface daily insolation in Denver, Colorado, if the collector array is tilted an angle the same as the value of latitude, 40 deg?
   a. 2108 BTU/ft$^2$-day
   b. 1810 BTU/ft$^2$-day
   c. 2182 BTU/ft$^2$-day
   d. 1777 BTU/ft$^2$-day

15. At what maximum altitude can an object be relative to a collector array at 40 deg north latitude so as not to interfere with the "solar window"?
   a. 10 deg
   b. 15 deg
   c. 20 deg
   d. 25 deg
   e. 30 deg

# System Installation: Roof Mounting and Ground Mounting Procedures

Siting the collector array has determined whether or not the collectors should be roof mounted or ground mounted. Mounting the collectors is perhaps the most arduous task in the installation process. It may not be the first thing to do especially if there is a possibility that the collectors may not be filled for several days, in which event the major part of the plumbing should be completed first. Although ground mounting is an easier task, roof mounting is preferable. It is not because the collectors are closer to the sun when roof mounted, but that (1) they are not as subject to breakage and vandalism, (2) there is normally less exposed pipe, thereby increasing system performance and reducing freeze problems, and (3) system installation is less costly in time and materials.

There are many factors to consider in mounting the collectors including safety, types of mounting, basic supporting rack designs, and associated loading considerations, as well as the actual step-by-step mounting procedures. Before installing the system, remember to investigate the need for a building permit at the local town/city office and consult the local building codes for material and installation regulations.

## ROOF MOUNTING

### General Information

Personal safety is of the utmost importance when installing the support rack and attaching the collectors. For people who do not work well at heights, a ⅜-in. safety rope can be used and/or a roof ladder can be constructed as illustrated in Appendix A (Construction of Useful Tools). A roof ladder can be invaluable when first establishing the collector racks. Collectors should be mounted near midheight of a pitched roof and should if possible be parallel with the ridge. (This can be accomplished by measuring the distance from the top of the collector rack to the ridge line at each end of the array, drawing a chalk line, and checking the parallel line with a bubble level.) Collectors should not be positioned near the ridge of the roof nor in a position in which they are susceptible to high winds. Collectors should also not be placed within 24 in. of the lower edge of a pitched roof, because ice dams will have a tendency to form. Consideration should be given to minimizing the length of pipe run to storage thus preventing unnecessary heat loss and reducing pipe friction to flow. One should also ensure that there is no potential danger with snow and ice sliding off the collectors, dropping onto pedestrian walkways, driveways, or building entrances. The roof surface should be in good repair prior to mounting the collectors. If there is a need for reshingling or other repair work, it should be done. The tools and materials necessary for roof mounting include:

| *Material* | *Tools* |
|---|---|
| Silicon caulking | Hammer |
| Solder (95/5) | Socket set |
| Flux | Caulking gun |
| Unions | Electric Drill and bits |
| ⅜ in. Lag screws or threaded rods | Extension cords |
| | Propane torch |
| Nails (8d) | Ladders |
| 2 × 4 lengths | Rope |
| Galvanized pipe | Level |
| or angle iron | Chalk line |
| | Allen Wrench (where required) |
| | Hacksaw |

The actual tilt of the roof must be determined to ensure the necessary tilt angle is within latitude ±10 deg, thereby meeting the requirements discussed in Chapter 6. If the roof angle is not within the required specifications, an additional angle must be added to the collector support rack to compensate. The angle of a typical gable roof can be obtained by determining either the ***pitch*** or ***slope*** of the roof. The terms *pitch* and *slope* are not synonymous, however, either of these terms can be used to determine approximate roof angle. Equations 7-1 and 7-2 explain the difference between pitch and slope.

Equation 7-1 represents the determination of pitch and the terms involved are illustrated in Figure 7-1.

***Equation 7-1***

$$\text{Pitch} = \frac{\text{rise}}{\text{span}}$$

*where* Rise = vertical distance from attic floor to roof peak
Span = width of the house

For example, if the rise was 12 ft and the span was 24 ft, then the pitch would be ½ or

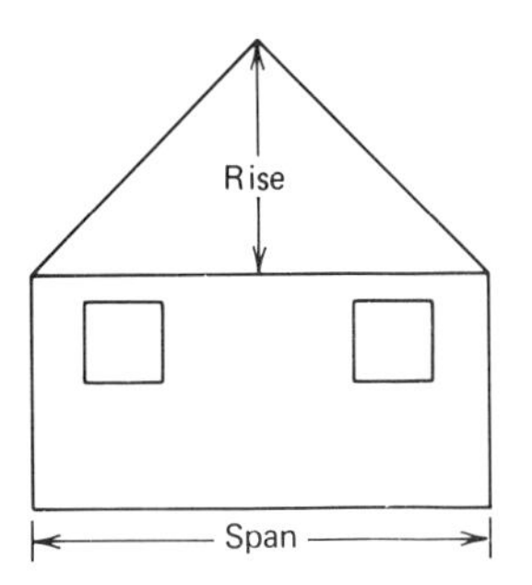

**Figure 7-1.** Determination of roof pitch.

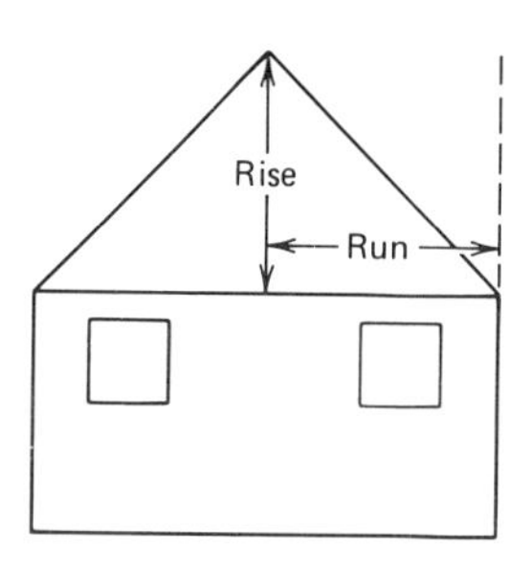

**Figure 7-2.** Determination of roof slope.

$$\text{Pitch} = \frac{\text{rise}}{\text{span}} = \frac{12 \text{ ft}}{24 \text{ ft}} = \frac{1}{2}$$

Equation 7-2 represents the determination of slope and the terms involved are illustrated in Figure 7-2.

***Equation 7-2***

$$\text{Slope} = \frac{\text{rise}}{\text{run}}$$

*where* Rise = vertical distance from attic floor to roof peak
Run = horizontal distance from the roof peak to the end of the roof section being measured.

For example, if the rise is 12 ft and the run is 12 ft, then the slope would be 1 or

$$\text{Slope} = \frac{\text{rise}}{\text{run}} = \frac{12}{12} = 1$$

The comparison of these two examples illustrates the same house to have a roof slope equal to 1 and a roof pitch equal to ½. These ratios measure different relationships and should not be thought of as equivalent fractions. The actual determination of roof angle in degrees can be obtained with a little background in geometry and trigonometry.

Slope is often included in architectual plans and is normally thought of in terms of a certain number of inches vertically for every 12 in. of run horizontally. This is illustrated in Figure 7-3. The angle $\theta_1$, which is the roof angle in units of degrees, can be determined by solving for angle $\theta_2$. From basic geometry it can be proved that $\theta_1 = \theta_2$ and the following geometric theorem can be used:

> If two parallel lines (lines $A$ and $B$) are cut by a transversal $T$, then alternate interior angles, $\theta_1$ and $\theta_2$, are congruent.

The angle $\theta_2$ can be determined by using the trigonometric identity for the tangent of an angle as defined in Appendix B:

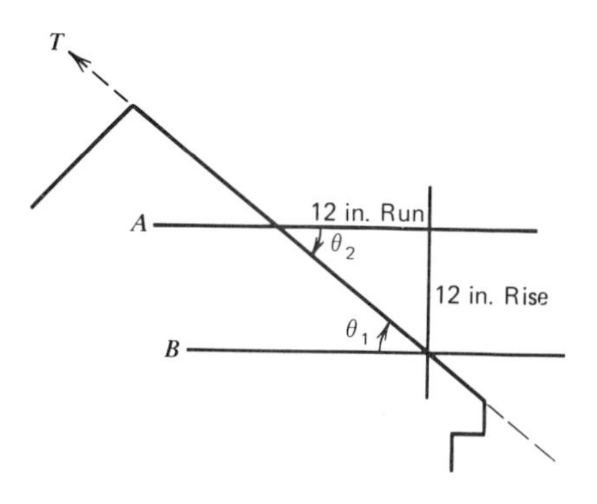


**Figure 7-3.** Determination of slope from fractions to degrees.

$$\text{Tan } \theta_2 = \frac{12'' \text{ rise}}{12'' \text{ run}} = 1$$

and

$$\theta_2 = \tan^{-1}(1)$$
$$\theta_2 = 45 \text{ deg}$$

Since $\theta_1 = \theta_2$ from our geometric theorem, then

$$\theta_1 = \theta_2 = 45 \text{ deg}$$

Now that we understand how the roof angle is derived we can effectively use the roof angle chart depicted in Table 7-1, which summarizes the more common relationships between pitch, slope, and roof angle.

For example, if a house was located at 40 deg north latitude and oriented within $\pm 15$ deg of true south, the angle of the roof should ideally be 30 to 50 deg for a direct mount collector array, which is equivalent to a slope of 7/12 to 14/12 and a pitch of 7/24 to 7/12. If the slope or pitch is other than within these ranges, then the support rack must be compensated by either an increase or decrease for the proper tilt angle.

There are four general categories in which flat-plate collectors can be mounted. These categories consist of (1) direct rack mounting to existing roof slope, (2) tilt rack mounting to existing roof slope, (3) tilt rack mounting to a flat roof, and (4) integral mounting.

### Direct Mounting

If a suitable tilt angle exists as determined by the methods of Chapter 6, the collectors can be mounted on a rack directly onto the roof with no additional tilt to the rack. This is the easiest type of roof mounting in that rack design may consist of a parallel array of galvanized piping or angle iron. There are no additional angles that must be considered in the design of the supporting rack. The bottom of the collector array should be located from 4 to 6 in. above the roof surface as illustrated in Figure 7-4. This height above the roof surface will tend to have a minimum effect on the fire resistance characteris-

**Table 7-1**
**Roof Angle Chart**

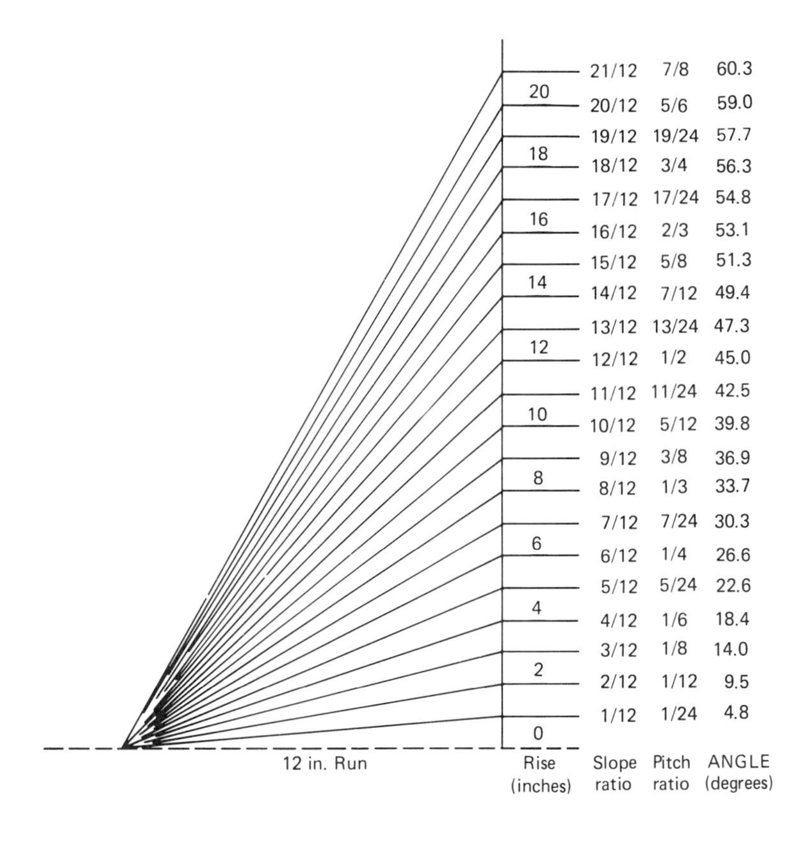


| Slope ratio | Pitch ratio | ANGLE (degrees) |
|---|---|---|
| 21/12 | 7/8 | 60.3 |
| 20/12 | 5/6 | 59.0 |
| 19/12 | 19/24 | 57.7 |
| 18/12 | 3/4 | 56.3 |
| 17/12 | 17/24 | 54.8 |
| 16/12 | 2/3 | 53.1 |
| 15/12 | 5/8 | 51.3 |
| 14/12 | 7/12 | 49.4 |
| 13/12 | 13/24 | 47.3 |
| 12/12 | 1/2 | 45.0 |
| 11/12 | 11/24 | 42.5 |
| 10/12 | 5/12 | 39.8 |
| 9/12 | 3/8 | 36.9 |
| 8/12 | 1/3 | 33.7 |
| 7/12 | 7/24 | 30.3 |
| 6/12 | 1/4 | 26.6 |
| 5/12 | 5/24 | 22.6 |
| 4/12 | 1/6 | 18.4 |
| 3/12 | 1/8 | 14.0 |
| 2/12 | 1/12 | 9.5 |
| 1/12 | 1/24 | 4.8 |

**Figure 7-4**

tics of the roof-covering materials and will allow water, snow, ice, and other debris to pass underneath. Mildew and leakage problems are also minimized.

## Tilt Mounting to Sloped Roof

If a suitable tilt angle does not exist as determined by the methods of Chapter 6, the collectors' supporting rack must be adjusted in angle in addition to the roof angle as illustrated in Figure 7-5.

The additional height of the rear standoff of the support rack can be calculated using basic trigonometry. For example, if the house was located 40 deg north latitude and oriented within ±15 deg of true south and the angle of the roof was determined to be only 20 deg ($\theta_1 = 20$ deg) we find the support rack should be tilted an additional minimum increase of 10 deg and ideally an additional tilt increase of 20 deg ($\theta_2 = 20$ deg) for a collector tilt equal to the actual latitude. As illustrated in Figure 7-6 we know the collector length (assume 96 in.) by simple measurement as well as the increase in angle needed ($\theta_2 = 20$ deg). The additional height ($h$) needed to raise the tilt of the collector another 20 deg can be obtained by using one of the trigonometric relations defined in Appendix B:

$$Sin\ \theta_2 = \frac{h}{96}$$

Solving for $h$ we have

$$h = 96 \sin \theta_2 = 96 \sin (20°) = 96(0.34)$$

And $h = 32.6$ in., which means the rear standoff should be 32.6 in. longer than the front standoff. If the front standoff is 4 in. longer, then the rear standoff should be 4 in. plus 32.6 in. or 36.6 in. Table 7-2 includes typical rear standoff heights for collectors of approximately 8 ft in length.

### Tilt Rack Mounting to a Flat Roof

As illustrated in Chapter 6, there is little benefit in mounting a collector in a horizontal position for solar domestic hot water. Because there is no roof slope or pitch, there is no roof angle to determine. The collector support

Figure 7-5

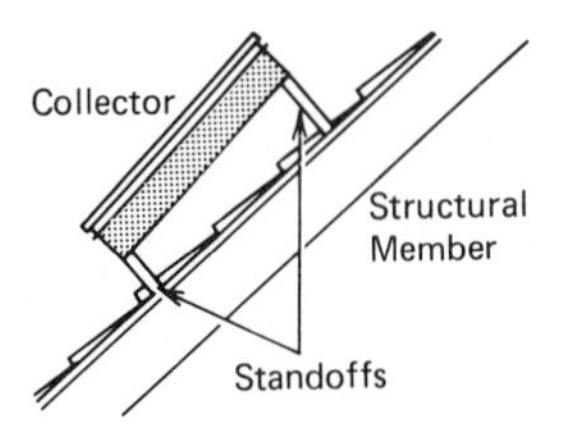


Figure 7-6

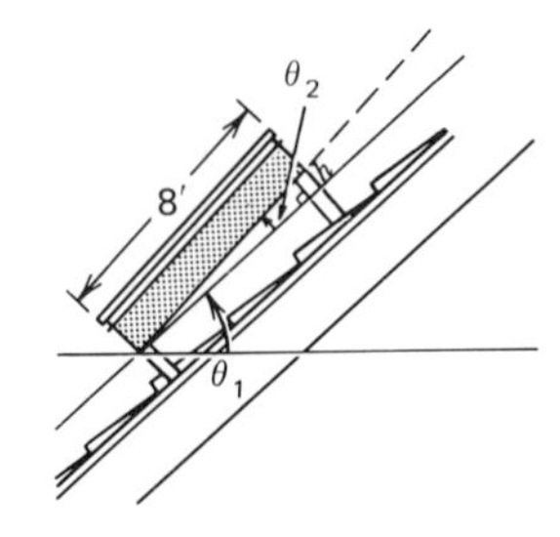

**Table 7-2**
**Additional Height for Rear Standoff**
**(Assume Standard Collector Length = 96 In.)**

| | North Latitude | | | | | | | | | |
|---|---|---|---|---|---|---|---|---|---|---|
| *Roof Angle* | *24°* | *28°* | *32°* | *36°* | *40°* | *44°* | *48°* | *52°* | *56°* | *60°* |
| 4.8 | 31.6 | 37.8 | 43.9 | 49.7 | 55.3 | 60.7 | 65.7 | 70.4 | 74.8 | 78.8 |
| 9.5 | 24.0 | 30.5 | 36.7 | 42.8 | 48.7 | 54.4 | 59.8 | 64.9 | 69.6 | 74.1 |
| 14.0 | 16.7 | 23.2 | 29.7 | 36.0 | 42.1 | 48.0 | 53.7 | 59.1 | 64.2 | 69.1 |
| 18.4 | 9.4 | 16.0 | 22.6 | 29.0 | 35.3 | 41.5 | 47.4 | 53.1 | 58.6 | 63.7 |
| 22.6 | 2.4 | 9.0 | 15.7 | 22.2 | 28.7 | 35.0 | 41.2 | 47.1 | 52.8 | 58.3 |
| 26.6 | — | 2.3 | 9.0 | 15.7 | 22.2 | 28.7 | 35.0 | 41.2 | 47.1 | 52.8 |
| 30.3 | — | — | 4.8 | 9.5 | 16.2 | 22.7 | 29.2 | 35.5 | 41.6 | 47.6 |
| 33.7 | — | — | — | 3.9 | 10.5 | 17.2 | 23.7 | 30.1 | 36.4 | 42.5 |
| 36.9 | — | — | — | — | 5.2 | 11.9 | 18.5 | 25.0 | 31.4 | 37.7 |
| 39.8 | — | — | — | — | 0.3 | 7.0 | 13.7 | 20.3 | 26.8 | 33.1 |
| 42.5 | — | — | — | — | — | 2.5 | 9.2 | 15.8 | 22.4 | 28.9 |
| 45.0 | — | — | — | — | — | — | 5.0 | 11.7 | 18.3 | 24.9 |
| 47.3 | — | — | — | — | — | — | 1.2 | 7.9 | 14.5 | 21.1 |
| 49.4 | — | — | — | — | — | — | — | 4.4 | 11.0 | 17.7 |
| 51.3 | — | — | — | — | — | — | — | 1.2 | 7.9 | 14.5 |
| 53.1 | — | — | — | — | — | — | — | — | 4.9 | 11.5 |
| 54.8 | — | — | — | — | — | — | — | — | 2.0 | 8.7 |
| 56.3 | — | — | — | — | — | — | — | — | — | 6.2 |
| 57.7 | — | — | — | — | — | — | — | — | — | 3.9 |
| 59.8 | — | — | — | — | — | — | — | — | — | 0.3 |

rack should be constructed to tilt the collectors at an angle equivalent to the latitude.

If air collectors are used, a separate structure should be built to house the ductwork as illustrated in Figures 7-7 and 7-8.

If liquid collectors are used, the collector support rack can be fabricated from galvanized pipe or angle iron as illustrated in Figure 7-9. Because ¾-in. copper pipe is used in most situations, an exterior housing is not a necessity for liquid collector systems.

## Integral Mounting

Unlike the three previous mounting categories, integral mounting places the collector within the roof construction itself. Thus, the collector is attached to and supported by the structural framing members. In addition, the top of the collector serves as the finished roof surface. Weather tightness is therefore critical to avoid problems of water damage and mildew. This type of mounting is typical for air collectors as illustrated in Figure 7-10 and the site-built collectors in new construction as illustrated in Figure 7-11.

**Figure 7-7.** Typical flat roof mounting support for air collectors (front view). (Courtesy of Granite State Solar Industries, Inc., Dover, N.H.)

**Figure 7-8.** Typical flat roof mounting support for air collectors (back view). (Courtesy of Granite State Solar Industries, Inc., Dover, N.H.)

**Figure 7-9.** Typical flat roof mounting support for liquid collectors (back view). (Courtesy of Applied Technologies, Inc., Kittery, Maine)

**Figure 7-10.** Integral mounting of air collectors. (Courtesy of Granite State Solar Industries, Inc., Dover, N.H.)

**Figure 7-11.** Integral mounting of site building collectors. (Courtesy of Reimann-Georger Hoist Co., Inc., Buffalo, N.Y.)

## Collector Support Racks

### Typical Types

There are several types of mounting bracket kits available from the various collector manufacturers for direct and tilt mounted applications. Supporting racks can be constructed from slotted angle rail (i.e., 12 gauge, 1½ in. by 1½ in.) as illustrated in Figures 7-12 and 7-13.

Direct mounting bracket kits (Figure 7-12) include T-brackets that lock the collector to commercially available Unistrut® channels which are attached to the roof or other substructure. The tilt mounting bracket kits (Figure 7-13) include a standard top and bottom bracket designed for use with transition struts. These bracket kits have been engineered to facilitate collector installation and to contribute to the reliability of the system.

Supporting racks can also be constructed from galvanized pipe and fittings. This type of support rack can be easily assembled by the homeowner because it can be constructed with only a hacksaw and an Allen wrench. Note that if it is necessary to cut any piece of galvanized metal, the exposed edge should be coated with zinc chromate-based paint for corrosion protection. Standard pipe dimensions (schedule 40) require 1 in. galvanized pipe for direct mounted applications to 1½ in. galvanized pipe for tilt mounted applications. Standard pipe dimensions are illustrated in Figure 7-14.

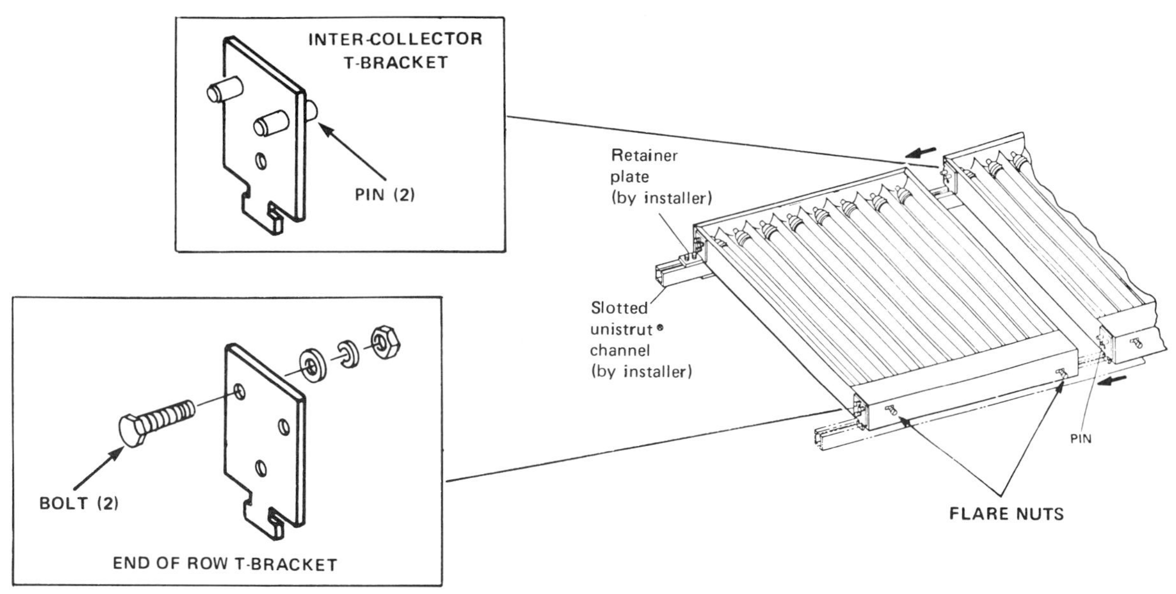


Figure 7-12. Direct mounting kits. (Courtesy of General Electric Co., Philadelphia, Penn.)

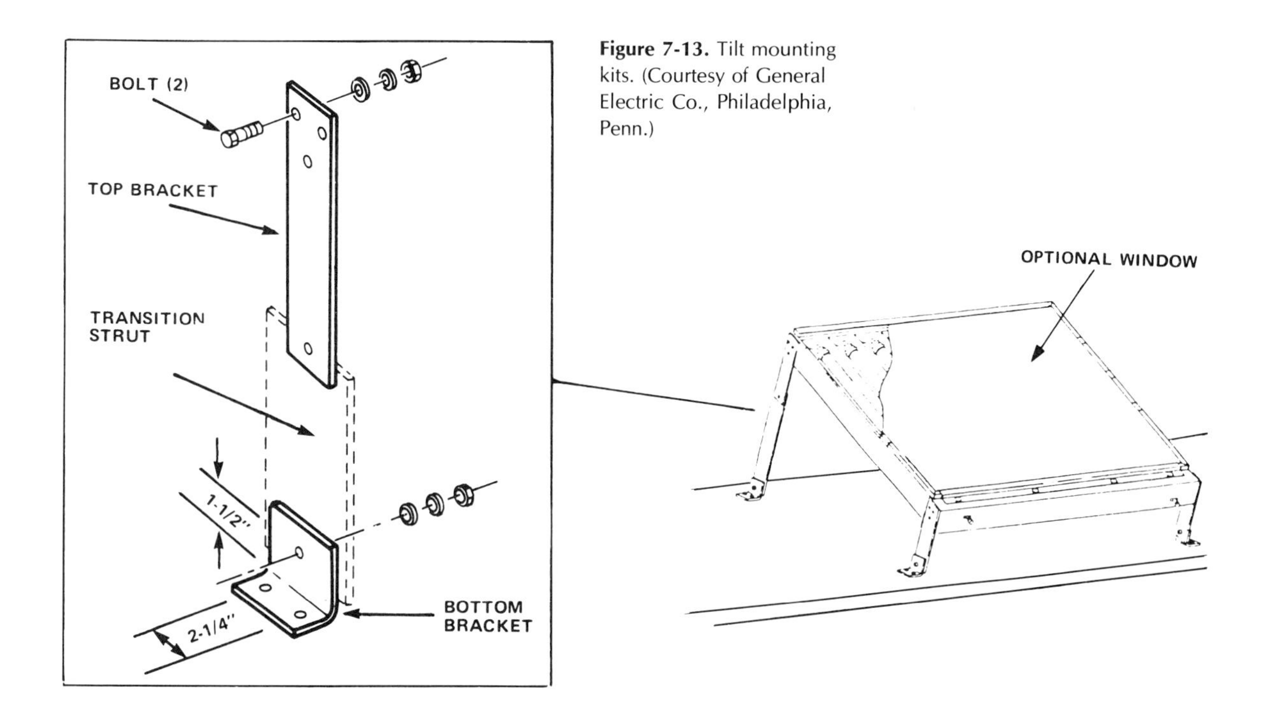


Figure 7-13. Tilt mounting kits. (Courtesy of General Electric Co., Philadelphia, Penn.)

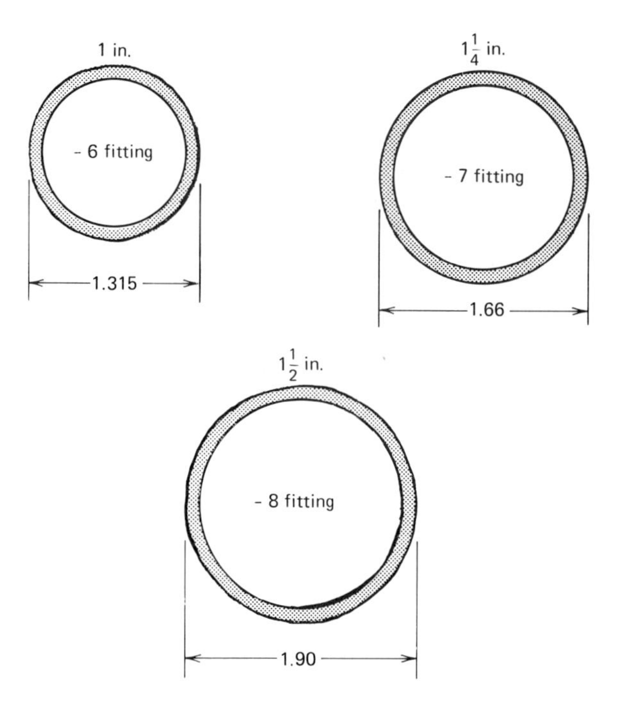


**Figure 7-14.** Standard pipe dimensions.

Galvanized pipe is one of the easiest materials to obtain and work with that can be used to construct the support rack. It is available in standard 21-ft lengths with threaded ends. These threaded ends can serve a very useful function as illustrated in Figure 7-15. If the collector system is ever ex-

**Figure 7-15.** Extension of support rack.

panded, additional collectors can easily be added by using threaded couplers and extending the original support rack to suit. This type of rack has a distinct advantage over a one-piece welded support rack.

Several standard pipe fittings are illustrated in Figures 7-16*a* through 7-16*e*. Only an Allen wrench is required to secure the pipe to these fittings. All mounting hardware (i.e., bolts, screws, washers, mounting angles, clips, etc.) should be protected against corrosion and should consist of galvanized steel, aluminum, or other non-corrosive metal. To reiterate, if it is necessary to cut any piece of galvanized metal, the exposed edge should be coated with zinc chromate-based paint for protection. In addition, to prevent electrolytic corrosion, it is essential not to have two different metals in contact with one another.

### Loading Considerations

If the collectors are part of the roofing system, they must be engineered to support service personnel as well as environmental loads such as wind and snow. The FHA (Federal Housing Administration) suggests that the integral roof system be engineered to withstand a concentrated load of 250 lb distributed over a 4 in.$^2$ area where maintenance may be required. Accessible roof-mounted collectors do not need to support workmen when being in-

**Figure 7-16*a*.** Rail support.

**Figure 7-16*b*.** Crossover.

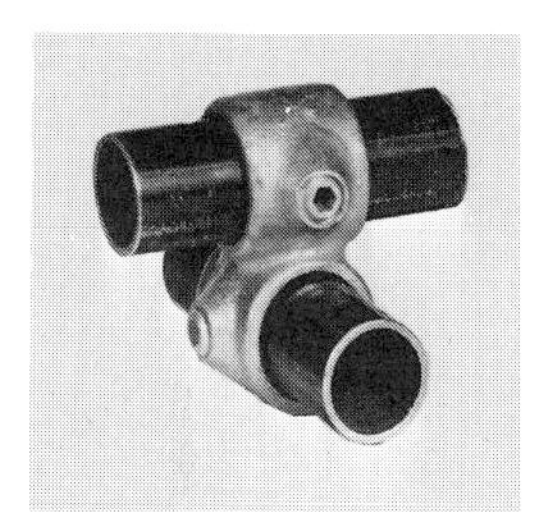

**Figure 7-16*c*.** Swivel flange.

**Figure 7-16*d*.** Combination socket tee and crossover.

**Figure 7-16*e*.** Swivel flange. (Photos for Figures 7-16*a* through 7-16*e* Courtesy of Kee Klamps, Gascoigne Industrial Products Ltd., Buffalo, N.Y.)

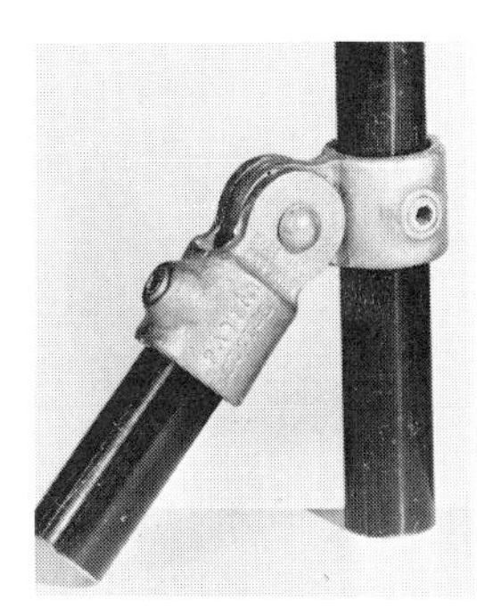

**Table 7-3**
**Load Capacity Approximations (lb) (Standard Iron Pipe, Schedule 40 Continuous Beam) Safety Factor 2½ to 1 Based on Ultimate Strength**

| | *Pipe Size (I.D. in.)* | | |
|---|---|---|---|
| *Span (ft)* | *1 in.* | *1¼ in.* | *1½ in.* |
| 3′-0 in. | 706 | 1249 | 1736 |
| 3′-6 in. | 605 | 1070 | 1488 |
| 4′-0 in. | 529 | 937 | 1306 |
| 4′-6 in. | 470 | 832 | 1120 |
| 5′-0 in. | 423 | 749 | 1042 |
| 5′-6 in. | 388 | 681 | 947 |
| 6′-0 in. | 353 | 624 | 868 |
| 6′-6 in. | — | 589 | 801 |
| 7′-0 in. | — | 535 | 744 |
| 7′-6 in. | — | 499 | 694 |
| 8′-0 in. | — | — | 651 |

*Source:* Reprinted courtesy of Kee Klamp. Gascoigne Industrial Products Ltd., Buffalo, New York.

stalled or repaired. Hence, they need only to sustain the required environmental loading such as wind, snow, and hail. In accordance with HUD's Intermediate Minimum Property Standards, the support rack for a direct mounting application should be built to withstand winds up to at least 100 mph. This imposes a wind load of 40 lb/ft$^2$ on a vertical surface and an average of 25 lb/ft$^2$ on a sloped roof. As the collector array is tilted, it becomes more susceptible to wind gusts, and the mounting support for a tilted support rack should be built to withstand winds of at least 150 mph. These loading considerations can be factored into the pipe diameter used and the spans between supports as illustrated in Table 7-3.

The capacities shown in Table 7-3 are for one piece of pipe supported by three or more uprights with uniform loading and is intended only as a guide for support rack design. Because most support racks are designed to use two or more longitudinal support members, the maximum safe load can be determined by multiplying the particular load capacity in Table 7-3 by the number of supporting members.

For example, assume a three collector array (direct mounting), as illustrated in Figure 7-17, has three supporting rails which must carry an average total load of 25 lb/ft$^2$ of surface area (wind load) and a total collector weight of 255 lb (dead load). The total load for collectors of 32 ft$^2$ surface area per

Figure 7-17

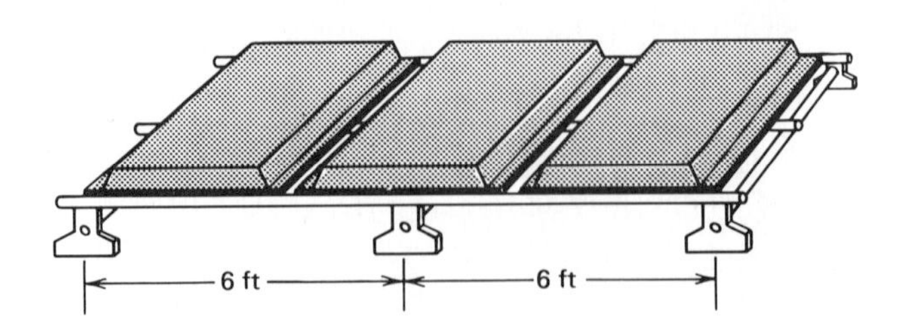

collector would thus be (32 $ft^2$) × (25#/$ft^2$) plus (255#) for a total load of 1055 lb. The spans are 6 ft between supports. Therefore, a 1-in. pipe size is chosen because three rails will support 1059 lb. (i.e, 3 × 353 lb, from Table 7-3). If this collector array is tilt mounted, then a minimum of 40 lb/$ft^2$ load should be assumed for capacity calculations. The cross bracing as illustrated in Figure 7-20 should also be included to resist sideways pressure.

There are several other types of loading conditions besides wind and dead weight. These include snow, hail, seismic, and thermal stress loading. For the most part, the manufactured collector design will take these parameters into consideration. A collector array that collapses during the first windstorm or snowstorm is not a good investment. The supporting structure must be built to withstand the rigors of nature.

### Typical Support Rack Designs

As mentioned previously, support racks can either be welded together as one-piece units or constructed using pipe fittings. There are basically five types of support designs as illustrated in Figures 7-18*a* through 7-18*e*. These

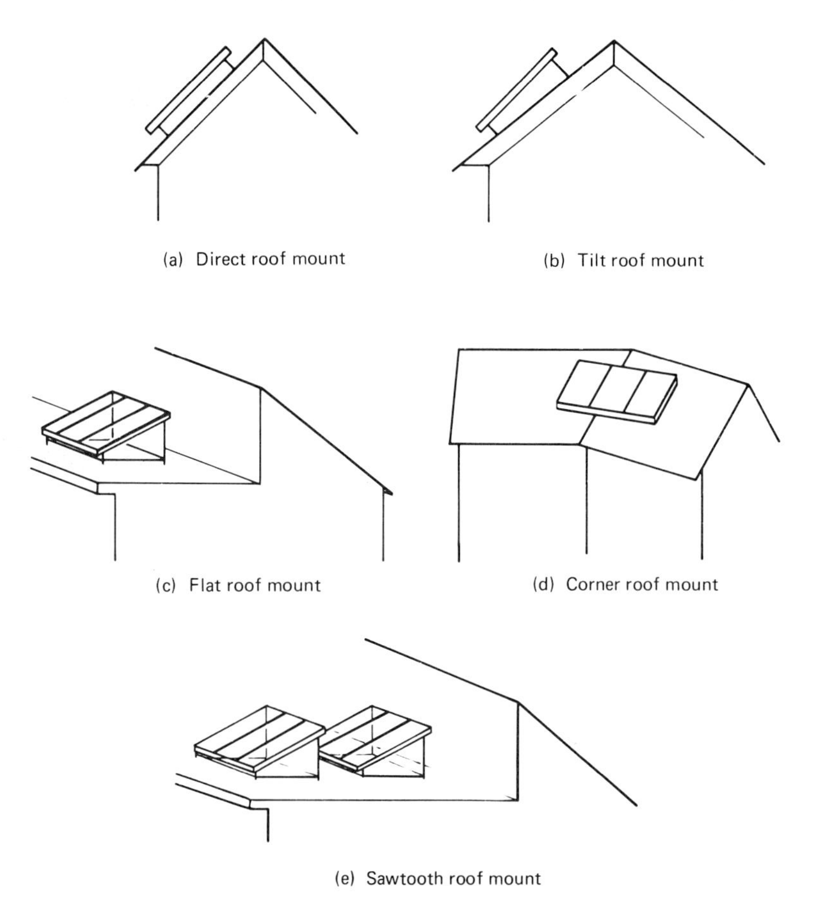


**Figure 7-18.** Support rack designs. (*a*) Direct roof mount. (*b*) Tilt roof mount. (*c*) Flat roof mount. (*d*) Corner roof mount. (*e*) Sawtooth roof mount.

include (a) direct roof mount, (b) tilt roof mount, (c) flat roof mount, (d) corner roof mount, and (e) sawtooth roof mount.

On-site support rack construction can be accomplished with Kee Klamps® (i.e., Figures 7-16*a* through 17-16*e*), and the following examples of rack designs are constructed using these types of pipe fittings. Dimensions of each support rack will vary depending on collector weight, size, and orientation. Pipe size can be determined upon consideration of loading conditions previously discussed. A typical direct roof mount support rack is illustrated in Figure 7-19. A typical tilt or flat roof support rack can be constructed in the same configuration as illustrated in Figure 7-20. A typical corner roof mount support rack is illustrated in Figure 7-21.

A typical sawtooth support rack is constructed the same as a flat roof support rack. If collectors' arrays must be stacked back to back due to unavailable roof area, spacing between collectors is required to avoid collector shading. Such an array is illustrated in Figure 7-22.

The required spacing between the collector is determined from the relationships between tilt angle and latitude as shown in Figure 7-23. The "local slope" is defined as latitude minus the tilt of the collector mounting

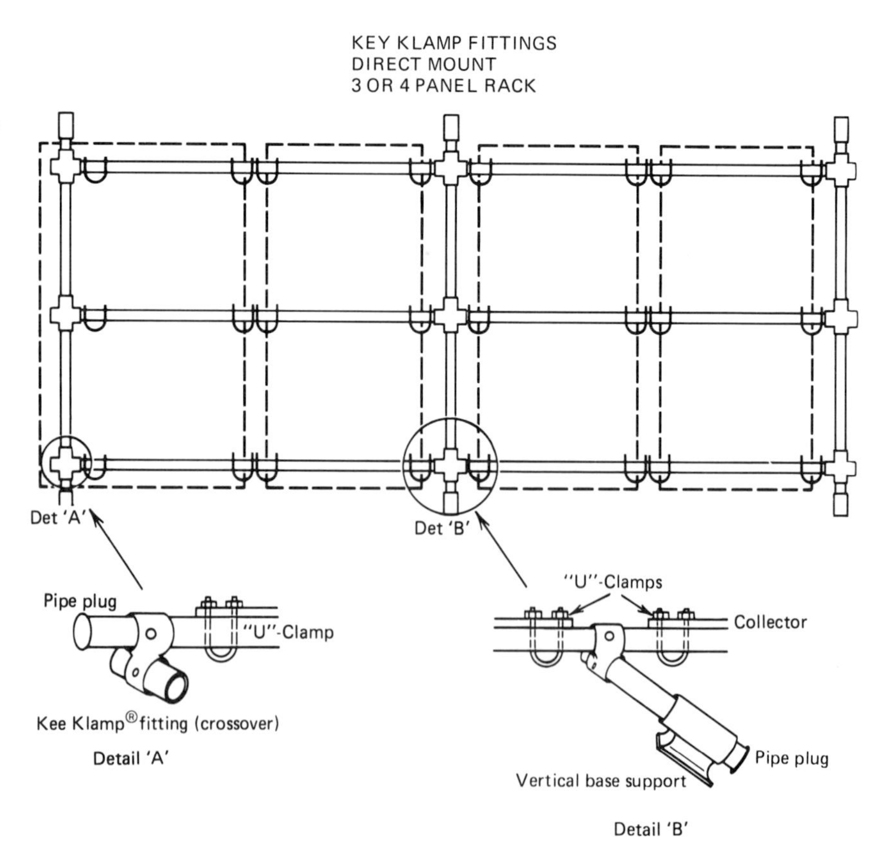


**Figure 7-19.** Direct roof mount support rack (Key Klamp® fittings, direct mount, 3- or 4-panel rack).

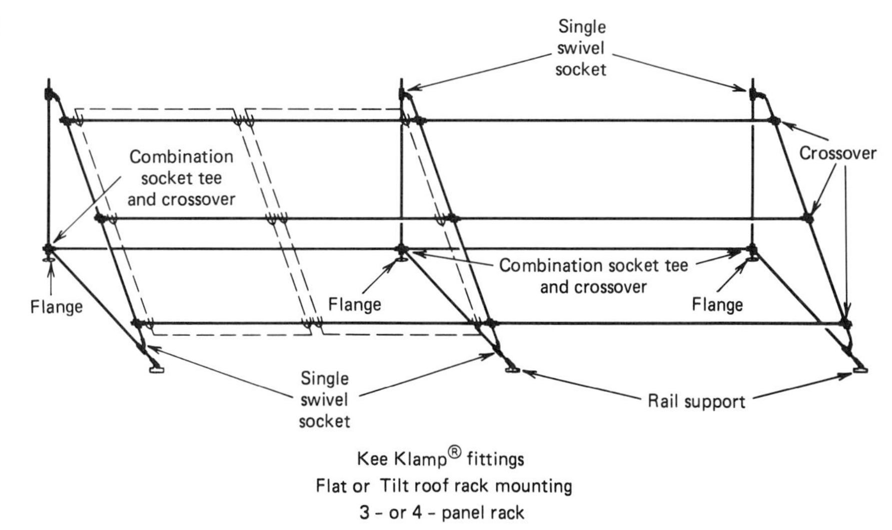


**Figure 7-20.** Flat or tilt roof mount support rack.

surface. For the collector tilt desired, the separation factor "D" is determined from Figure 7-23. The required spacing can then be determined by multiplying the spacing factor "D" by the effective collector height.

## Ground-to-Roof Transfer of Collectors

Before discussing physical attachment of the collector and support rack to the roof structure we should briefly illustrate three basic ways of lifting the collectors onto the roof.

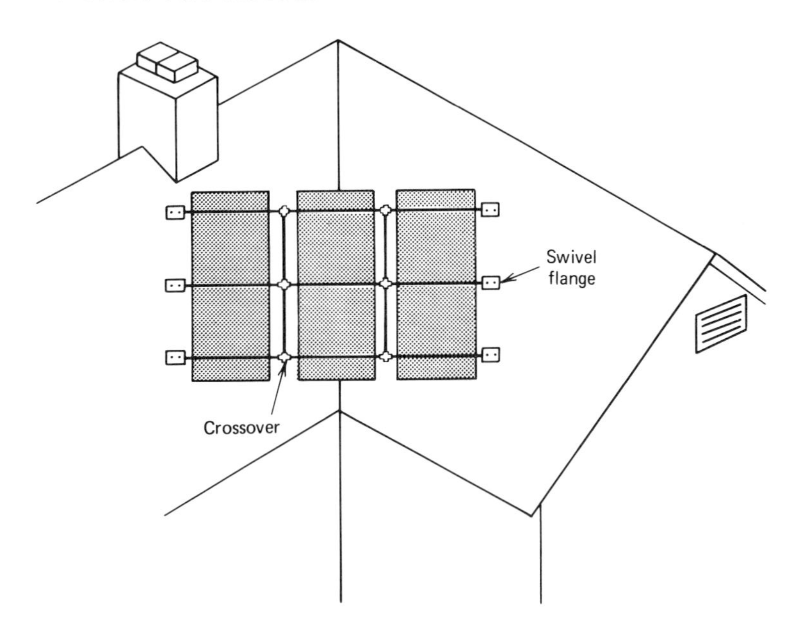


**Figure 7-21.** Corner roof mount support rack.

**Figure 7-22.** Sawtooth roof mount. (Courtesy of General Electric Co., Solar Heating & Cooling Division, Philadelphia, Penn.)

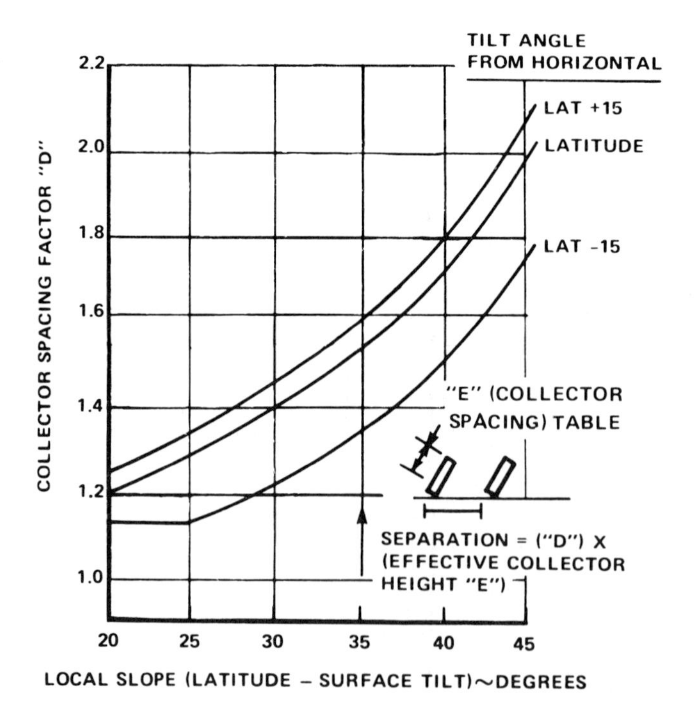


**Figure 7-23.** Distance between collector rows for minimal performance degradation. (Reprinted with permission from General Electric Co., Solar Heating & Cooling Division, Philadelphia, Penn.)

The collectors can be lifted manually using two ladders in parallel and three or four people to "push" and "pull" the collectors onto the roof. The rope provides not only a means of pulling the collector but also a means of guidance. For one- and two-story houses this method is the most widely used and is normally designated as the Parallel Ladder Method as illustrated in Figure 7-24. We could also call this procedure the "grunt and tug" method if the collector weighs upward of 200 lb.

For situations in which normal extension ladders cannot reach the roof, a movable pulley system can be used as illustrated in Figure 7-25. This method provides a greater distance in moving ability as well as a mechanical advantage in lifting. Figure 7-25 shows there are four cords on the movable pulley, thus providing a theoretical mechanical advantage of four. The actual mechanical advantage will be somewhat less depending on the friction of the pulley system. For instance, for every 25 lb of force applied to the cord, we can theoretically lift 100 lb by the cord.

Mechanical advantage is the ratio of the force exerted by the pulley system to the force applied to it, and is achieved by the sacrifice of distance. From basic physics it is known that work is a product of force and distance. A small force multiplied by a large displacement can be made equivalent to a large force multiplied by a small displacement. The movable pulley has a mechanical advantage because the cord running over the pulley travels

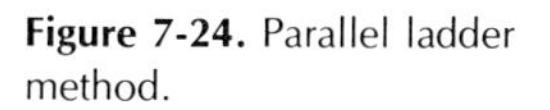

**Figure 7-24.** Parallel ladder method.

**Figure 7-25*a* and *b*.** Movable pulley method. (Courtesy of Granite State Solar Industries, Inc., Dover, N.H.)

farther than the pulley during a displacement. The pulley itself can therefore exert a force greater than the force applied to it via the cord. The advantage of such a lifting method is readily observed in Figure 7-25*b*.

For lifting heavy collectors or for individuals who install collectors on a day-to-day basis, some type of mechanized load lifting device can be considered as a third method of lifting the collectors onto the roof. Such load-lifting devices can include "cherry picker," cranes as depicted in Figures 7-26*a* and *b*, fork lifts and motorized units for ladders as depicted in Figures 7-26*c* and *d*.

## Support Rack and Collector to Roof Attachment

Essentially the collector or mounting racks can be attached to the roof by either spanner mounting or lag bolt mounting. Each type of mounting is one of individual preference. All mounting hardware used (bolts, screws, washers, mounting angles, etc.) for either type should be protected against corrosion and should consist of galvanized steel, aluminum, or other noncorrosive metal. In addition, to prevent electrolytic corrosion, it is essential not

**Figure 7-26a.** (Courtesy of B.H. Hoist Co., Tyler, Texas.

**Figure 7-26b.** (Courtesy of B. H. Hoist Co., Tyler, Texas)

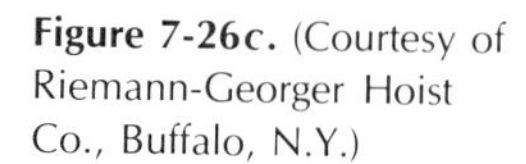

**Figure 7-26c.** (Courtesy of Riemann-Georger Hoist Co., Buffalo, N.Y.)

**Figure 7-26d.** Mechanized load lifting. (Courtesy of Riemann-Georger Hoist Co., Buffalo, N.Y.)

to have dissimilar metals in contact with one another. An example of such a situation with a high-corrosion potential could be the use of steel or galvanized (zinc-coated) mounting hardware in contact with an aluminum collector case.

### Spanner Mounting

Spanner mounting can be accomplished as illustrated in Figure 7-27 for an attic that is accessible and for which there is no finished ceiling flush to the rafters. Holes should be predrilled through the roof and through the mounting spacers (if required) just large enough to accept the bolts or threaded rods (which should be at least ⅜ in. in diameter). The bolts or threaded rods should then be inserted before securing the spanners to assure proper alignment of the holes. The 2 × 4 spanners can then be screwed or nailed perpendicular, and directly to the rafters inside the attic. Wood spacer blocks can be sized to fit snugly between the spanner and roof sheathing prior to tightening the two nuts as shown in Figure 7-27, or sleepers can be installed between rafters as depicted in Figure 7-28.

Spanner mounting can also be accomplished as illustrated in Figure 7-29 for an attic that is inaccessible and for which there is to be a finished ceiling flush to the rafters. In this situation, wooden blocks are nailed to the inside of the rafters at least 3 in. short of the ceiling edge, and the spanners are screwed or nailed to these blocks. In all cases, silicon sealant should be applied generously to prevent leakage. The sealant should be applied until it literally oozes from around bolt heads when they are tightened.

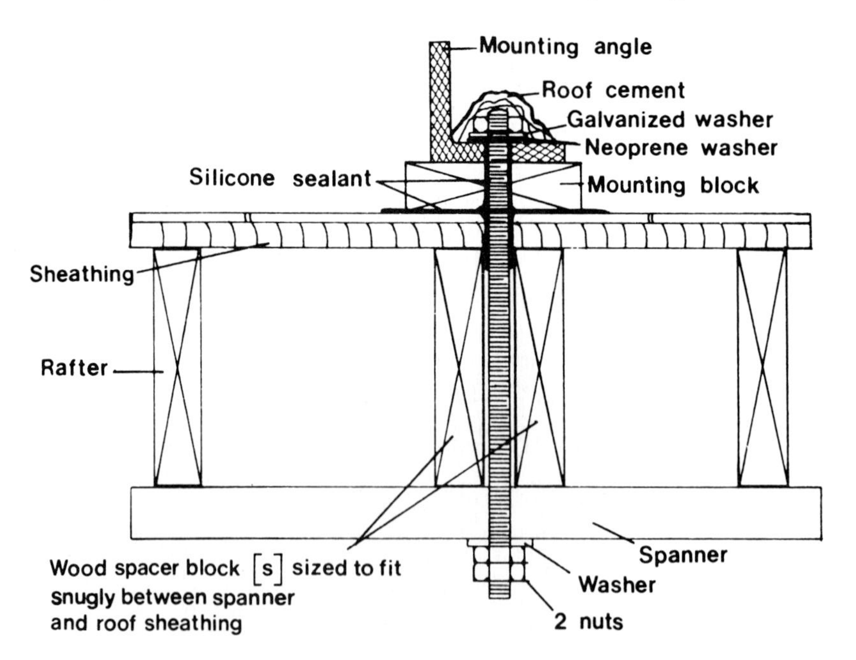


**Figure 7-27.** Spanner mounting with spacer blocks—accessible attic. (*Source. Installation Guidelines for Solar DHW Systems in One and Two Family Dwellings*, U.S. Dept. of Housing & Urban Development)

**Figure 7-28.** Spanner mounting with sleepers—accessible attic.

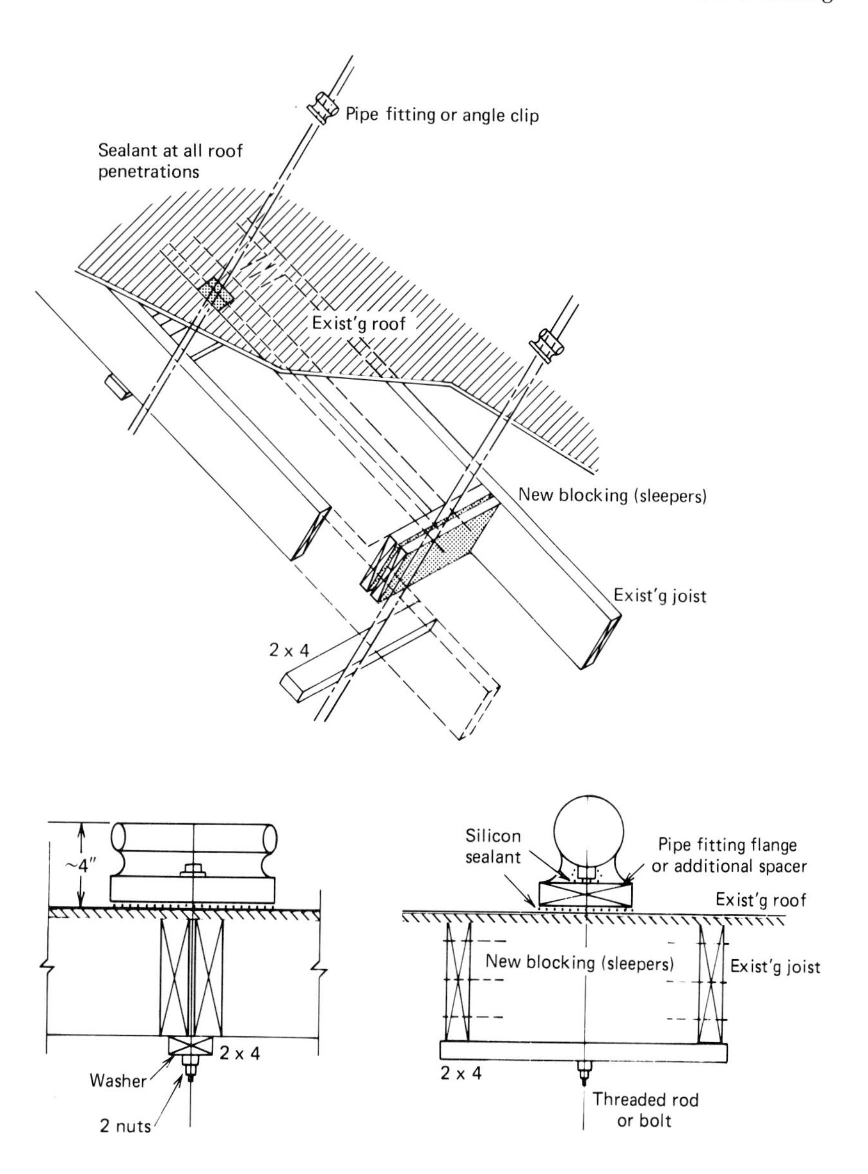


## Lag Bolt Mounting

Lag bolt mounting can be accomplished regardless of whether or not the attic is accessible, by lag bolting directly to the rafters from the outside, as illustrated in Figure 7-30. Rafters or trusses can be located by tapping the roof with a hammer and using a small nail as a guide to locate one side of the rafter. Rafters or trusses will normally be on 16 or 24 in. centers. If the attic is accessible, then sleepers can be installed between rafters so the lag bolts do not have to fall exactly on the center of each rafter. Lag bolts should be a

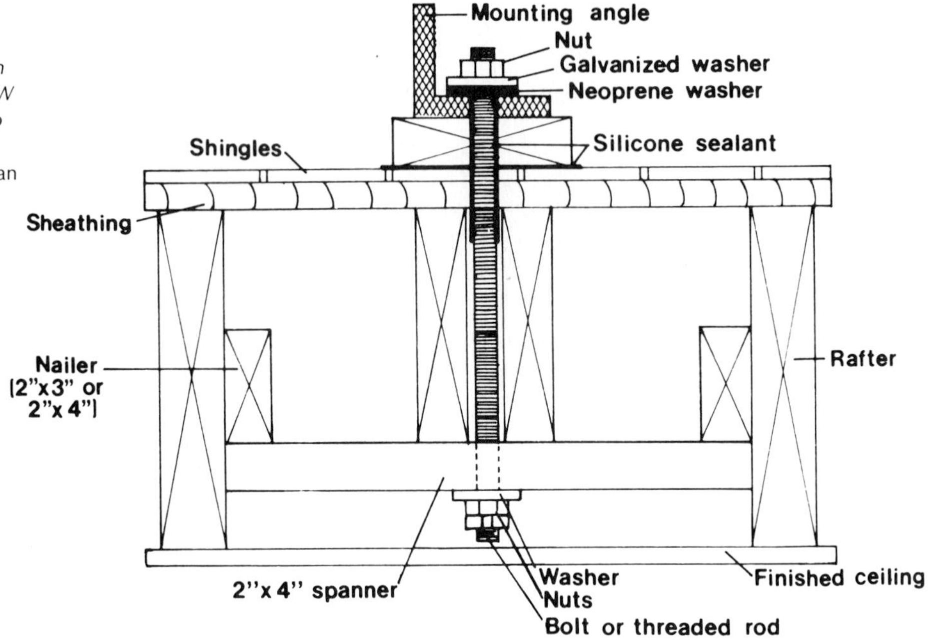


**Figure 7-29.** Spanner mounting—inaccessible attic. (*Source: Installation Guidelines for Solar DHW Systems in One and Two Family Dwellings,* U.S. Dept. of Housing & Urban Development)

minimum of ⅜ in. in diameter by 3½ in. in length and should penetrate the rafters at least 2 in. Holes, several sizes smaller than the lag bolts, should be predrilled through the roof to assure a tight fit. The lag bolt should be inserted through a galvanized washer and then through a neoprene washer before going through the collector flange and mounting block or pipe flange.

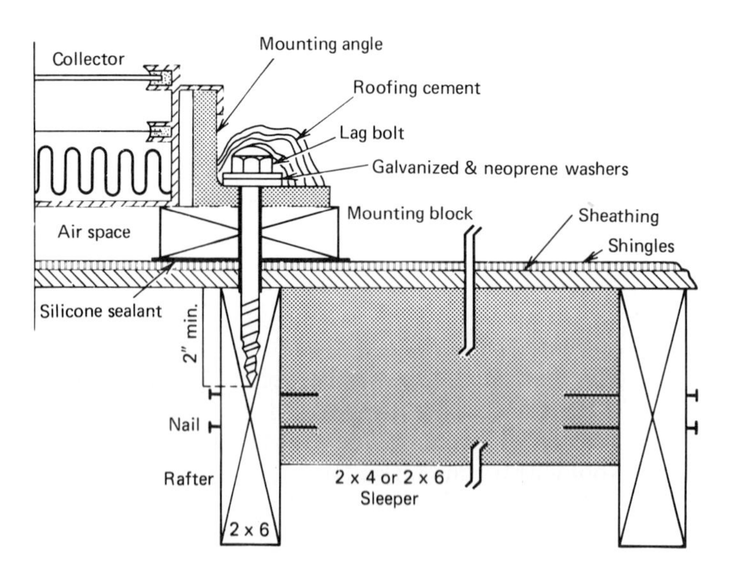


**Figure 7-30.** Lag bolt mounting.

Always apply a sufficient amount of silicone sealant between mounting blocks, washers, and the roof surface before tightening lag bolts. After the bolts have been tightened, more sealant should be applied over the entire assembly.

The following photographs sequence a typical direct roof mounting for a closed loop freeze resistance system. Bear in mind that there may be many variations of this sequence depending on the type of system installed and on the particular roof situation, as previously discussed. This step-by-step sequence provides a descriptive "hands-on" view of collector mounting.

***Step 1 (Figure 7-31).*** Solder unions and other collector to collector fittings on the ground to save time and also working difficulties on the roof. Many collectors have some type of external gasket that may have to be protected by an escutcheon plate (i.e., asbestos) when soldering the fittings. Remember to use high temperature 95/5 (tin/antimony) solder for all collector header fittings.

Figure 7-31

***Step 2 (Figure 7-32).*** Align the collector headers, position the collectors on the support rack, and premount the collector array on the ground as a "trial run." Any problems associated with the rack and collector attachment can then be corrected on the ground rather than from a precarious position on a sloped roof. Before handling the collectors, surface temperature of the headers should be checked because temperatures of 150° F can burn unprotected skin. The collectors may have to be covered prior to handling.

Figure 7-32

*Step 3 (Figure 7-33).* Remove the preassembled collectors from the support rack and lift the support rack onto the roof.

Figure 7-33

***Step 4 (Figures 7-34 and 7-35).*** Holding the support rack in place, mark the positions of the rail supports for installation of sleepers or spanners.

Figure 7-34

Figure 7-35

***Step 5 (Figures 7-36 and 7-37).*** Install sleepers or spanners as needed.

Figure 7-36

Figure 7-37

***Step 6 (Figures 7-38 and 7-39).*** Predrill holes for lag bolts or threaded rods.

Figure 7-38

Figure 7-39

***Step 7 (Figure 7-40).*** Ensure all support fittings are tight.

**Figure 7-40**

***Step 8 (Figure 7-41).*** Apply a generous amount of silicone sealant between the mounting fitting and the roof surface.

**Figure 7-41**

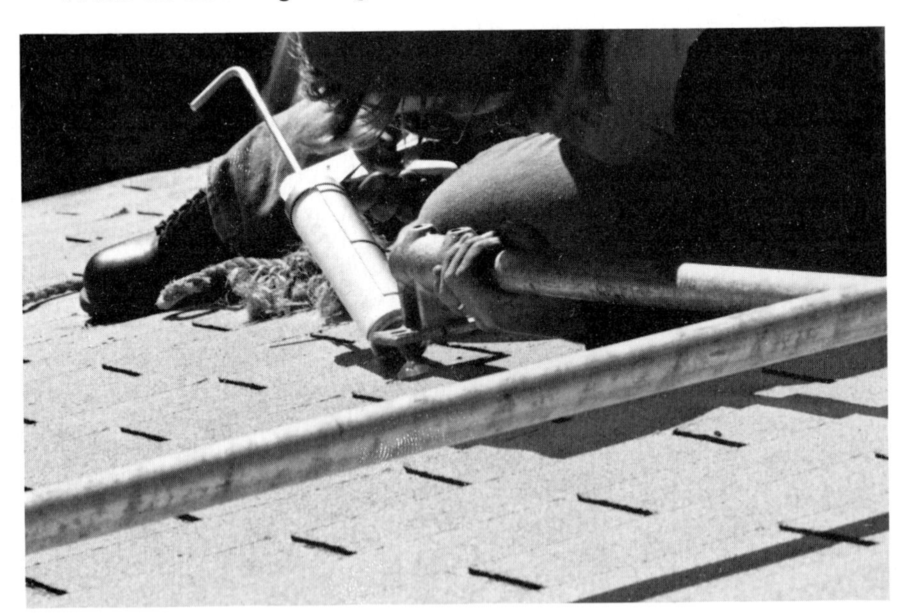

***Step 9 (Figures 7-42 and 7-43).*** Fasten the support rack to the roof using lag bolts or threaded rods.

**Figure 7-42**

**Figure 7-43**

***Step 10 (Figure 7-44).*** Apply silicon sealant over the entire fastened rail support assembly.

Figure 7-44

***Step 11 (Figures 7-45, 7-46, 7-47, 7-48, and 7-49).*** Lift the collectors onto the roof.

Figure 7-45

Figure 7-46

Figure 7-47

Figure 7-48

Figure 7-49

***Step 12 (Figures 7-50, 7-51, and 7-52).*** Attach the collectors to the support rack. Ensure that weep holes, if any, are located at the bottom.

**Figure** 7-50

**Figure** 7-51

Figure 7-52

*Step 13 (Figure 7-53).* Connect the collector manifolds to each other. Remember to ensure at least ⅛ in. between collectors to allow for thermal expansion.

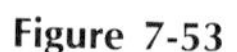

Figure 7-53

This sequence completes the roof-mounting procedure. The only work that remains on the roof involves the penetration for the inlet and outlet headers (discussed in Chapter 8), lightning protection of the collector array (dis-

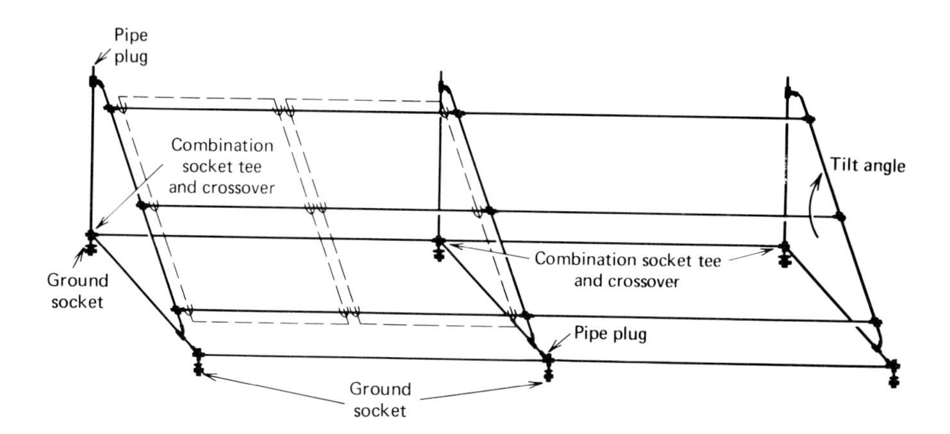


**Figure 7-54.** Ground mount support rack.

cussed in Chapter 8), attachment of the sensor to the collector header for the differential controller (discussed in Chapter 8), the collector loop fill procedure (discussed in Chapter 9), and the insulation of the exterior piping (discussed in Chapter 9).

## GROUND MOUNTING

### General Information

Much of the preceding information associated with roof mounting also pertains to ground mounting. For ground mounting, however, the support rack structure is essentially definitive in that collectors should be mounted on a rack similar to the one designed for a flat roof, as illustrated in Figure 7-20. A typical ground-mounting support rack is illustrated in Figure 7-54. The only difference in this support rack design is the ground socket, as shown in Figure 7-55.

Before starting construction of a ground-mounted collector array, the

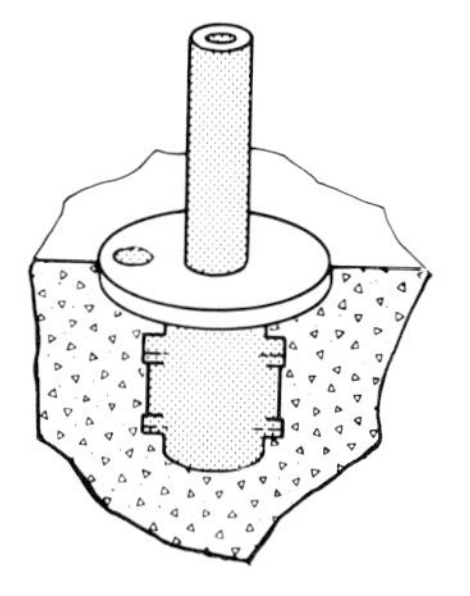

**Figure 7-55.** Ground socket pipe fitting. (Photo courtesy of Kee Klamps Division, Gascoigne Industrial Products, Ltd., Buffalo, N.Y.)

local zoning ordinance should be checked for such conditions as setback and classification of structure. Safety is also an important consideration. The support racks should be located where they are not likely to be used as playthings by children. All exposed piping must be insulated so that no one is inadvertently burned.

## Installation Considerations

Collectors are mounted on a rack or frame that must be attached securely to footings extending below the local frost line. Locations for the footings can be established by "sounding" the ground with an iron bar. All footings should be dug at the same time, and the necessary amount of concrete should be poured at the same time. (A typical concrete mix normally consists of 1 part concrete, 2 parts sand, and 3 parts gravel.) There should be at least 4 pier footings, 8 in. × 8 in. square or 8 in. diameter round for a two-panel system and two additional footings for every collector thereafter. Footing holes should be dug below the frost line and the bottoms filled with an inch or two of dry, washed pebbles. The footings should be poured with a frame member or a threaded rod protruding from the concrete as illustrated in Figure 7-56.

If the frame is embedded, a temporary jig should be made so that the rack holds its shape while the concrete sets. If the frame member is wood, it must be treated to slow deterioration. If a threaded rod is used, it must be at least ⅜ in. in diameter. Warping of the collector array caused from soil freezing can be prevented by bracing with a continuous strong member around the base of the framework or by building the footings as two walls, as illustrated in Figure 7-57.

The collector array should be mounted as close to the house as possible,

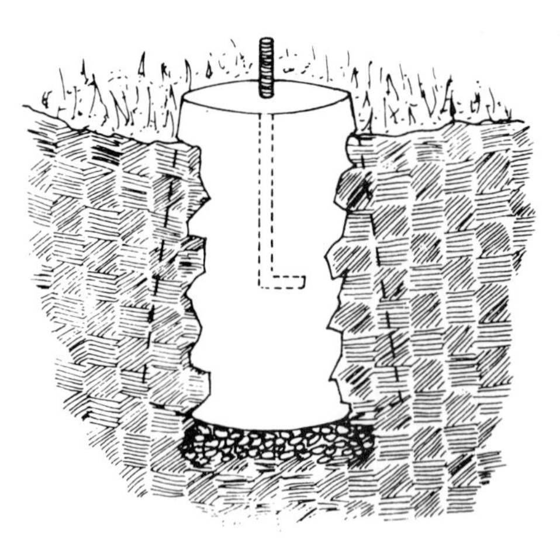

**Figure 7-56.** Pier footing. (*Source. Installation Guidelines for Solar DHW Systems in One and Two Family Dwellings*, U.S. Dept. of Housing & Urban Development)

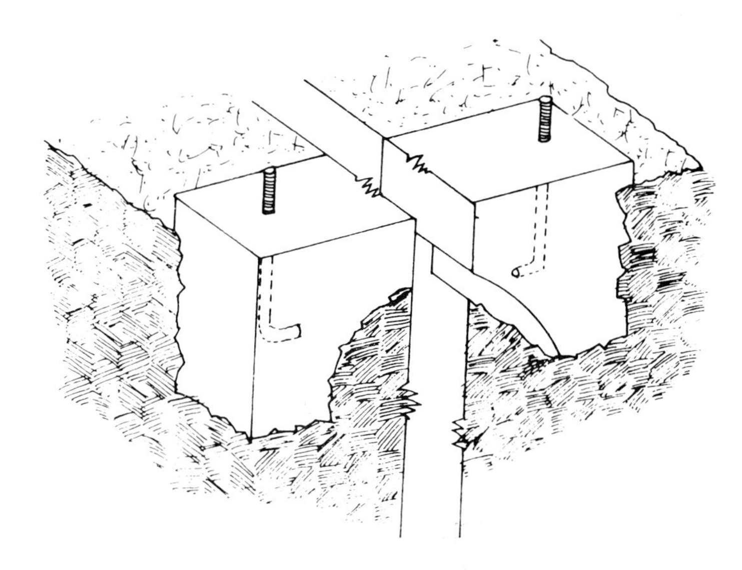

**Figure 7-57.** Wall footing. (*Source. Installation Guidelines for Solar DHW Systems in One and Two Family Dwellings*, U.S. Dept. of Housing & Urban Development)

as illustrated in Figure 7-58. If there is no danger of snow falling directly off the roof onto the collectors, the collector array can be mounted against the house with the exterior pipes insulated above ground.

If there is a potential danger of snow fall as illustrated in Figure 7-59, then the pipes should be run underground below the frost line, insulated, and weatherproofed. Insulated piping should be encased with PVC plastic pipe and sealed shut. Then, 6 in. of earth shoud be dumped into the trench and carefully tamped around the pipe to assure proper support. The remaining depth should be filled and tamped at 6-in. intervals to prevent unsightly settling of the lawn.

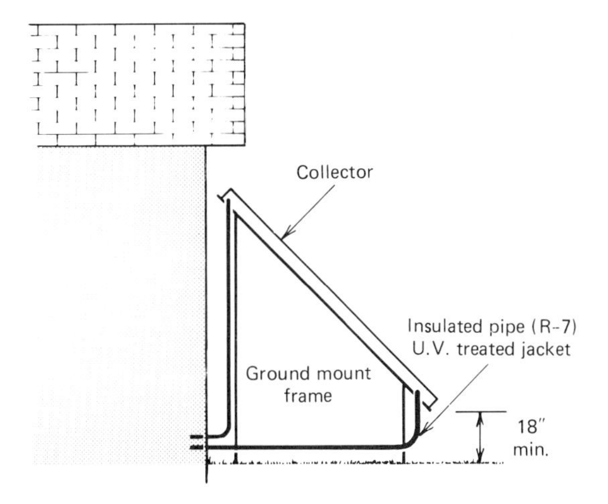


**Figure 7-58**

Figure 7-59

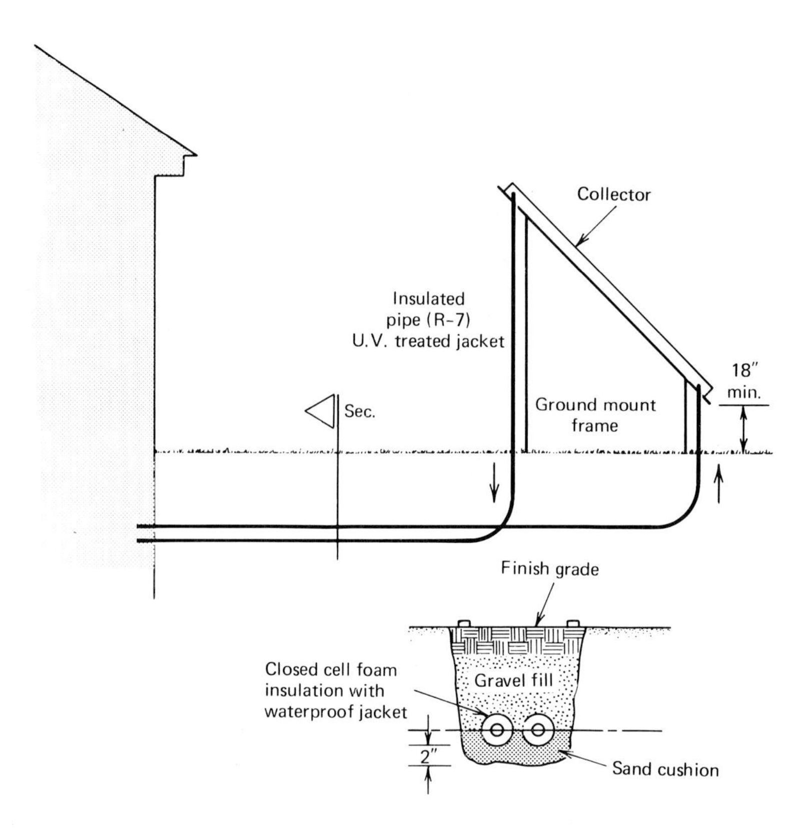


## REVIEW QUESTIONS

1. Which of the following considerations does not apply for proper collector roof mounting?
   a. Mount the collector near midheight and parallel with the ridge of a pitched roof.
   b. Mount the collector within 24 in. of the lower edge of a pitched roof.
   c. Reshingle the roof prior to collector mounting
   d. Ensure that the collector tilt is within latitude $\pm 10°$.
   e. None of the above

2. If the pitch of a roof has been determined to be ½ and the span is 24 ft, what is the equivalent slope?
   a. 1 b. 1/4
   c. 3/2 d. 2/3

3. How many inches above a roof's surface should a direct mounted collector array be established?
   a. 1 in. b. 2 in.
   c. 3 in. d. 4 in.

4. Air collectors are more likely to be mounted in which of the following mounting configurations?
   a. Direct mount
   b. Tilt mount to a flat roof
   c. Integral mount
   d. Tilt mount to a sloped roof
5. What wind speed should a direct mount collector array be able to withstand?
   a. 100 mph b. 125 mph
   c. 150 mph d. 175 mph
   e. None of the above
6. What wind speed should a tilt-mounted collector array be able to withstand?
   a. 100 mph b. 125 mph
   c. 150 mph d. 175 mph
   e. None of the above
7. Suppose a direct mounted three-collector array has a total dead load weight of 300 lb and a total surface area of 66 $ft^2$. If the pipe diameter used to support 3 rails is 1¼ in., what is the recommended span between supports?
   a. 6 ft 6 in. b. 6 ft 0 in. c. 5 ft 6 in.
   d. 4 ft 6 in. e. 3 ft 0 in.
8. What is the recommended span for review question 7 if the collector array is tilt mounted?
   a. 3 ft 6 in. b. 4 ft 0 in.
   c. 4 ft 6 in. d. 5 ft 0 in.
   e. None of the above
9. What is the theoretical mechanical advantage of a pulley system capable of lifting three collectors weighing 400 lb over a distance of 10 ft with a 50 lb force?
   a. 3 b. 5
   c. 8 d. 16
10. Spanner mounting is subject to which of the following conditions?
    a. Accessible attic
    b. Mounting locations in the center of rafters
    c. Rafter on 24-in. centers
    d. Existing ceiling not flush to the roof rafters
    e. Sleepers

# System Installation: Guidelines

8

Ideally, the system components should be plumbed and wired prior to mounting the collectors. Plumbing and wiring can be accomplished by one person; mounting the collectors, however, depends on the availability of human resources. If help is only available at a particular time and the collectors are mounted first, they must be covered to prevent stagnation conditions from causing possible damage to the absorber plate.

A solar DHW system installer using these guidelines should also refer to the installation manual normally provided by each collector manufacturer and/or dealer. The instructions from each manufacturer may require specific procedures/methods that must be followed to retain collector warranty. This chapter will provide the fundamentals necessary to help ensure a complete and successful system installation.

## ELECTRICAL INSTALLATION

### General Considerations

There are three possible electrical wiring aspects to consider during the installation of the solar DHW system. These aspects include: (1) wiring to the electrical back-up elements in the storage tank, (2) wiring to the differential controller and sensors, and (3) wiring to the circulator/pump. We will discuss these wiring details under each respective topic in this chapter. One must remember to check National Electric Codes and local codes to ensure proper electrical installation. Health and safety are essential considerations during installation. Always disconnect the electrical power while the electrical associated components are being installed, serviced, repaired, or replaced. If one is totally unfamiliar with electrical wiring procedures, a licensed electrician should be called. The volt-ohm-milliammeter (VOM) is an important tool used in the electrical wiring and troubleshooting process.

### VOM Circuit Theory

The fundamental relationship between current $I$, voltage $E$, and resistance $R$ is expressed mathematically as Ohm's law, $E = IR$. The unit of resistance is the ohm, which is one volt per ampere and is customarily abbreviated by the Greek capital letter, omega—$\Omega$. The volt-ohm-milliammeter (VOM), is an electronic device that is used to measure these three parameters of Ohm's law. This instrument comprises a voltmeter, ohmmeter, and milliammeter (VOM) into a single compact unit as illustrated in Figure 8-1.

**Figure 8-1.** Typical volt-OHM Millammeter. (Courtesy of Radio Shack, a Division of Tandy Corp., Fort Worth, Texas)

It is not absolutely necessary that we understand the fundamental circuit theory of a VOM in order to use the instrument properly. It may, however, be helpful for us to better understand the parameters we are measuring if we describe the three basic circuits of the VOM that are used to measure current, voltage, and resistance. These three basic circuits are illustrated in Figure 8-2.

***Current measurement*** is achieved as illustrated in Figure 8-2*a*. Shunt-switched resistors, R1, R2, and R3, resistance of the meter movement, source resistance $Rs$, and source voltage $V$ are considered. The accuracy of the current reading depends (1) on whether or not the meter resistance remains the same over its entire range, (2) on the accuracy of the shunting resistors, (R1, R2, and R3) and, (3) the effect of the measuring circuit itself upon the circuit under test. The resistance in the measuring circuit should be much smaller than the resistance of the circuit we are measuring in order to provide maximum accuracy. This situation is depicted by the equation in Figure 8-2*a*.

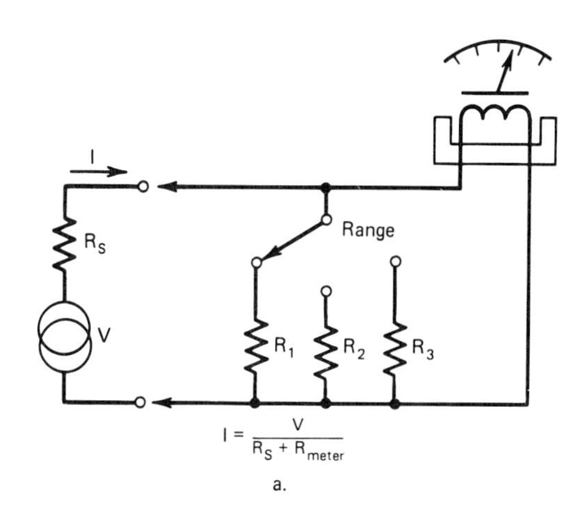


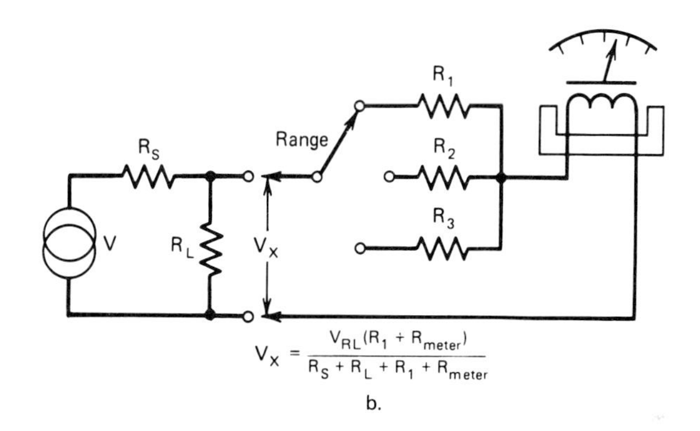


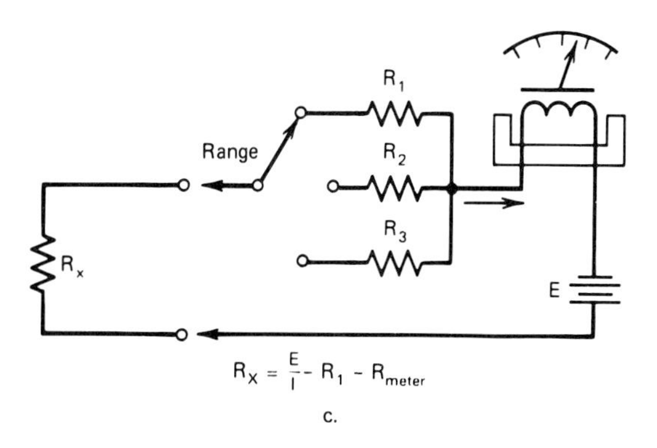


**Figure 8-2.** Basic VOM circuits. (*a*) Current measurement. (*b*) Voltage measurement. (*c*) Resistance measurement. (*Source. Tested Electronics Trouble-Shooting Methods* by Walter H. Buchbaum, reprinted with permission by Prentice-Hall, Inc., Englewood Cliffs, N.J. 07632)

***Voltage measurement*** is achieved as illustrated in Figure 8-2*b*. The voltage under test, $Vx$, is determined by the original source voltage $V$, the source resistance, $Rs$, and the load resistance $RL$. In this case it is important to have a very high resistance value for the shunt-switched resistors R1, R2, and R3. These values should be high so that the amount of current flowing in the meter is less than one-tenth of the current flowing through $RL$; this will determine the accuracy of the voltage measurement. The equation depicted in Figure 8-2$b$ illustrates why VOM's are normally rated in "ohms per volt." A typical value of 10,000 Ω per volt means that when $Vx$ is indicated as 1 V on the meter, the combined shunt resistance and meter resistance would be 10,000 Ω.

***Resistance measurement*** is achieved as illustrated in Figure 8-2$c$. The main difference between the resistance circuit and the current and voltage circuits is the addition of an internal battery voltage $E$. Because the resistor under test is a passive device, it is necessary to drive a current $I$ through the meter with the use of the internal battery. The amount of current driven is determined by the combined resistance $Rx$, R1, and the meter resistance. The unknown resistance $Rx$ can then be determined as depicted by the equation in Figure 8-2$c$.

## VOM Operation

Operation of a volt-ohm-milliammeter is quite easy to understand, and an instruction booklet that usually accompanies each VOM upon purchase will provide the information necessary to understand the different measurement settings. The resistance scale of a VOM, as illustrated in Figure 8-1, expresses "Ø" ohms at the right-hand end and infinitely large ohms (∞) at the left-hand end. The voltage and current scales read opposite to that of the resistance scale and read increasingly from left to right. Only a mechanical zero adjustment may be necessary prior to voltage and current readings. For resistance measurements, the pointer must be zeroed by a separate adjustment on the VOM. For this adjustment, the VOM is selected to the desired range, the test leads are shorted together creating a series circuit within the meter, and the VOM is adjusted to read zero ohms. This operation should be done each time a different resistance range is selected. This adjustment determines the largest test current possible within the VOM causing a full-scale deflection. With the test leads separated, the series circuit is opened, the VOM current becomes zero, and the deflection is infinity, indicating a very high resistance or open circuit.

In most of our troubleshooting situations involving electrical checks, we will be measuring voltage and resistance. Examples of these measurements are illustrated in Figure 8-3.

Before taking ***voltage measurements***, (1) the mechanical zero should be checked and set to zero if required, (2) the instrument range must be set higher than the maximum value of any voltage quantity to be measured, and (3) the type of current to be read should be switched accordingly to ac (alternating current) or dc (direct current). (For input and output voltages to most controllers and pumps, we shall be measuring ac voltage.) Input and output voltage measurements can be taken without removing any wiring and by simply observing the deflection of the voltmeter across each load and noting the range selected on the VOM.

Before taking ***resistance measurements*** (1) the mechanical zero should be

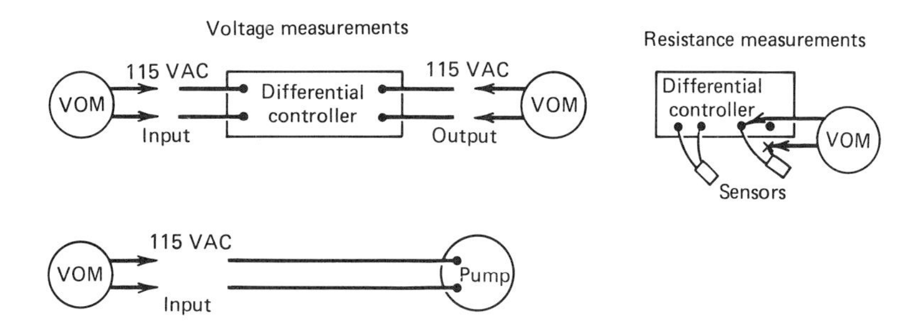


**Figure 8-3.** Voltage and resistance measurements with a VOM.

checked and set to zero if required, and (2) with the test leads shorted together, the VOM should be adjusted to read zero ohms at the selected resistance range. The circuits to be checked must be de-energized, and one lead must be disconnected from the circuit so that the resistance reading is not affected by other resistance loads. The VOM generates the test current and measures the resistance. The three types of resistance readings observed are depicted in Figure 8-4.

Resistance readings on a VOM are corrected by a multiplying factor (i.e., R × 1, R × 10, R × 100, R × 1000, R × 10,000). Therefore, to obtain a true reading, the displayed value must be multiplied by the multiplying factor indicated next to the range switch. For example, if the display reads 5.0 ohms and the range switch is selected to the "R × 1000" scale, the VOM is actually reading 5000 ohms. To increase one's confidence in taking resistance measurements, several resistors of known values can be purchased from a local electronics store and used for practice.

# PIPING INSTALLATION

## General Considerations

Reread Chapter 5, (the section on piping and miscellaneous hardware) for a discussion of the type and size of piping and for a discussion of the miscellaneous hardware necessary for system installation. Piping arrangements to and from the collector should be configured for parallel flow-direct return as described previously in Chapter 4, and the piping run should be as short

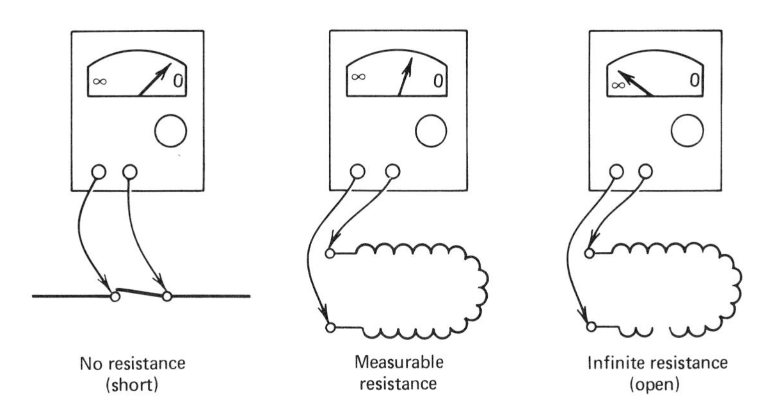


**Figure 8-4.** Resistance measurements.

as possible. Unions rather than sweat fittings should be used between collectors and at the collector inlet and outlet. (There is no need to use Teflon® tape with unions because they have press-fit surfaces.) Soft copper tubing with compression fittings should be avoided. Teflon tape should be used on all threaded joints. Pressure and temperature gauges must be installed; they are essential in determining satisfactory system operation. Keep outside pipe runs to an absolute minimum, and avoid making joints in pipe runs that are inside wall partitions or in underground conduits. Sharp bends should also be avoided. Do not install insulation on pipes unless joints are accessible or until the system has been pressure tested and filled as will be discussed in Chapter 9, "Charge-Up, Start-Up, Maintenance, and Troubleshooting."

## Soldering Techniques

The solar DHW system is comprised of many components and fittings that must be soldered together. (The types of solders have been discussed previously in Chapter 5.) For the nonskilled plumber, this section should provide the techniques needed for proper system piping installation. Remember, all ***collector connections*** should be soldered with 95/5 high-temperature solder. In ***Freon*** systems, if lines are brazed, nitrogen must be pumped through the system to inhibit system corrosion. Refrigerant grade fittings must be used for these systems.

There are eight basic steps necessary to achieve a good solder joint. We shall illustrate each of these steps as follows:

***Step 1 (Figure 8-5a).*** *Measure* and *mark* the length of copper tubing needed.

***Step 2 (Figure 8-5b).*** *Cut* the copper tubing with a standard tube/pipe cutter (available from most hardware stores).

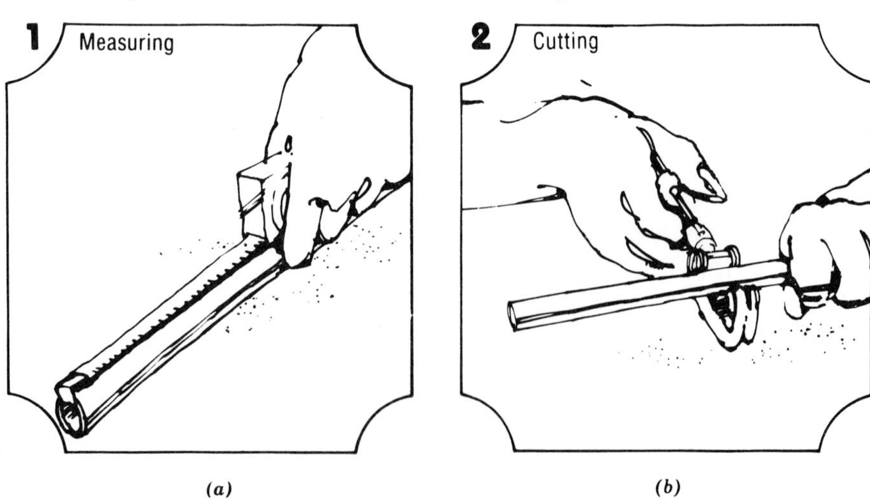


(a) (b)

**Figure 8-5.** (Reprinted with permission from *The Solar Decision Book*, John Wiley & Sons, Inc., © 1979)

*Step 3 (Figure 8-5c). Ream* out both ends of the copper tubing once it has been cut. Cutting the pipe can leave small burrs that can cause poor soldering joints and localized turbulence, resulting in corrosion problems.

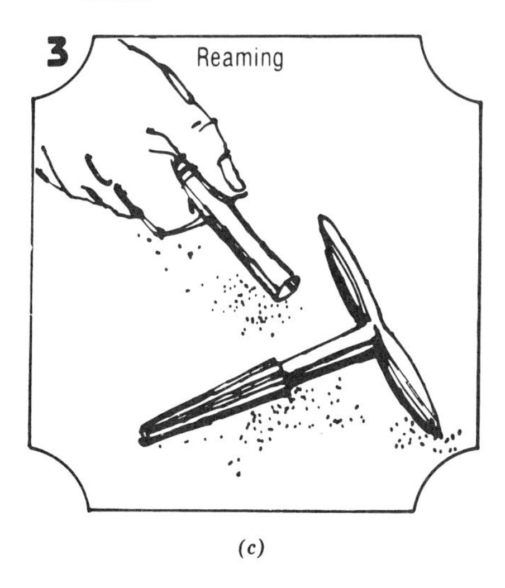


*(c)*

*Step 4 (Figures 8-5d and 8-5e). Clean* both the tube and socket pipe connections using a fine crocus cloth (#0 or #00) or emery paper 200–400 grit). The surfaces must be free from oil, grease, and heavy oxide. Do not handle the clean portions of the pipes and sockets with your hands after cleaning.

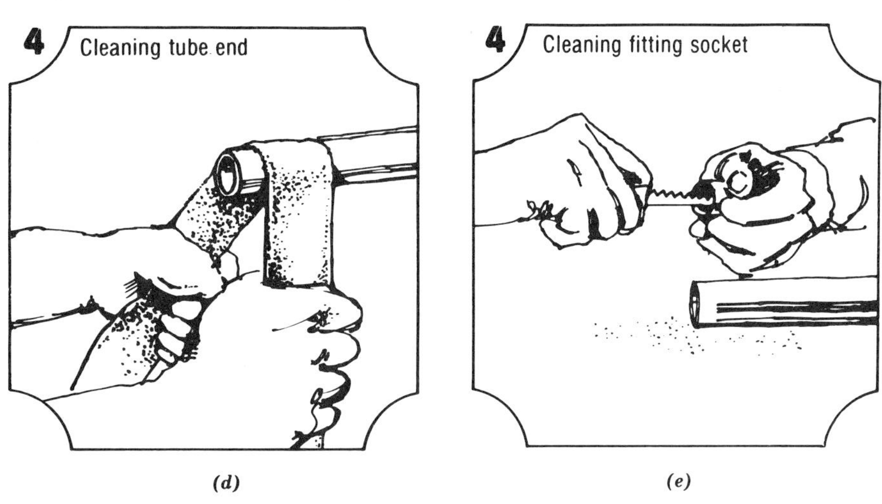


*(d)* *(e)*

*Step 5 (Figures 8-5f and 8-5g).* After cleaning, apply a thin film of self-cleaning flux onto the tube surface and onto the inside of the socket fitting. The flux can be applied with a small brush or a rag. Note that flux accidentally carried to the eyes can be very harmful.

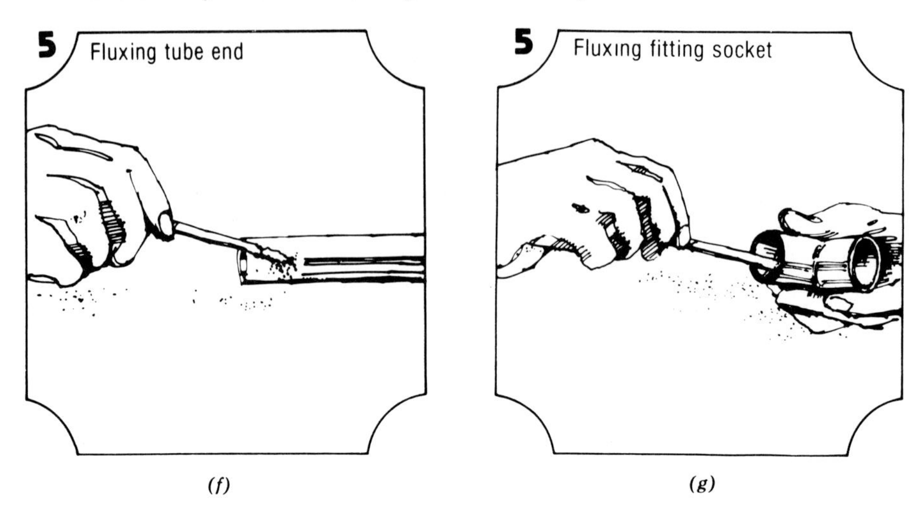


(f) (g)

*Step 6 (Figures 8-5h and 8-5i).* Assemble the tube and socket fitting. Ensure the tube is firmly against the end of the socket. All excess flux should be removed with a rag.

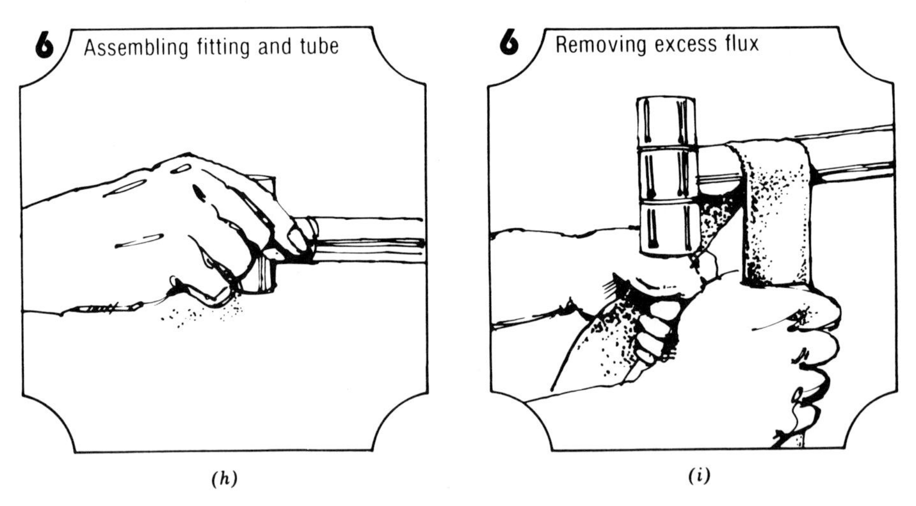


(h) (i)

*Step 7 (Figures 8-5j and 8-5k).* Apply the heat from a butane, propane, or acetylene torch onto the fitting rather than the tube. Don't overheat the joint. Once the solder is applied, it will be drawn into the fitting by capillary action much as ink is drawn into a blotter. Excess solder should be wiped quickly with a clean rag. Be careful not to touch the hot solder and the immediate vicinity of the soldered joint. When soldering valves, be careful to leave the valve in the open position so as not to damage the seating surface.

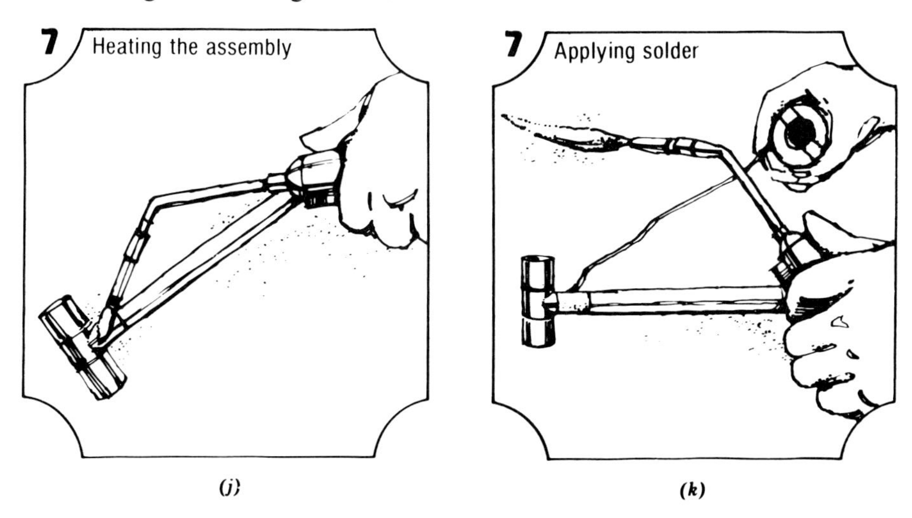


(j) (k)

*Step 8 (Figure 8-5l).* Allow the solder joint to cool naturally. Too-rapid cooling contributes to poor solder joints. If solder joints are made carefully, there should be no leaks during the final system pressure testing (to be discussed in Chapter 9). If the joint leaks during system testing, the liquid (if used) must be drained, and the joint resoldered. If the leak still persists, the fitting should be disassembled and Step 4 through 8 repeated.

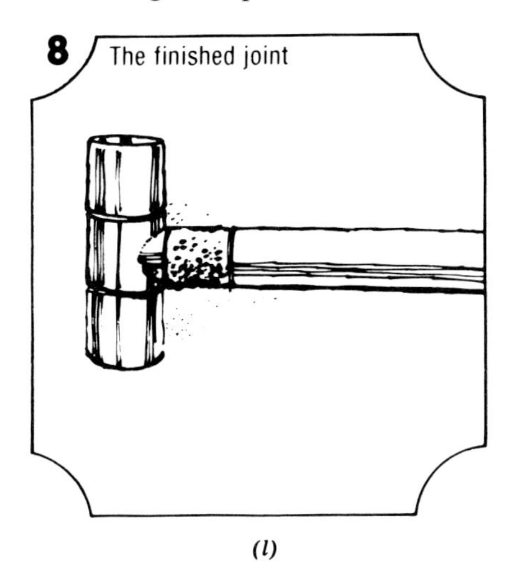


(l)

## Roof Penetrations

The number of penetrations through the roof should be kept to a minimum to avoid leakage problems. Penetrating the roof for inlet and outlet connections should be the last step in plumbing the system. A slightly larger diameter hole should be drilled than the diameter of the system piping. For instance, if ¾-inch copper inlet and outlet pipes are used, then a 1-in. hole should be drilled for the roof penetrations. If the inlet or outlet pipes are to enter perpendicular to the roof, the drilling locations can be found by simply cutting a piece of pipe to length, loosely attaching a 90 deg elbow onto the collector pipe, fitting the length of pipe into the elbow, and positioning the elbow and pipe so the length of pipe touches the shingles 90 deg to the roof's surface. A circle can then be traced around the circumference of the pipe for the inlet and outlet, thus marking the drilling location. The pipe and elbow can then be removed and the penetrations can then be drilled. These penetrations can be flashed much the same way as vent pipes or soil stacks.

Neoprene boots are normally available from most lumberyards with sizes ranging from 1¼ to 4 in. It is not recommended that the insulation be inserted through the roof, but rather it should be butted up against the roof surface due to possible damage to the insulation, and in turn leakage through the roof. The boot opening can be applied over a short piece of copper pipe and reduced at both ends with reducing couplers to match the diameter of the system piping, as illustrated in Figure 8-6.

A sufficient length of copper pipe should then be cut to fit from the elbow at the collectors through the boot to several inches below the interior of the roof's surface to meet the interior piping. The bottom of the flange as well as the penetration, should be coated with silicone sealant, and the top of the flange should be under the shingles as far as it can go. All solder joints should then be completed and all exposed seams should be sealed with

**Figure 8-6.** Reduced neoprene boot penetration.

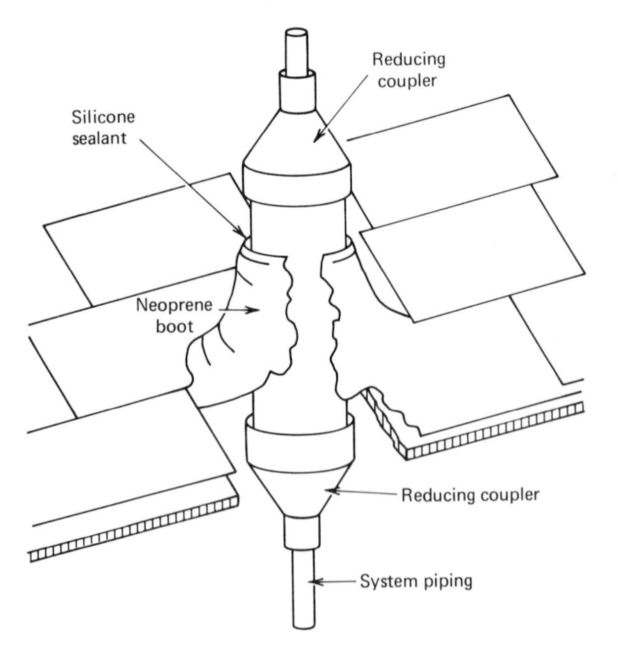

silicone. A completed roof penetration is illustrated in Figure 8-7. Installation of the pipe insulation will be discussed in Chapter 9.

Sensor wiring can simply be brought through the roof by carefully lifting a shingle, drilling a ⅛-in. hole, insetting a ⅛ in. diameter piece of soft copper (1-in. length) flush to the roof's exterior surface, inserting the sensor wiring, sealing the entire area with silicone sealant, and laying the shingle back down over the wire.

## COMPONENT INSTALLATION

### Differential Controller Installation

**Electrical Considerations**

Proper installation of the differential controller is crucial to the safe and economical operation of the solar DHW system. In most cases, this device is accompanied by a complete set of wiring instructions provided by the manufacturer. A typical wiring diagram is illustrated in Figure 8-8.

Wire size and type depend on load size, pump or circulator horsepower requirements, and state and local electrical codes. The controller can be hardwired or softwired depending on the controller model purchased. Hardwiring refers to point to point direct wiring (i.e., #14-2 Romex with ground). Softwiring refers to controllers ordered with a line cord that can be plugged into a convenience outlet. Extension cords should not be used unless a proper ground is maintained. Prior to any electrical work, one should ensure that circuit breakers are opened or fuses are removed. The power leads to the controller should be checked with a VOM to ensure there is no voltage to these lines during installation. Once input leads have been established the output connections can be made to an optional relay coil (Figure 8-19) and/or the circulator. It is *important to remember* not to energize the differential controller until a load (i.e., circulator) has been connected. Some controllers can be damaged if they are energized under no-load conditions.

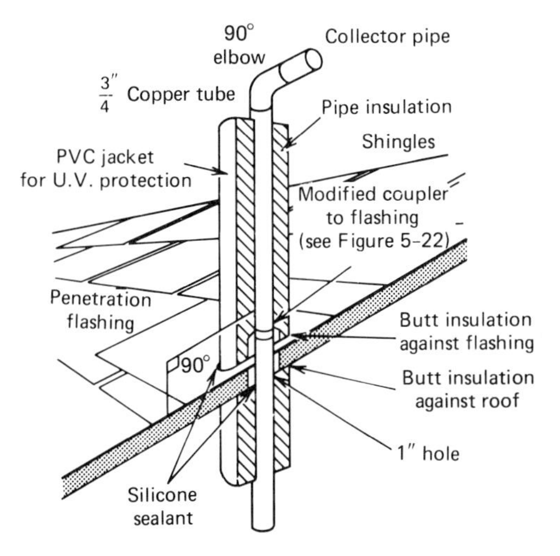


**Figure 8-7.** Completed roof penetration.

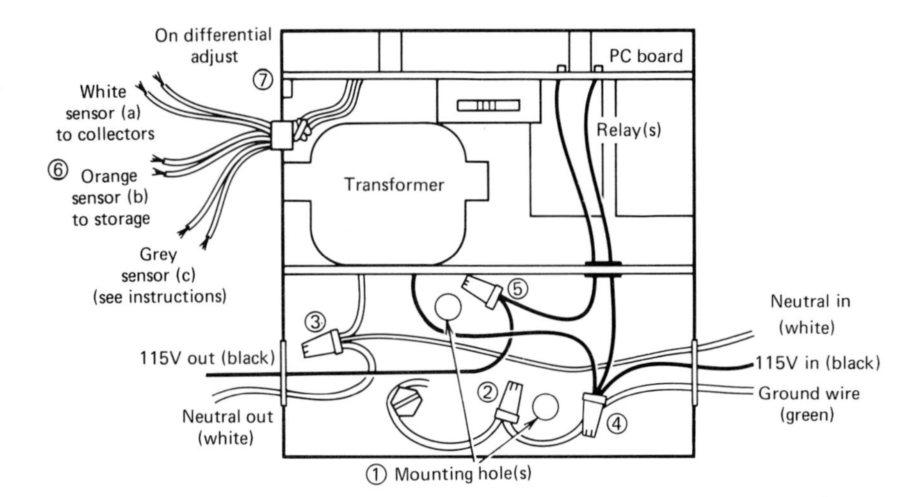


**Figure 8-8.** Typical differential controller wiring diagram. (Courtesy of Heliotrope General)

① Mount box on inside wall close to an electrical source and/or the motor to be controlled. Two holes are provided inside the box for mounting.

② Ground Delta-T® using the green wire attached to inside of box.

③ Connect white (neutral) from electrical source and from pump, etc., to the white lead from the transformer in the Delta-T® using a wire-nut.

④ Connect black (power-in) from electrical source to one of the black leads from the PC board and the black lead from transformer.

⑤ Connect black (power out) from pump, etc., to the other lead from the PC board as shown using a wire-nut.

⑥ Connect sensors: A—White to collector B—Orange (some models have clean leads) to storage. C—Grey or brown, refer to reverse side for specific mounting applications, and for specific mounting instructions for auxiliary sensors such as FS-4, FS-5 and OPB-6-80. If grey leads are not used, separate conductors and insulate with tape. Use 18-24 GA. Two-conductor wire for connecting sensor to Delta-T® up to 1,000.

⑦ By changing screw slot setting, the differentials are changed.

General rules of thumb to consider include the following:

1. Install the controller as close as possible to the circulator to minimize wire runs.
2. Always ensure accessibility to the controller by a screwdriver or other tools.
3. Ensure a proper wire gauge for the length of wire run and load size.
4. Ensure there is sufficient voltage at the control.
5. Do not install the controller on the same AC power line as high current drawing appliances such as a refrigerator or an air conditioner. Activation of such appliances may result in insufficient voltage to the controller and therefore improper controller operation.

6. Ensure that the manufacturer's power output specifications are compatible with the rated load requirements.
7. Locate the controller away from electrical interference devices such as intercoms and CB's (citizen band radios).
8. Ensure that the controller is properly grounded.

### Sensor Location and Attachment

The differential controller can only function appropriately in response to its sensors. Therefore, the location and attachment of the sensor are critical to proper system operation. Before locating and attaching each sensor, the resistance at ambient air temperature should be checked with a volt-ohm-multimeter and compared with the manufacturers' resistance versus temperature chart. Table 8-1 illustrates a typical chart for a 3000 ohm and 10,000 ohm thermistor sensor. Checking the sensors prior to installation may reveal open or short circuits which could result in lost time searching for a fault after the system is installed.

For instance, if the thermistor to be mounted at the collector was defective and indicated zero resistance (short), the controller could assume the collector outlet to be over 200° F and the circulator would run continuously. If the same thermistor indicated an infinite resistance (open), the controller would assume the collector outlet to be below freezing and the circulator would never run. On the other hand, if the thermistor to be mounted at the storage tank was defective and indicated zero resistance (short), the controller could assume the storage to be over 200° F and the circulator would never run. If the same thermistor indicated an infinite resistance (open), the controller would assume the storage to be below freezing and the circulator would run continuously.

**Table 8-1**
**Temperature versus Thermistor Resistance Chart**

| | | *Resistance (ohms)* | |
|---|---|---|---|
| *°F* | *°C* | *3,000 ohms at 25° C* | *10,000 ohms at 25° C* |
| 32 | 0 | 9,810 | 32,660 |
| 50 | 10 | 5,970 | 19,900 |
| 68 | 20 | 3,750 | 12,490 |
| 86 | 30 | 2,420 | 8,058 |
| 104 | 40 | 1,600 | 5,327 |
| 122 | 50 | 1,080 | 3,602 |
| 140 | 60 | 747 | 2,488 |
| 158 | 70 | 525 | 1,751 |
| 178 | 80 | 378 | 1,256 |
| 194 | 90 | 275 | 916 |
| 212 | 100 | 204 | 679 |

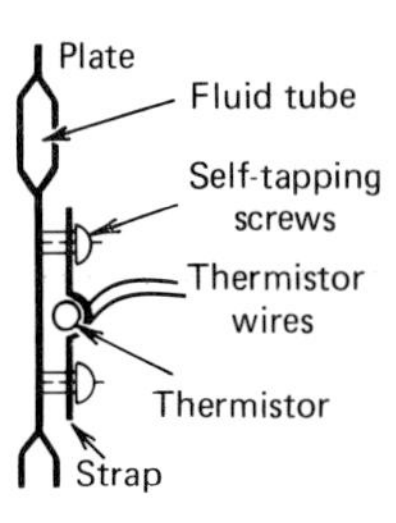


**Figure 8-9.** Pre-attached sensor to absorber plate.

The ***collector sensor*** should be mounted either onto the first 3 in. of the collector outlet pipe at the top of the collectors or onto the internal manifold pipe between the last two collectors just prior to the outlet pipe. Some solar collectors have a sensor already attached to the collector plate as illustrated in Figure 8-9. This situation can sometimes limit the types of controllers that can be used in a system.

The collector sensor can be clamped mechanically onto a manifold pipe with a copper clamp. With a roughed and cleaned manifold surface, a thermal cement (i.e., Devcon®) can be applied to provide a good thermal contact between the sensor and pipe as illustrated in Figure 8-10.

The thermistor wire entrance into the copper lug should be covered with silicone sealant to prevent water from entering the thermistor during inclement weather, thus shorting the device before it is protected by exterior insulation. The sensor should not touch the wall of a metallic collector housing because the housing will act as a heat sink, resulting in lower temperature readings from thermistor to controller. If connections between collectors are soldered and unions are not used, then the sensors must be attached after the soldering joints have been made. The sensors should not

**Figure 8-10.** Collector sensor attachment.

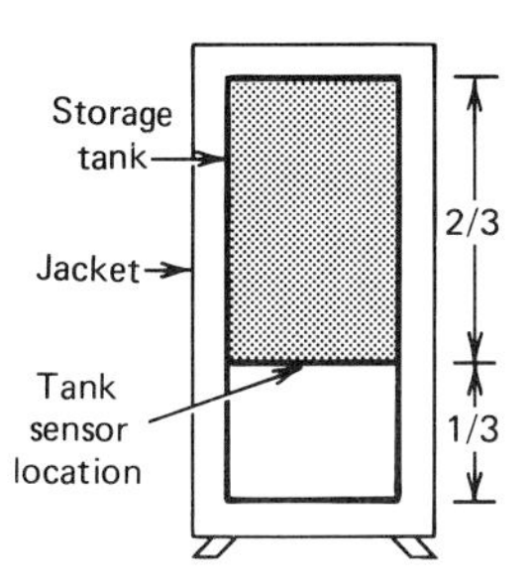


**Figure 8-11.** Tank sensor location.

be subjected to open flame or soldering. Thermal shock may permanently damage the sensors and cause false readings.

The location of the ***tank sensor*** should be at the bottom third of the storage tank as illustrated in Figure 8-11.

Attachment of the sensor to the tank can be accomplished by inserting the sensor mechanically into a clip that has been welded to the side of some solar storage tanks. Attachment can also be accomplished by roughing the surface with steel wool and attaching the sensor with a thermal-conducting epoxy such as Devcon®, as illustrated in Figure 8-12. The thermistor can be held in place with electrical tape until the epoxy hardens and then the tape should be removed.

Another type of sensor commonly used with various controllers is the ***frost sensor***. This sensor is generally located in the collector housing or on

**Figure 8-12.** Tank sensor attachment.

the outlet and inlet connections of the collector. The sensor should not be insulated. This type of sensor is common in air-to-liquid exchangers for freeze-recirculation. Once a preset temperature is reached (i.e., 37° F), the water from storage is circulated back into the exchanger using heat from the storage tank to keep the water in the exchanger from freezing. The frost sensor is used in many open loop systems to activate solenoid valves allowing the water to drain down or drain back thus preventing freezing conditions in the collector array. It is preferable to have this sensor attached to the absorber plate, because the collector, analogous to a blackbody, will always be cooler than the connecting pipes during noninsolation periods.

Most sensors come available with 6- to 10-in. leads with Teflon-insulated wire. This high temperature sleeving is necessary to withstand the rigors of 300° F temperatures during possible stagnation conditions. If the cable run from collector to storage is in contact with the hot outlet return, it must also use the high temperature sleeving. In most situations a high temperature sleeving is not needed for the entire length of run, and #18 AWG (American Wire Gauge) two conductor ''bellwire'' can be used for runs under 100 ft in length. Shielded cable such as Belden 8408 or 8759 can be used for runs greater than 100 ft in length. Remember that sensor wiring should not be run in parallel with AC lines unless shielded wire is used. Noise interference could cause false triggering of the controller resulting in premature start-up or shut-down of the system. At slightly added expense, four conductor cables can be run, thus providing spare leads for other sensors or for backup capabilities. Lead connections can be made by stripping wire leads and twisting together the wires to be connected. This twisted portion of wire should be approximately ⅜ in. A wire nut should be screwed over this twisted pair and a silicone sealant should then be injected. Strain relief for this connection can be provided with an overhand knot as illustrated in Figure 8-13.

## Pump Installation

### Piping Considerations

Prior to pump installation, ensure that the correct type of pump has been selected for the job. Reread portions of Chapter 5 for a discussion of pumps.

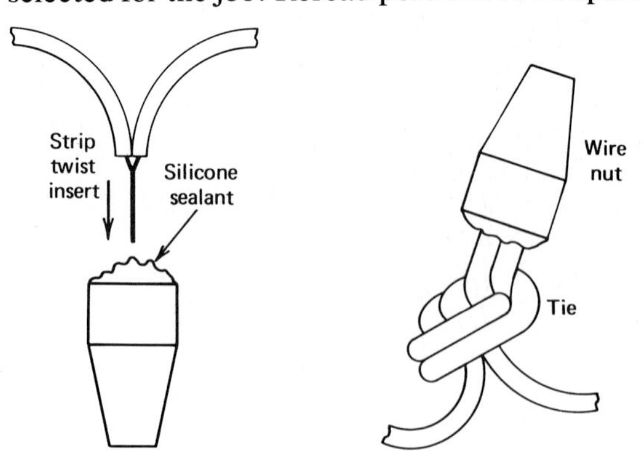


**Figure 8-13.** Weather tight thermistor connections.

**Figure 8-14.** Mounting the pump. (Courtesy of Grundfos Pump Corp., Clovis, Calif.)

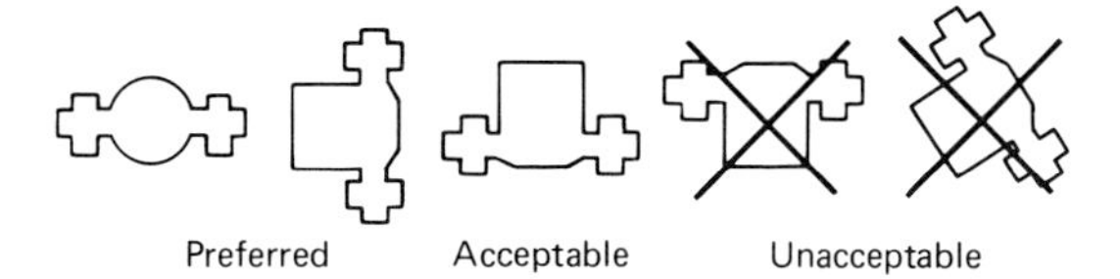


Remember, closed loop systems use a circulator with an iron housing, whereas open loop systems must incorporate a pump in which all water-touched parts are manufactured of stainless steel or bronze.

Arrows on the side or bottom of the pump housing as depicted in Figure 5-43 normally indicate the proper direction of flow. Most pumps can be installed with the motor shaft horizontal or vertical as shown in Figure 8-14. One must be careful not to install a pump in a position where the motor shaft falls below the horizontal plane causing possible wear-out of bearings. Manufacturer's installation sheets should be reviewed carefully.

Mounting flanges are normally included with the pump as shown in Figure 8-15. A ¾ × ¾ in. threaded male adapter to sweat fitting can be screwed into the flange.

The pump is then installed as shown in Figure 8-16. A gasket fits between the flange and the pump body, and the pump can be bolted securely to the flange fittings.

Isolation valves (i.e., one ball valve and one gate valve) should be installed on either side of the pump for flow control and for isolation if repair/replacement is necessary. Whenever possible, avoid high pressure/loss fittings such as elbows and tees directly on either side of the pump. The pump and/or piping on each side of the pump should be adequately

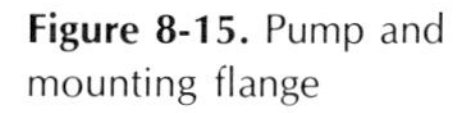

**Figure 8-15.** Pump and mounting flange

**Figure 8-16.** Attachment of pump to flange.

supported to reduce thermal and mechanical stresses on the pump. The pump should not be installed at the lowest point of the system due to the natural accumulation of dirt and sediment. The pump should always be installed in the supply line to the collectors because the transfer fluid will be coolest at this point. This location is of particular importance in open loop systems so that the suction side of the pump is always flooded with the fluid being pumped. Minimum static head requirements in this situation require approximately 3 ft of water column. This requirement will prevent noise and pump cavitation. *Never* insulate the motor housing, because a pump cools itself by radiating heat through the housing.

### Electrical Considerations

The pump should be installed on a separate service line to prevent any possibility of overload and pump failure. The proper operating voltage and current requirements can normally be found on a nameplate attached to the motor. Safe operation requires that the pump be grounded in accordance with the National Electric Code and local governing codes. The ground wires should be copper conductor of at least the size of the circuit conductor supplying power to the pump. Minimum ground wire is #14 AWG (American Wire Gauge). Do *not* ground the pump to a gas supply line. Most pump motors have a built-in, automatic resetting thermal protection, which does not require additional external protection.

In ***open loop systems***, pumps must be selected with higher static head capabilities. Unfortunately, these higher head capabilities also include un-

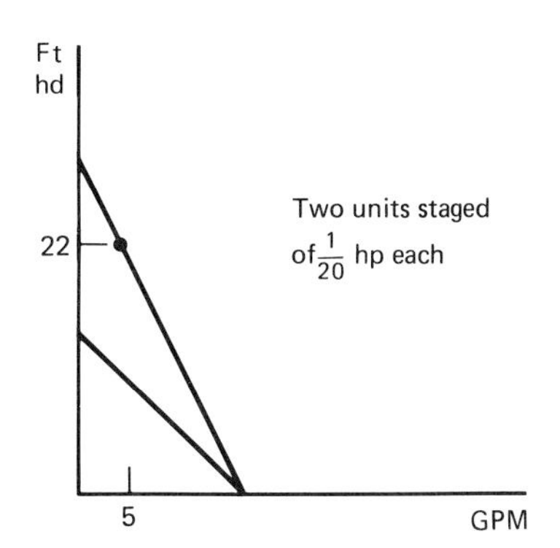


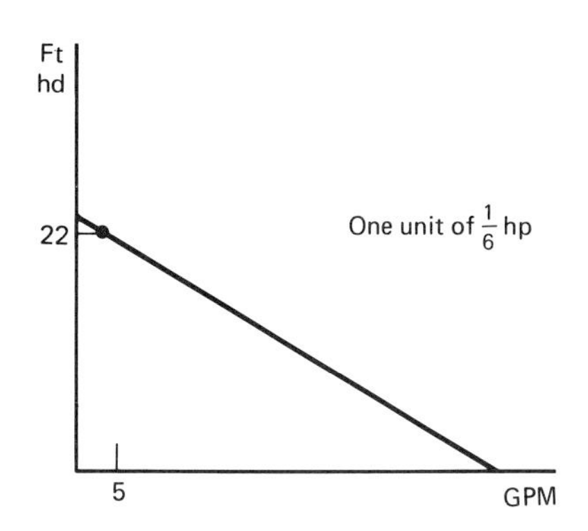


**Figure 8-17.** Circulator staging comparison. (Reprinted with permission from *Solar Engineering Magazine*, 2636 Walnut Hill Lane, Suite 257, Dallas, Texas)

necessary increased flow capacities. Once the solar system is started, the larger horsepower and high head capability is not needed. We can eliminate the use of a higher horsepower pump by staging two or more smaller circulators in series. For two circulators staged in series, the result is twice the pressure and a more efficient pumping system rather than using one large pump. For example, Figure 8-17 shows a comparison using two circulators in series or one larger pump for a requirement of 5 gal/min at 22 ft of head. The total horsepower of the staged pumps is 1/10 hp in comparison with the 1/6 hp required by the larger pump. Normally the cost of two smaller units is comparable to the one larger unit.

By now you may be wondering what all this has to do with wiring the circulator. Understandably, if two circulators are necessary to fill a system initially, and then only one is required to maintain and operate the system, we must have some means of electrically de-energizing the unnecessary circulator(s). This can be accomplished using the wiring diagram of Figure 8-18. We can meet the initial high head requirements by using a time delay relay wired to de-energize the second circulator after the time needed to fill the system has been met.

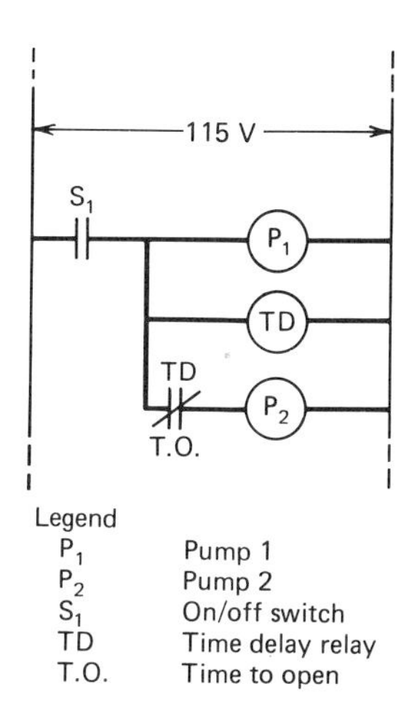


**Figure 8-18.** Time delay wiring for series circulators. (Reprinted with permission from *Solar Engineering Magazine*, Dallas, Texas)

## Storage Tank Installation

Solar hot water storage tanks are extremely heavy in comparison with the conventional glass-lined water heaters. Typical shipping weights for 65, 80, 100, and 120 gal stone-lined tanks are 370, 450, 475, and 525 lb, respectively. This tank weight, when added with the weight of water at 8.33 lb/gal, results in a large amount of weight in a concentrated area. Prospective tank locations, where adequate draining is available, should be checked to ensure this substantial weight factor can be sustained. Because of their weight, care should be taken in moving these tanks to avoid snapping heat exchanger connections and cracking tank linings. As a *minimum*, two workers and a hand truck are necessary. Electric elements are typically 12 in. in length and access must be allowed for repair and replacement. The storage tank should be level and should be at least 2 in. off the floor to avoid water damage and heat loss.

### Piping Considerations

When plumbing solar storage tanks in series with auxiliary conventional fired backup tanks, the cold water supply should always be brought into the solar storage tank first. This, of course, will prevent the water from first being heated conventionally, thus defeating the purpose of the solar system. Two tanks in tandem are best in high ambient surroundings because there is greater storage capacity. One tank systems, which serve as solar storage as well as provide for backup hot water storage, are best in low-ambient surroundings because they loose less stored heat and use the solar storage more effectively than two tank systems. In this situation, the old hot water tank can be stripped of its outer casing and insulation, and can be used as a tempering tank preceding the solar hot water tank.

A vacuum relief should be installed above the tank on the cold water supply line. This will protect the tank walls from possible collapse should there be a sudden large demand on the main water supply. A tempering/mixing valve can be installed 12 in. below the hot water outlet with the cold water entering the bottom. After locating the tank, the shipping nipple can be removed from the top of the tank where the temperature/pressure relief valve is installed. The necessary elbows and male adapters should be soldered together and then all fittings should be screwed into the tank. Do not sweat (solder) any fittings directly to the tank. Soldering at the tank could cause premature failures of the heat exchanger connections or melting of plastic dip tubes.

Flow and temperature parameters of storage tanks under dynamic conditions can be very complex. Because thermal stratification is desirable and should be maintained as much as possible, the transfer media in a closed loop system should be introduced to the storage tank at the highest point of the exchanger. The pump (circulator) should be connected to the bottom of the heat exchanger to ensure pump inlet head under all conditions. This is of particular importance to open loop systems.

### Electrical Considerations

The water heater should be on its own circuit with its own circuit breaker or fuse. The storage tank must be filled with water before the electric element is energized. Otherwise, the element will burn out if allowed to run dry. Wiring diagrams for the electric backup element(s) are normally provided by the manufacturer as previously discussed in Chapter 5. One must take care to observe all National and Local Electrical Codes and be particularly safety conscious when working with 240 VAC imput to electric heating elements. If a solar hot water tank with an internal heat exchanger is supplied with two heating elements, only the top element need be wired for backup heating. To service or replace an electric element (cal rod), the tank must first be drained (see Chapter 9, the section on Operation and Periodic Maintenance—Storage Tanks). The following steps can then be taken:

*Step 1.* Remove the electrical access panel to the heating element.

*Step 2.* Check the electrical terminals with a VOM to ensure they are de-energized. Remove the electrical connection.

*Step 3.* Take out the flange nuts and pull out the heating element. Be careful not to loosen the flange bolts.

*Step 4.* Replace the unit. (It is recommended that a new gasket be used.) Tighten the flange nuts.

*Step 5.* Replace the electrical connections. Refer to the tank's wiring diagram and ensure that wire reconnection is in agreement with the proper terminals.

*Step 6.* Refill the tank (see Chapter 9, the section on Operation and Periodic Maintenance—Storage Tanks).

Sensor location and attachment to the tank are accomplished as discussed during the wiring of the differential controller. The thermostat setting on the single tank solar hot water heater will be determined primarily by whether or not you own a dishwasher. If a dishwasher is used without a preheater as mentioned previously in Chapter 6, then the storage temperature should be maintained at approximately 135° F. Otherwise storage tank temperature need be maintained at only 120° F depending of course, on personal preference. Adjusting the water heater's thermostat is as simple as removing an access panel on the tank's jacket and turning a knob or a screw.

To increase the solar system's efficiency a relay should be added to the wiring arrangement as illustrated in Figure 8-19. This relay prevents the electric heating element (backup) from being energized while the solar system is in operation. Once the differential controller energizes the circulator, the relay is also energized and opens the 240 VAC contacts to the heating element. System operation can be further improved by adding an automatic timer in series just prior to the relay as illustrated in Figure 8-20.

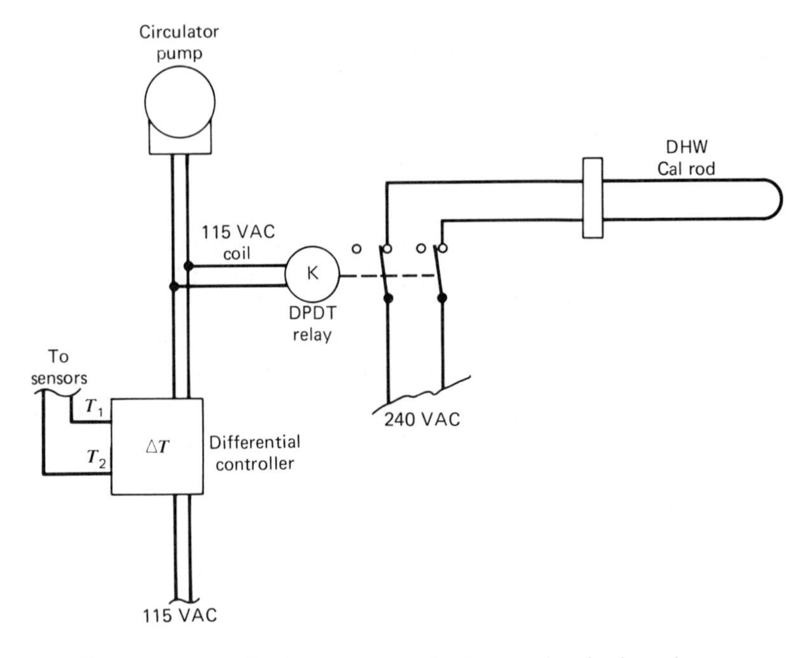


**Figure 8-19.** Relay addition for increased system efficiency. (Courtesy of Applied Technologies, Inc., Kittery, Maine)

These automatic timers are electromechanical and use a small clock motor to turn the heating elements on and off at selected intervals during the day. Timers are easy to install with the instructions supplied by the manufacturer. One must ensure that the unit is rated for the wattage of the heating element. This information is normally stamped on the water heater's identification plate. The time intervals selected will depend upon a person's living habits. Typically the timer can be set to turn on a heating element the first thing in the morning to bring the tank up to a preselected thermostat setting in case the solar system has not completely heated the tank on the previous day. It can also be set to turn the heating element on again from late afternoon to early evening if there was insufficient insolation during the day. If the timer attempts to energize the heating element while the solar system is operating (circulator energized), the relay depicted in Figure 8-19 will prevent the tank from being electrically heated until the differential controller either turns off automatically or is turned off manually. A timer, relay, and differential controller arrangement is illustrated in Figure 8-21.

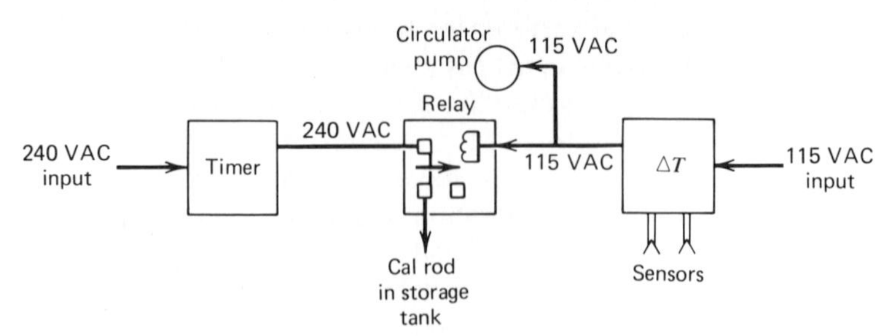


**Figure 8-20.** Wiring block diagram.

**Figure 8-21.** Hot water timer, relay, and differential controller.

### Insulating the Tank

Hot water tanks should be insulated to a minimum R-11. This insulation can be added by using commercially available kits or by simply wrapping the tank with insulation. One must observe safety precautions when adding insulation. The pressure temperature relief valve at the top of the tank should be left unobstructed. Access should also be left to the water heater's controls and sensor attachment. On gas heaters, at least a 2-in. clear space should be left around the exhaust vent at the top of the tank and at least a 3-in. air space should be left around the base of the unit so the main burner and pilot have a sufficient air supply.

## LIGHTNING PROTECTION INSTALLATION

Lightning protection can be an important addition to the solar collector array installation in order to provide a degree of protection to the structure, to the differential controller and, above all, to personal safety. Primary charges, secondary (induced) charges, and electrostatic charges can severely damage or destroy portions of the solar collector system that are not designed to conduct large electrical currents.

***Primary*** charges are created when a large group of electrons seeks a difference in potential from a cloud to the earth. These electrons pulse forward every 50 millionths of a second, which is the start of the lightning bolt. As this bolt nears the ground, electrons are drawn from the air near the earth's surface, creating an ionized stroke that rises from the earth to meet the lightning bolt from the overhead cloud. Once the ionized path is complete, the air around the path is charged, creating the main or return stroke that illuminates the sky.

When lightning strikes an object, it can side flash to another object. This ***secondary*** discharge between one metal conductor and another metal body of conductance or inductance is caused by a large potential difference build-up between the two bodies. Interconnection of metallic objects to ground will eliminate this potential difference. ***Electrostatic*** charges can be generated by windswept particles being drawn over plastic FRP (Fiberglass Reinforced Polyester) or glass glazing materials. When the glazing becomes charged, an ungrounded metal absorber plate can aquire an equal but opposite charge.

Roof-mounted panels are more susceptible to lightning than ground-mounted panels, simply because the roof-mounted panels are closer to the source of discharge. Because this gap is closer than the earth, lightning can strike the solar system, conduct through the water return system down into the hot water storage tank, through the differential controller, through various plumbing parts, and finally out the main water supply pipe to the earthen ground.

Providing a good metallic path to earth for roof- or ground-mounted collectors can prevent possible damage from lightning except in the event of a direct lightning strike. All metal solar collector and support structure parts should be grounded to a cold water supply pipe or to a 3-ft copper clad steel shaft driven into the earth. Steps to ground a collector array are illustrated in Figure 8-22.

***Step 1.*** Ground all inlet and outlet piping or ductwork.

***Step 2.*** If panels are isolated, provide jumpers between frames. Each end of each jumper should have an 8 sq in. contact area.

***Step 3.*** If panels are isolated from the metal support frame, provide a minimum of two jumpers between the panels and frame. If the frame is

**Figure 8-22.** Lightning protection installation. (Reprinted courtesy of Vanderbeck Lightning Rod Co., Hohokus, N.J.)

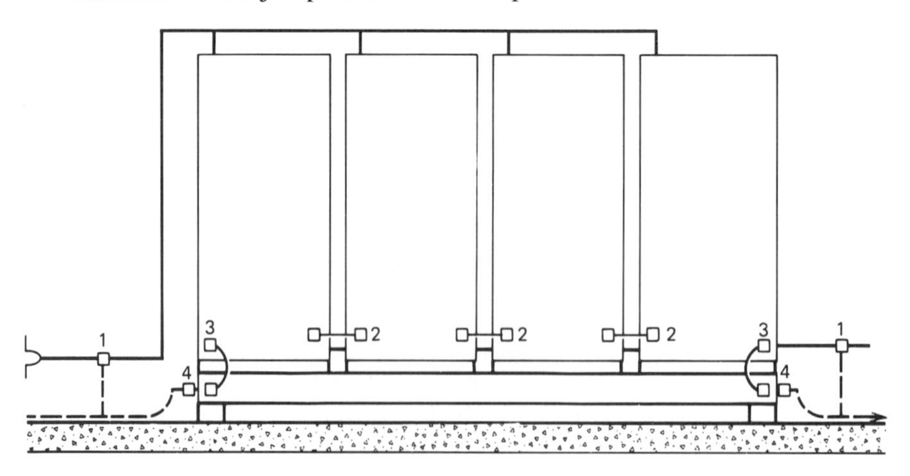

nonmetallic, these connections should be from the ground to the continuous panel network.

*Step 4.* Provide a minimum of two connections from the collector array to ground.

A lightning arrester is a simple but important addition to the installation and affords another degree of protection to the system. A 10-in. piece of #12 copper wire can be used as illustrated in Figure 8-23. One end of the wire should be sharpened to a point and directed skyward, and the other end should be bent at a right angle and clamped to the grounded collector housing or pipe.

This pointed metal rod will discharge the charge-laden thunderclouds more efficiently and quietly than the building or collector array. Because it is easier to discharge a charged body by means of a pointed object rather than a blunt object, the clouds are discharged before they have time to build up enough charge to ionize the air and overcome the resistance (potential difference between the cloud and the earth) violently by means of a lightning bolt.

Lightning protection will lessen the chances of lightning being intercepted, conducted, and grounded through the solar system; it can prevent

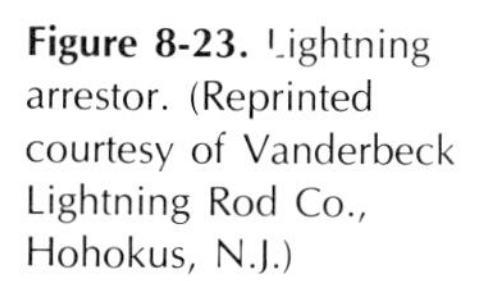

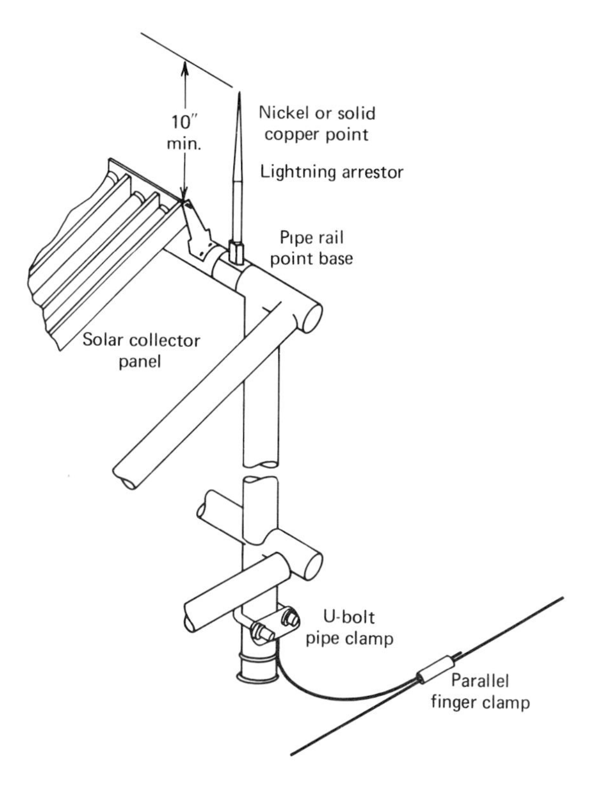


**Figure 8-23.** Lightning arrestor. (Reprinted courtesy of Vanderbeck Lightning Rod Co., Hohokus, N.J.)

the destruction of the differential controller; and it can protect other bodies from secondary discharges.

## REVIEW QUESTIONS

1. What proportions of tin/antimony solder should be used at the collector connections?

   a. 50/50 b. 45/55 c. 90/10 d. 95/5

2. In Freon systems, what gas should be used to purge the lines, if brazed, to prevent corrosion?

   a. Oxygen b. Nitrogen c. Freon d. Argon

3. What is considered to be the last step in plumbing the solar loop?

   a. Connections to the storage tank
   b. Header connections to the collectors
   c. Roof penetrations for inlet and outlet connections
   d. Connections to the pump or circulator if required
   e. Installation of pipe insulation

4. Which of the following is not a significant step in installing a differential controller?

   a. Not to energize the differential controller until a load has been established
   b. Ensure accessibility
   c. Ensure that output leads are connected to a relay prior to the pump connection
   d. Ensure sufficient voltage at the controller
   e. Ensure that the controller is properly grounded

5. If a typical 3000 ohm (@ 25° C) sensor (Table 8-1) indicates a resistance reading of 400 ohms, what is the temperature at its location?

   a. 180° F b. 176° F c. 188° F d. 173° F e. 160° F

6. The collector sensor should be mounted at which of the following locations?

   a. Within the first 3 in. from the collector outlet
   b. Within the first 3 in. from the roof penetration
   c. Between the first two collectors prior to the outlet pipe
   d. Inside the glazing on the absorber plate
   e. On the collector housing prior to the outlet pipe

7. The storage tank sensor should be mounted at which of the following locations?

a. On the outlet pipe away from the tank
b. On the inlet pipe away from the tank
c. At the bottom third of the storage tank surface
d. Two-thirds up from the bottom of the storage tank surface
e. Directly on the bottom of the tank

8. A frost sensor used in open loop systems should be well insulated and can be mounted directly to the collector housing.
   True or False

9. A frost sensor is best attached when located at one of the following points.
   a. On the absorber plate
   b. On the outlet connection from the collectors
   c. On the inlet connection to the collectors
   d. On the collector housing

10. Which of the following steps are not suggested for pump installation?
    a. Install the pump at the lowest point of the system
    b. Insulate the pump housing after installation
    c. Install isolation valves at the pump inlet and outlet
    d. Support the piping on each side of the pump

11. What is the weight of an 80-gal stone-lined tank when filled with water if its weight empy is 450 lb?
    a. 1,116 lb
    b. 666 lb
    c. 754 lb
    d. None of the above

12. Two tanks in tandem are best in low ambient surroundings because there is greater storage capacity.
    True or False

13. When lightning strikes an object, it can side flash to another object. This discharge is considered to be which of the following?
    a. Primary discharge
    b. Secondary discharge
    c. Electrostatic
    d. Induced

14. If the collectors are isolated from each other, jumpers can be connected from one to another. What should the contact area be at the end of each jumper?

a. 3 in.$^2$ b. 5 in.$^2$
c. 6 in.$^2$ d. 8 in.$^2$

15. Which component is most likely to be damaged by an induced charge?
    a. The pump
    b. The controller
    c. The collector
    d. The storage tank

# System Installation: Charge-Up, Start-Up, Maintenance, and Troubleshooting

Once the plumbing and wiring of the required system components have been completed, there are four final steps that remain to be accomplished to complete the installation. We must (1) leak check the system, (2) flush the system, (3) fill and start the system and, (4) insulate the pipes. These steps must be completed in their repective order. Once the system is operational, we shall discuss the importance of periodic maintenance, and troubleshooting procedures to preclude typical possible system failure.

## CHARGE UP THE SYSTEM

Before filling the system and insulating the piping we should first check the system for leaks. This pressurized charge-up of the system is extremely important. Even the best professional plumbers can occasionally find a pinhole leak in a soldering joint thought to be good. It can be very costly to find such a leak if the system has been filled with a heat transfer fluid other than water. Damage to construction materials can also be avoided by initially charging the system prior to fill. A leak test of the collector loop can be performed with either an air pressure test or hydrostatic test with water. The type of leak test to be used depends on the type of heat transfer fluid to be used. If the heat transfer fluid is nonmiscible (not capable of being mixed in any ratio) with water, an air pressure test is recommended. If a system was hydrostatically charged with water when a system was to be filled with a nonmiscible transfer fluid (i.e., synthetic hydrocarbon), it would be extremely difficult to remove all the water droplets from the collector loop to avoid potential freezing, boiling, and corrosion problems once the system was filled.

### Air Pressure Test

In warm, ambient surroundings and during conditions of high insolation the collectors should be covered to maintain a stable temperature when testing the collector to storage loop. The system can be pressurized with a portable air compressor as shown in Figure 9-1.

Normally, systems are pressure tested from 1½ to 2 times their normal operating pressures. For instance, if a closed loop system were to normally operate at 15 psi, then the recommended test pressure would be 30 psi. If higher test pressures are used, system components subject to damage (such as the expansion tank) should be removed. Manufacturers' specifications should be consulted for pressure limitations. All gauges and valves should be left in line to check for leaks through their various threaded and soldered adapters while checking for leaks in the general piping.

Testing can be accomplished by attaching the portable air compressor or a bicycle pump in place of the outlet temperature gauge. The boiler drain valves should be shut (see Figure 5-10 for closed loop detail), and all remaining valves (isolation valves) should remain open. Ensure that the automatic air vents are either removed from the system or that their caps are securely tightened. System pressure can be monitored with a pressure gauge on the portable air compressor or on the "in-line" closed loop pressure gauge. Once the loop has been filled with the desired air pressure, leaks can be found by using the "soap-bubble" test. By brushing a soapy solution around each solder or threaded joint, a leak can be detected if the soap applied bubbles and froths as illustrated in Figure 9-2.

Once the leaks have been located, the air pressure can be released by opening the boiler drains, and the joints can then be resoldered or tightened. The air pressure test should then be performed again until all leaks have been fixed. The collector loop should remain pressurized for at least 1 hour

**Figure 9-1.** Portable air compressor and threaded test adapter. (Courtesy of Applied Technologies, Inc., Kittery, Maine)

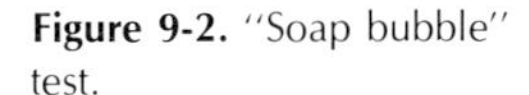

**Figure 9-2.** "Soap bubble" test.

to ensure that there is no change in pressure. Successful completion of the "soap-bubble" test is indicated by the absence of continuous bubbling at all joints. Once the air pressure test is complete, the soap film should be removed by wiping with a clean cloth moistened with demineralized water.

## Hydrostatic Water Test

If the heat transfer fluid is water or a gycol-based fluid that is mixed with a percentage of water, a hydrostatic test can be performed. The same criteria apply for the hydrostatic test as for the air pressure test. The collectors should be covered to maintain a stable temperature during the test. Excessive pressures to system components should be avoided. The system can be pressurized at 1½ to 2 times its operating pressure by using a high head self-priming pump or town/city well water of sufficient pressure. A ¾-in. washing machine hose (female connection) can be screwed onto each of the boiler drains as depicted in Figure 9-3. Three hoses will be needed and are attached (1) from boiler drain no. 1 to the transfer pump, (2) from the transfer pump to the fill bucket, and (3) from boiler drain no. 2 to the fill bucket.

A check valve can be used between the boiler drains to force the water to circulate in one direction in a complete loop toward the bottom of the collector. Using a check valve instead of a gate valve at this point is a matter of economics. Because a check valve must be used in the system to prevent thermosiphoning of heat from storage to the collectors during periods of

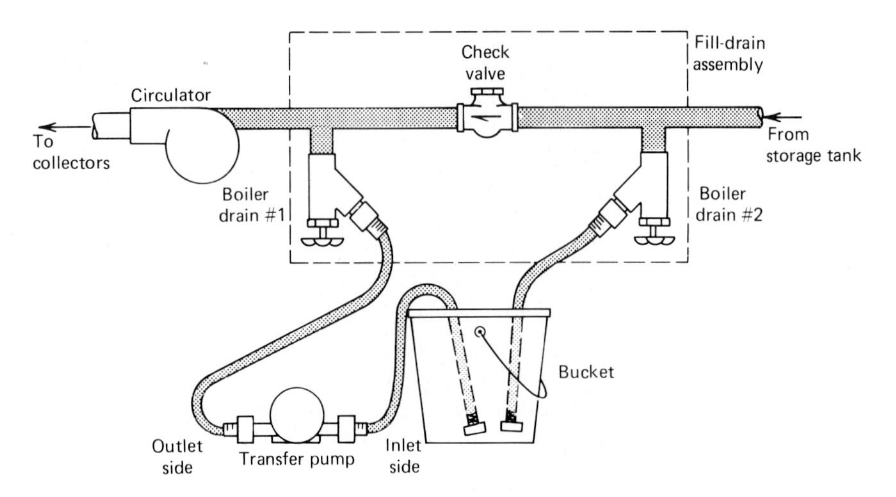


**Figure 9-3.** Hydrostatic and fill procedure arrangement.

noninsolation, it can be placed between the two boiler drains thereby serving the same functions of a gate valve and eliminating the extra cost. The following steps apply to the hydrostatic test.

***Step 1.*** Open boiler drains nos. 1 and 2. Ensure that the differential controller is off.

***Step 2.*** Fill a large bucket with clean water. Ensure that additional water is available if needed.

***Step 3.*** Connect the outlet side of the transfer pump with a washing machine type hose and fitting to boiler drain no. 1.

***Step 4.*** Connect the inlet side of the transfer pump with a washing machine type hose and fitting to the bucket of water.

***Step 5.*** Connect a washing machine type hose and fitting from boiler drain no. 2 to the bucket of water.

***Step 6.*** Ensure that all isolation valves are open and that all drains are secured. Start the transfer pump.

***Step 7.*** Pump the water through the collector to storage loop until water exits boiler drain no. 2.

***Step 8.*** Close boiler drain no. 2 and watch the system pressure until it is 1½ to twice the normal operating pressure.

***Step 9.*** Once the test pressure is reached, close boiler drain no. 1 and quickly shut off the transfer pump.

***Step 10.*** Check all soldered and threaded joints for leaks. The collector loop should remain pressurized for at least 1 hour to ensure that there is no change in pressure.

*Step 11.* Disconnect the outlet side of the transfer pump and place each boiler drain hose connection in the bucket. Open boiler drains nos. 1 and 2 to drain the system.

*Step 12.* Resolder all joints and tighten all connections that were found to leak. Repeat Steps 1 to 12 as necessary.

## FLUSH THE SYSTEM

Once all leaks have been corrected, the collector to storage loop should be flushed of all metal filings, flux, solder, and any other contaminants. This can be accomplished by continuously running the heat transfer fluid with the transfer pump from the containment illustrated in Figure 9-3, through the collector loop and back to the containment. Loose solder can sometimes be freed by "burping" the system. This is accomplished by closing boiler drain no. 2, allowing the pressure to build up, and then opening the same boiler drain. A strainer upstream of the circulator can be useful during this flushing procedure. The strainer should have a bronze body and stainless steel 20 mesh screen and should be cleaned once flushing is complete. This procedure should be run for approximately 10 minutes. It is important to remember *not* to use water to flush out any system or parts of systems that use hydrocarbons or silicone heat transfer fluids. After flushing, hardware peculiar to each generic system should be checked as recommended by the manufacturer. Such items as satisfactory draining from pitched headers for drain back and drain down systems should be checked as well as solenoid and motorized valve operation.

## FILL AND START UP THE SYSTEM

The fill and start-up procedure is perhaps the most gratifying of all steps in the installation process in that all the work done in assembling an alternative energy system will come to fruition in actual system operation. We shall illustrate a typical fill and start-up procedure for a closed loop freeze-resistant system. Other generic types of systems will have varying idiosyncrasies in the fill procedure, and many collector manufacturers provide a recommended fill procedure for their particular system. The amount of heat transfer fluid needed for system fill can be approximated by using Tables 5-2, 5-8, and manufacturers' data. An example is depicted in Table 9-1.

Approximate system capacity can also be determined during a hydrostatic test by measuring the actual amount of fluid when the system is drained. The fill procedure is very similar to the hydrostatic test. Figure 9-4 illustrates the equipment hookup used during this procedure.

The following steps apply to fill and start-up of a closed loop freeze-resistant system.

**Figure 9-4.** Fill procedure.

**Table 9-1**
**Gallons of Transfer Fluid Needed for Fill**
**(Glycol Solution)**

| | *Quantity Required (gal)* |
|---|---|
| Length of pipe<br>(Table 5-8)<br>(i.e., ¾ in. pipe @ 40 ft) | 4.09 |
| Heat Exchanger<br>Table 5-2<br>(i.e., 80 gal tank) | 0.25 |
| Solar Collector<br>(manufacturers' data)<br>(i.e., 3 collectors @ 0.39 gal/panel) | 1.17 |
| Expansion Tank<br>(dependent on system press)<br>(i.e., 0.25 gal) | 0.25 |
| Total gallons needed for fill | 5.76 gal |

***Step 1.*** Open boiler drains nos. 1 and 2 (Figure 9-3). Ensure that the differential controller is off.

***Step 2.*** Fill a large bucket with a predetermined amount of heat transfer fluid. Ensure additional tranfer fluid is available if needed.

***Step 3.*** Connect the outlet side of the transfer pump with a washing machine type hose and fitting to boiler drain no. 1 (Figure 9-3).

***Step 4.*** Connect the inlet side of the transfer pump with a washing machine type hose and fitting to the bucket of heat transfer fluid.

***Step 5.*** Connect a washing machine type hose and fitting from boiler drain no. 2 to the bucket of heat transfer fluid. Do *not* extend the hose into the fluid.

***Step 6.*** Turn the screw caps on the automatic air vents located at the air purger and at the highest outlet point on the collector array one to two turns counterclockwise to help bleed off air trapped in the collector loop during system fill. A coin vent installed at the collector outlet as shown in Figure 9-5 is often used in systems that require few changes in transfer fluids. This type of air vent is less subject to freeze because it can be completely insulated.

One person must be positioned on the roof with a screwdriver and small container to catch the initial fluid exiting the coin vent. The vent should be open before the transfer pump is started.

**Figure 9-5.** Coin vent at collector outlet.

***Step 7.*** Start the transfer pump at the fill and drain assembly. Air in the pipe will be forced through the highest air vent as well as boiler drain no. 2 (Figure 9-3). A small container should be placed under the coin vent as illustrated in Figure 9-6 to catch the initial fluid expelled. The vent should then be closed.

***Step 8.*** Once the fluid exits boiler drain no. 2 (Figure 9-3), the hose should be immersed into the transfer fluid. If the hose had been previously immersed while air was exiting the piping, more air bubbles would have been entrapped in the transfer fluid.

***Step 9.*** Pump the transfer fluid through the collector loop until most of the air bubbles in the container have been removed.

***Step 10.*** Close boiler drain no. 2 (Figure 9-3) and watch the system pressure gauge until the desired system pressure is attained. Then close boiler drain no. 1 and quickly shut off the transfer pump. The system should now be free of most air and is now pressurized. Record the pressure.

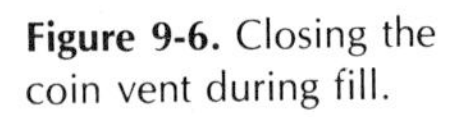

**Figure 9-6.** Closing the coin vent during fill.

It should be remembered that to enable the expansion tank to maintain system pressure, it is necessary at the initial fill to pressurize the system fluid to a point slightly higher than the air charge in the expansion tank. For instance, if we have an expansion tank charge at 12 psi, a fluid fill at 14 to 16 psi is suggested. The effect of this slightly higher pressure is to force some of the heat transfer fluid into the expansion tank at the initial fill.

*Step 11.* Disconnect the hoses. (You may also wish to remove the handles from the boiler drain valves so children won't be subject to inadvertently opening the valves and possibly scalding themselves with the heat transfer fluid.) Uncover the collectors if previously covered during leak testing. Turn the differential controller on manual and listen for fluid circulation. System pressure may increase slightly when the circulator is activated. Check the pressure gauge to ensure system pressure is holding.

*Step 12.* Turn the differential controller to automatic, and the system is now operational. Only insulation of the pipes remains. Watch the pressure for several days before installing the pipe insulation. Note that as the transfer fluid is heated, the dissolved air is gradually vented to the atmosphere via the air purger and air vent. The fluid in the expansion tank is forced into the system by the diaphragm to displace the volume of vented air. After a few sunny days the system pressure should stabilize at some point between the initial fill and the expansion tank air charge.

## INSULATED EXTERIOR AND INTERIOR PIPING

Insulating the solar DWH plumbing is the last important step in the installation squence. The various types of insulation that should be used are discussed in Chapter 5. All exterior pipes must be insulated. Remember *not* to use elastomeric insulation in exposed weather conditions because it will crack and deteriorate rapidly unless jacketed or coated with a paint approved by the manufacturer. Only the indoor collector outlet to storage pipe needs to be insulated with the elastomeric type of pipe insulation. In some instances it may not be economical to insulate the indoor storage to collector return pipe because the collector will operate at a higher efficiency with a cooler inlet fluid. (This is only true if the return pipe run to the collectors is short and few BTU are lost.) If the indoor pipe insulation has not been added by this point, it can now be slit, installed, and glued with the manufacturer's recommended adhesive. Outside pipe insulation can be installed as illustrated in the following pictorial sequence.

### Insulation Between Collectors

*Step 1 (Figure 9-7).* Exterior pipe insulation is normally available in 3-ft lengths. Mark the desired length to be cut.

Figure 9-7

*Step 2 (Figures 9-8 and 9-9).* Cut the insulation to fit with a hacksaw. Contour the shape as necessary to accommodate collector sides, attached sensors, and unions.

Figure 9-8

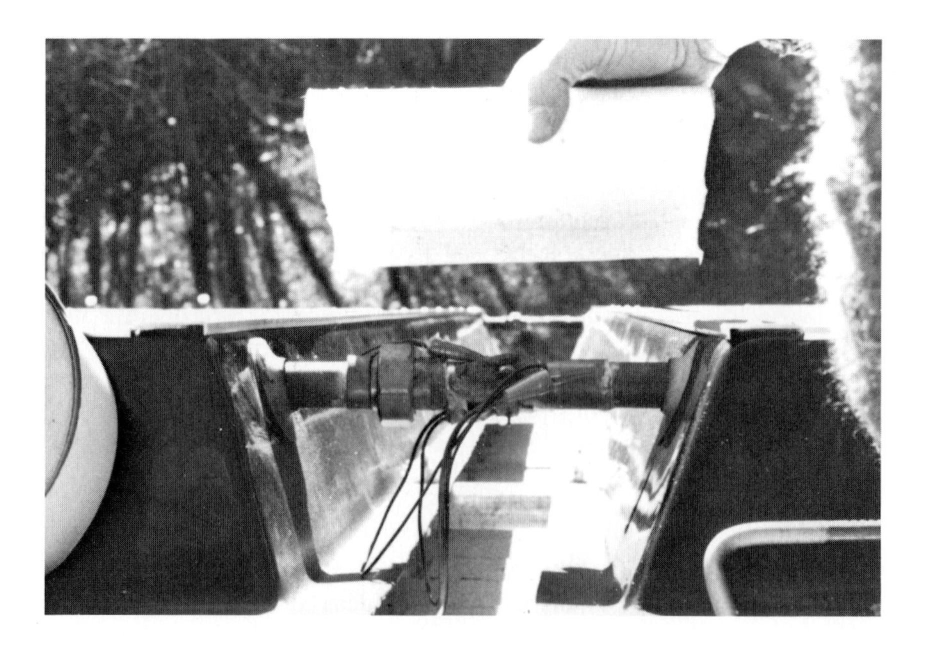

Figure 9-9

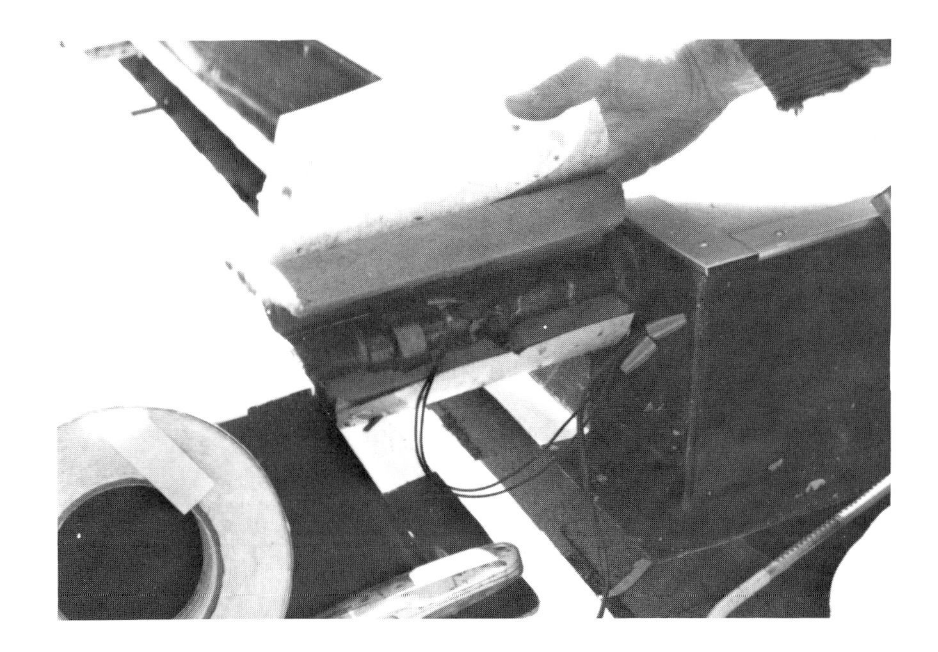

***Step 3 (Figures 9-10 and 9-11).*** The slits in the jacket should be positioned on the downward side of the pipes, sealed with adhesive, and secured with Tedlar® or U.V. resistant PVC tape.

Figure 9-10

Figure 9-11

*Step 4 (Figure 9-12).* Seal any exposed insulation with silicone sealant to prevent water penetration.

Figure 9-12

## Insulation at Collector Outlet

*Step 1 (Figure 9-13).* Cut the insulation to fit. Cut a 45 deg miter for 90 deg joints or use preformed fittings.

*Step 2 (Figures 9-14 and 9-15).* Seal the vinyl jacket with a brush-on or spray adhesive and use Tedlar tape or PVC to seal all seams.

**Figure 9-13**

**Figure 9-14**

**Figure 9-15**

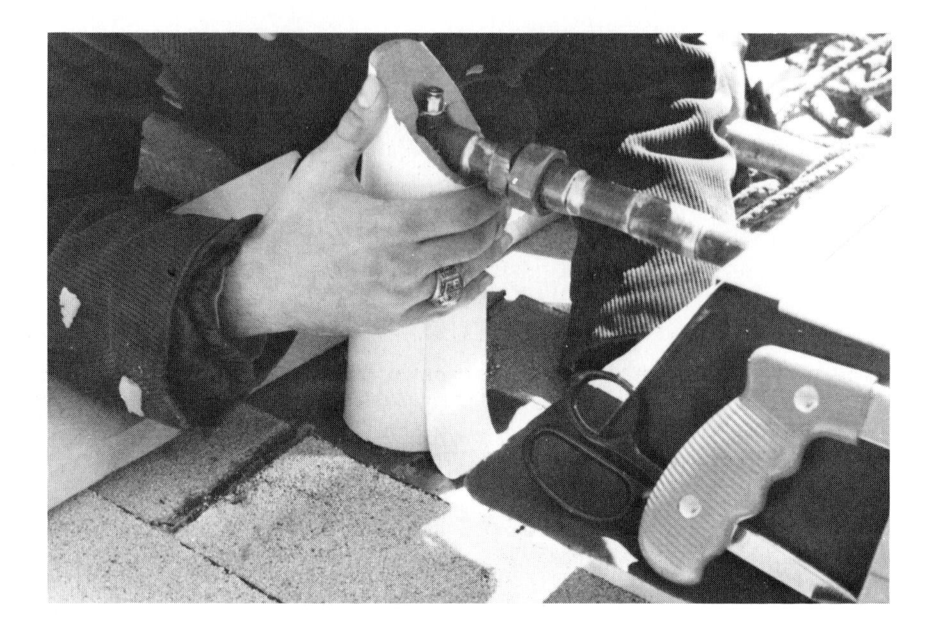

*Step 3 (Figures 9-16 and 9-17).* Seal exposed insulation at abutting surfaces with silicone sealant to prevent water penetration. Completely insulate over the manual coin vent. If an automatic air vent is used, do not insulate over the end of the vent cap.

**Figure 9-16**

Figure 9-17

## Insulation at Collector Inlet

***Step 1 (Figures 9-18 and 9-19).*** Cut the insulation to fit. Cut a 45 deg miter for 90 deg joints or use preformed fittings.

Figure 9-18

**Figure 9-19**

*Step 2 (Figure 9-20).* Seal the vinyl jacket with adhesive and tape the seams with Tedlar or PVC tape.

*Step 3 (Figures 9-21 and 9-22).* Seal exposed insulation at abutting surfaces with silicone sealant to prevent water penetration. The exterior piping insulation is now complete.

## OPERATION AND PERIODIC MAINTENANCE

If a solar system is not maintained properly, as in any other system, efficiency and effective system life cannot be maximized. As in all types of energy conversion systems, a much greater return can be realized on the money invested if the proper servicing steps are taken at regular intervals. Every system must be monitored periodically by the owner/installer, and minimum maintenance schedules should be established. An owner normally will not know whether or not the solar system has malfunctioned unless the temperature and pressure gauges are checked daily by personal observation, or until the monthly utility bill is received. Systems can also be monitored

Figure 9-20

Figure 9-21

Figure 9-22

constantly electronically for system pressure and reverse temperature cycling (caused by a faulty controller or sensor) as illustrated in Figure 9-23. Such a device can audibly warn the homeowner of system malfunction much like a smoke detector. It is this use of controlled monitoring, personal observation, and periodic servicing that will optimize the usage of the solar DHW system and, in consequence, its economic viability.

Periodic maintenance should be performed on the following components as suggested by the system manufacturer or system installer. Typical service intervals and methods may vary depending on the type of climate and system. In alphabetical order we shall discuss operation and service of the following:

- Collectors
- Controllers and sensors
- Heat transfer fluids
- Miscellaneous plumbing parts
  - Automatic air vents
  - Gauges, pressure, and temperature
  - Pipe insulation
  - Strainer
  - Valves
- Pumps and blowers
- Storage tanks

**Figure 9-23.** Solar system electronic monitor. (Courtesy of Applied Technologies, Inc., Kittery, Maine)

## Collectors

Collector glazing should be cleaned every 6 months. Tempered glass can be cleaned with commercially available products. Fiberglass reinforced polyester and acrylic materials should be cleaned with mild soap and water. Care should be taken not to scratch the glazing. Do *not* wash collectors in direct sunlight. Early morning or cloudy days are better times for this chore. If absorber plates are accessible, inspect for surface peeling and loose tubing (depending on the type of absorber plate). The temperature difference across each collector should be relatively the same.

## Controllers and Sensors

There is virtually no maintenance involved with the differential controller. Failure of a component within the device requires correction by the manufacturer or replacement.

Inspect the sensors at the collector every 6 months (if accessible), as well as at the storage tank to ensure good thermal contact. A loose or detached ***tank sensor*** would cause inefficient operation and could actually cause the storage tank to expel the energy stored during a sunny day (reverse temperature cycling) by making the controller assume the tank to be at a constant ambient temperature condition rather than the higher actual storage temperature attained during the day. The controller would therefore continue the collection cycle until the storage temperature was within the preset temperature limit (i.e., 5° F) of ambient. If the ***collector sensor*** were loose or detached, the sensor would cause inefficient operation and could shorten the daily collection ability of the solar DHW system. Proper resistance values of sensors should also be verified. If the collector sensor were short circuited, a differential controller would interpret the collectors to be hot even at night, thereby extracting heat from storage back through the collectors (reverse temperature cycling). Only an electronic monitor such as the Solar Sentry(TM) shown in Figure 9-23 could detect these sorts of malfunctions on a continual basis.

## Heat Transfer Fluids

Heat transfer fluids must be checked every fall for safe pH levels (i.e., measurement of solution acidity; $-\log_{10}[H_30^+]$) and for safe freeze protection levels. Commercial test kits are available as illustrated in Figure 9-24 to

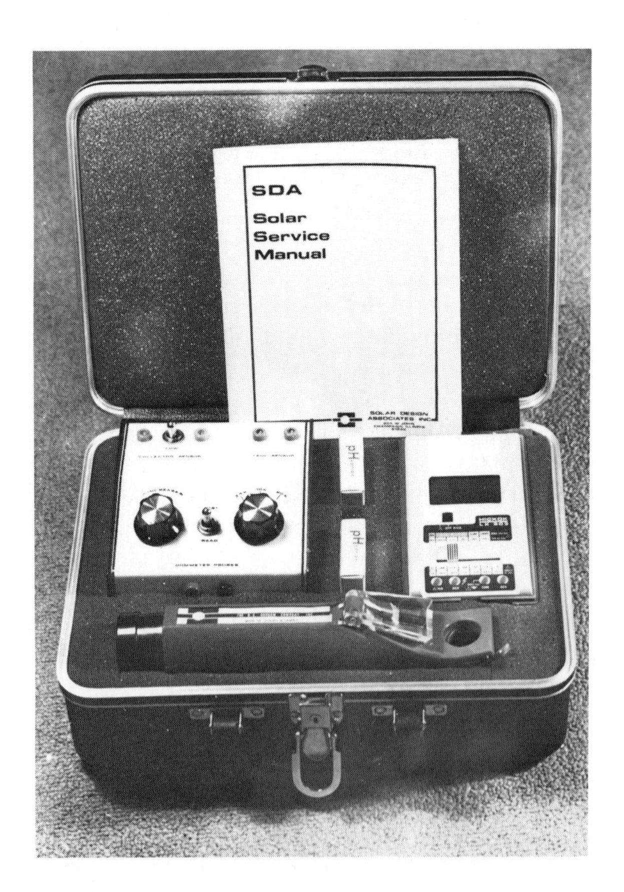


**Figure 9-24.** Typical service kit. (Courtesy of Solar Design Associates, Inc., Champaign, Ill.)

make fluid condition analysis easy. Such kits can include controller testers, Hydrion® papers for pH level checks, VOM's for troubleshooting electrical problems, and refractometers for freeze resistance checks.

Freeze resistance can be checked by either a refractometer, which requires only a few drops of fluid and uses the index of refraction for comparison, or by the determination of the specific gravity of the fluid. ***Specific gravity*** of material is the ratio of its density to that of water and can be determined with an instrument called a hydrometer. By determining the relative density to that of water, the freezing point of the fluid can be determined. This method requires the use of more transfer fluid than does the refractometer method. If too large a sample is taken, the system may have to be refilled. Using the index of refraction method avoids the large sampling. Figure 9-25 represents a typical index of refraction chart used to determine freeze potential of ethylene and propylene glycol. The antifreeze protection point is observed through the refractometer where the dividing line between light and dark (edge of shadow) crosses the scale.

The pH test can be performed using the same fluid withdrawn for the freeze concentration test. A piece of wide-range Hydrion paper can be dipped into the solution. The color of the paper is then matched to the color chart on the dispenser roll. A more accurate pH reading can be then obtained using a shorter range pH paper. The pH condition should be compared with the minimum standards specified by the collector manufacturer. If the pH or freeze requirements are not met, the fluid should be changed or buffers should be added, depending on the type of fluid used. Normally, glycol solutions should be changed every 2 years, hydrocarbons every 5 to

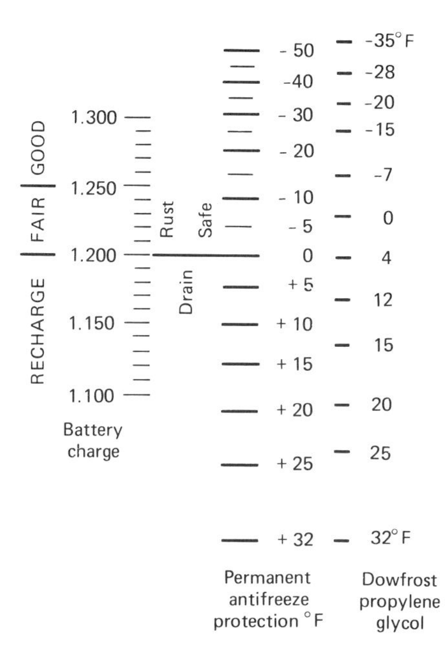


**Figure 9-25.** Index of refraction chart to determine freeze potentials. (Reprinted with permission from Solar Design Associates, Inc., Champaign, Ill.)

10 years, and silicones every 15 to 20 years as recommended by their manufacturers.

## Miscellaneous Plumbing Parts

### Automatic Air Vents

Check air vent caps to ensure they are at least two counterclockwise turns loose to allow air to escape from the system. An air vent will sputter audibly if dirt or debris is caught in the float mechanism.

### Gauges, Pressure, and Temperature

System pressure of a closed loop should be checked daily. (Initial system pressure should be recorded on a tag on the system fill line for comparison.) Temperatures should be checked daily for proper system operation. A differential temperature of 30° F between storage inlet and outlet is typical.

### Pipe Insulation

Inspect exterior pipe insulation once a year. Note any areas exposing insulation to weather. Seal those areas with silicone sealant. Ensure all jacket material is sealed. Look for ultraviolet damage.

### Strainer

Clean the strainer in an open loop system once a year as follows:

1. Drain the solar collector array.
2. Isolate the strainer from the tank (i.e., by closing a gate valve, Figure 3-6).
3. Open the strainer cap with an adjustable wrench.
4. Remove the stainless steel screen and clean it out with water.
5. Reinstall the screen, and tighten the strainer cap.
6. Open the tank to collector loop filling the system.

### Valves

Inspect all fittings and seals once a year from visible leaks. Check gaskets and valve seats. These seals are usually designed for resistance to water and not to all transfer fluids. Solenoid and automatic valves should be jumpered to simulate valve operation. Ensure that such valves are not getting wet, thus having the potential to short out.

## Pumps and Blowers

Some types of pumps and blowers are sealed and permanently lubricated, some are lubricated by the transfer fluid, and some require periodic lubrication. The manufacturer's literature should be checked for possible lubrication schedules. Some pumps and blowers have variable speed ranges that can be set by the homeowner. Flow may be varied depending on the season and latitude of the system. High temperatures must be maintained for practical household use while maximizing the total BTU collection. The pump speed should be varied accordingly. Using Equation 5-2 we can demonstrate this optimization of heat collection.

We shall assume two sets of conditions and note the results.

***Condition I***

*Assume:*

Inlet temperature to storage = 180° F }
Outlet temperature from storage = 83° F } Flow rate at 2 gal/min
Specific Heat $C_p$ = .56 BTU/lb-°F
Weight per unit $m$ = 7 lb/gal

Because $Q = mC_p \Delta T \times \text{(Flow Rate)}$

$$Q = (7 \text{ lb/gal})\left(\frac{0.56 \text{ BTU}}{\text{lb-°F}}\right) (180° \text{ F} - 83° \text{ F}) \times [(2 \text{ gal/min})(60 \text{ min/hr})]$$

And
$Q = 45{,}630$ BTU/hr

***Condition II***

*Assume:*

Inlet temperature to storage = 158° F }
Outlet temperature to storage = 133° F } Flow rate at 9.5 gal/min
Specific Heat $C_p$ = 0.56 BTU/lb-°F
Weight per unit $m$ = 7 lb/gal

Because $Q = C_p\Delta T \times \text{(Flow Rate)}$

$$Q = (7 \text{ lb/gal})\left(\frac{0.56 \text{ BTU}}{\text{lbs-°F}}\right) (158° \text{ F} - 133° \text{ F}) \times [(9.5 \text{ gal/min})(60 \text{ min/hr})]$$

And
$Q = 55{,}860$ BTU/hr

Because the outlet temperature from storage is sufficient for practical household functions, the working temperature difference is approximately 26° F across the collector, and there is more heat accumulated, Condition II would be the preferred operating condition. Typical pump operation will demonstrate increased pump speeds during summer use and slightly lower pump speeds during nonsummer use. Conditions will vary from one locale to another, and the installer/owner can vary the pump speeds while observing the corresponding change in system operation.

## Storage Tanks

To prolong the life of the water heater/storage tank it can be drained once each year to remove scale and sediment from the bottom. It is important to drain the tank completely because the last few gallons of water will draw the sediment out with it. If the tank is directly in the solar collector loop, then it should be drained at night or on a cloudy or rainy day. The following steps can be followed to drain the storage tank:

1. Shut off all power such as electric elements, oil, and gas burners prior to draining.
2. Shut off the cold water supply to the bottom of the tank. If the collector loop is open to the tank, it also must be drained in order to isolate and drain the tank.
3. If necessary, attach a hose to the tank outlet drain and run the hose to a suitable drain. Open the tank drain valve.
4. Lift the lever on the automatic pressure and temperature relief valve to allow air into the tank to assist in draining.
5. To reverse the process, close the drain valve and open the valve between the cold water inlet and the water heater. Open the nearest hot water faucet to facilitate filling, and close it once filling is complete.
6. Refill the solar collector loop if it was previously emptied, and turn on the backup power.

## TROUBLESHOOTING PROCEDURES

Anyone who understands the principle operation of a device can also learn to troubleshoot problems associated with that device. A person should be familiar with the fundamentals of electricity such as Ohm's law, be able to read manufacturers' specification sheets and instructions, and be able to use a standard volt-ohm-multimeter. Prior to performing troubleshooting procedures, ensure that there are no blown fuses or open breakers, and power input to each electrical component is available. Defective switches, connections, and wiring termination are potential sources of trouble and should be checked thoroughly during troubleshooting. Intermittent problems are always the worst to resolve, and many troubleshooting procedures will depend on the type of system involved. For example, if a ***leak*** is detected by either personal observation of the pressure gauge or by electronic monitoring in the collector loop of a closed loop system (i.e., Figure 5-10), the following steps should be followed immediately:

***Step 1.*** Shut off the circulator.

***Step 2.*** Close the isolation valves separating the collector loop from the storage loop.

***Step 3.*** Watch the pressure gauge. If pressure is still decreasing, the leak is on the collector side and not at the storage side.

***Step 4.*** Open the boiler drain and drain the remaining fluid into a container.

***Step 5.*** Air test the system to find the leak.

***Step 6.*** Resolder the defective joint and air test to ensure that the source of the problem has been corrected.

***Step 7.*** Refill and start-up the system.

We can troubleshoot each problem using a process of elimination. We should determine the symptom and then check the cause for each symptom and remedy the problem accordingly. Common problems frequent to solar domestic hot water systems are illustrated in Table 9-2. This table will not resolve all problems for all solar DHW systems, but it will provide an important starting point.

**Table 9-2**
**Common Troubleshooting Problems**

| *Symptom* | *Probable Cause* | *Steps to Remedy* |
|---|---|---|
| *Potable water is dye colored from tap* | Heat exchanger failure in storage tank | 1. Replace tank or heat exchanger if external heat exchanger is used. |
| *Pump runs continuously at night or sporadically* | Sensor or sensor wire failure | 1. Disconnect the sensor wires.<br>2. Measure the resistance of the sensors with a volt-ohm-multimeter. An open or short measurement indicates a defective sensor.<br>3. Measure the temperature at the sensors, and check the resistance values with the manufacturers' temperature versus resistance chart.<br>4. Ensure sensor location and attachment are adequate. Check sensor wires for damage (i.e., staples).<br>5. Replace the sensor if defective. |
| | Sensor reversal | 1. Trace sensor leads to collectors and storage. |
| | Frost sensor for freeze recirculation operational (i.e., air to liquid system) | 1. Normal operation. |

| *Symptom* | *Probable Cause* | *Steps to Remedy* |
|---|---|---|
| *Pump will not run* | Controller/Wiring or Sensor failure | 1. Verify temperature difference is sufficient to activate the controller.<br>2. Test controller for positive start-up in automatic by shorting the collector sensor terminals.<br>3. Check voltage output from the controller (i.e., 120 VAC) using a VOM with the controller manually on. Replace the controller if defective. |
| | Pump | 1. De-energize the system and check wiring continuity with a VOM between pump and controller<br>2. Re-energize the system and check the voltage input to the pump at the connector housing. Replace the pump if defective.<br>3. For the experienced electrician, wire jumpers circumventing the pump start switch to determine if the pump is the actual problem. Replace the pump if defective. |
| | Pump overheating | 1. Check for foreign matter in the impeller. |
| *Pump runs continuously* | Controller | 1. Test the controller for positive shut-off in automatic by shorting the tank sensor terminals. Replace the controller if defective. |
| | Open in tank sensor or short in collector sensor | 1. Disconnect the sensor wires.<br>2. Measure the resistance of the sensors with a VOM. Replace the defective sensor. |
| *System operates with poor performance* | Flow restriction | 1. Check isolation valve positions.<br>2. Check pump inlet and outlet. |
| | Air in system | 1. Check automatic air vent caps to ensure they are loose.<br>2. Drain and refill the system. |
| | System sizing | 1. Determine proper system sizing. If system is undersized, add more collector area. |

| *Symptom* | *Probable Cause* | *Steps to Remedy* |
|---|---|---|
| *Sudden pressure drop* | Leakage from pressurized collector to storage loop | 1. Determine location of leak with isolation valve if possible.<br>2. Drain the system.<br>3. Perform an air or hydrostatic test as appropriate.<br>4. Locate the leak and resolder or tighten the leaking joint(s).<br>5. Refill the system. |
| *Sudden pressure drop; no sign of leaks* | Adjustable pressure relief valve | 1. Check for transfer fluid discharge.<br>2. Reset the relief valve or replace it if defective. |
| | Expansion tank | 1. Check expansion tank pressure to determine a possible diaphragm failure.<br>2. Resize expansion tank to ensure it is not too small.<br>3. Replace if undersized or defective. |
| *Pressure/temperature valve on storage tank releasing hot water* | Stem on relief valve is not far enough inside the tank | 1. Drain the tank and replace the pressure/temperature relief valve with a longer stem. |
| | Thermostat for electric element is set too high or is defective | 1. Turn down the thermostat or replace the defective unit. |
| | Solar system sized improperly | 1. Resize the system. |
| *System won't drain (open loop systems)* | Valves | 1. Check electric valve operation. |
| | Air vents | 1. Ensure air vent caps are loose and not clogged.<br>2. Ensure a vacuum breaker is installed where necessary. |
| | Collector piping | 1. Ensure pipe slope is sufficient. |
| *System won't fill* | Cold water supply valve | 1. Valve must be in open position. |
| | Isolation valves | 1. Check for correct valve position. |

| *Symptom* | *Probable Cause* | *Steps to Remedy* |
|---|---|---|
| *No hot water* | System operation | 1. Search for other symptoms to define the problem. |
| | Tempering valve | 1. Valve is sticking or clogged. It should be cleaned. |
| | Backup system failure | 1. Check fuses and or circuit breakers and replace or reset.<br>2. Check backup thermostat.<br>3. Check electric element or other backup source. |

## REVIEW QUESTIONS

1. Which of the following is the last step to be performed once the plumbing and wiring of the solar system components have been completed?
   a. Leak check the system
   b. Flush the system
   c. Insulate the pipes
   d. Fill and start the system
2. In which order should the steps in Review Question #1 be completed?
   a. a, b, c, d
   b. a, c, b, d
   c. b, a, d, c
   d. a, b, d, c
   e. b, a, c, d
3. At which pressure range should pressure testing be accomplished?
   a. 1½ to twice the minimum expected operating pressure
   b. 1½ to twice the normal operating pressure
   c. 1½ to twice the maximum expected operating pressure
   d. Pressure test at the normal operating pressure
4. Hydrostatic leak testing uses which one of the following fluids?
   a. Water
   b. Air
   c. Propylene glycol
   d. Ethylene glycol
5. The hydrostatic test and fill procedure uses three hoses. Which of the following hose connections are improper?

a. Connection from boiler drain to inlet side of transfer pump.
b. Connection from boiler drain to outlet side of transfer pump
c. Connection from boiler drain to fill bucket
d. Connection from inlet side of transfer to fill bucket
e. Connection from outlet side of transfer pump to fill bucket

6. Which type of valve is the most economical to use between the boiler drain valves in the fill drain assembly?
   a. Globe valve
   b. Gate valve
   c. Ball valve
   d. Pressure reducing valve
   e. Check valve
7. At what point should a closed loop system be pressurized to enable the expansion tank to maintain system pressure?
   a. Slightly lower than the air charged in the expansion tank
   b. Slightly higher than the air charge in the expansion tank
   c. The same as the air charge in the expansion tank
   d. Atmospheric pressure
8. Which of the following portions of piping may not need to be insulated?
   a. Outside pipe between the collectors
   b. Inside return pipe from the collectors
   c. Outside pipe return pipe from the collectors
   d. Inside return pipe to the collectors
9. Which instruments can determine the freeze resistance of a fluid?
   a. Hydrometer
   b. Glycol meter
   c. VOM
   d. Refractometer
   e. Pyrheliometer
10. By varying the pump's speed, high temperatures can be maintained while maximizing the total BTU collection. Which parameter is not significant in determining this optimization of heat collection?
    a. Inlet temperature to storage
    b. Insolation available
    c. Outlet temperature from storage
    d. Specific heat
    e. Mass or weight per unit of transfer fluid

# The Economics of Solar Domestic Hot Water

The question may have already arisen as to why solar economics was not first discussed early in the text rather than as the last Chapter in the text. This question can philosophically be answered by quoting the poet, Robert Bridges, who once wrote:

> Wisdom wil repudiate thee,
> if thou think to enquire
> Why things are as they are
> or whence they came:
> thy task is first to learn
> What Is . . .

Now that there is an understanding of solar energy and its application to heat domestic water, it is now time to explain and illustrate the cost effectiveness of investing in a solar domestic hot water system. The previous chapters have provided the information necessary to understand and use solar energy in its most economical "first" application. Why should conventional methods of heating water remain to be used when there is a more economical way by using the sun? The technology is an old one; the present day ***use*** of the technology is a new one.

A solar DHW system simply must generate energy savings greater than its cost. Unless it lowers the cost of living, it becomes a burden rather than an investment. During the initial years, the system will have a negative impact on available monthly income from savings or from interest on money borrowed due to initial investment costs. As years go by and energy prices rise, however, the system will show an ever-increasing positive impact on ***cash flow***. This is what makes a system such an attractive investment. It must be understood that solar DHW systems are more expensive initially than conventional hot water systems simply because there are more components required. Labor is also more expensive than the installation of a conventional self-sufficient heating/storage tank because a solar domestic hot water system involves more facets of construction and installation including carpentry, plumbing, and electrical wiring. Solar system costs must be kept low, but one should not try to save money by purchasing inferior components. The system must be durable to maintain good economics, and it *must not be undersized*. System life is extremely important and a 20 year or more design life should be sought. This chapter could partake in an economic dissertation on payback and capital recovery from investments, but that is not its intention. The intention is simply to explain and illustrate the economic advantage to using solar energy for heating domestic hot water.

---

## ENERGY EQUIVALENCY AND DETERMINATION OF SYSTEM OUTPUT

A direct comparison of conventional fuel costs (i.e., gas, oil, electricity) versus solar energy costs can be established by determining the cost of fuel per unit energy. Oil, gas, electricity, coal, and wood have different terms of quantity (i.e., gallons, cubic feet, kilowatt-hours, etc.) which must be related to one common denominator: cost per BTU (British Thermal Unit; Chapter 2). The cost of each fuel per million BTU (MBTU's) can be compared with one another using the nomograph of Figure 10-1. Energy equivalencies are developed under the assumptions as prescribed.

For example, if a source of mixed hardwood is available at \$120 per cord, and fuel oil is available at \$1.60 per gallon, a cost per unit energy comparison can be derived using the assumptions in Figure 10-1 as follows.

*Mixed hardwood*

$$24\,\frac{\text{MBTU}}{\text{cord}} \times 0.5\text{ (efficiency)} = 12\,\frac{\text{MBTU}}{\text{cord}}$$

$$\frac{\$120}{\text{cord}} \times \frac{1\text{ cord}}{12\text{ MBTU}} = \frac{\$10.00}{\text{MBTU}}$$

*Fuel oil*

$$138{,}000\,\frac{\text{BTU}}{\text{gal}} \times 0.65\text{ (efficiency)} = 0.089\,\frac{\text{MBTU}}{\text{gal}}$$

$$\frac{\$1.60}{\text{gal}} \times \frac{1\text{ gal}}{0.089\text{ MBTU}} = \frac{\$17.84}{\text{MBTU}}$$

The equivalent fuel cost of the mixed hardwood in comparison with the fuel oil is \$10.00 per MBTU versus \$17.84 per MBTU, respectively. The mixed hardwood would be the better energy purchase in this situation.

To determine the "worth" of a solar domestic hot water system, the energy output of that system must first be determined. The solar energy output can be determined by systematically following the worksheet lines of Table 10-1a. Actual BTU requirements and available solar radiation were previously determined in Chapter 6 using Table 6-3. Table 6-3 can now be used in conjunction with Table 10-1a to determine the BTU output. Because the sun does not shine at all locations 100 percent of the time, the monthly insolation available must be adjusted statistically by a percentage of possible

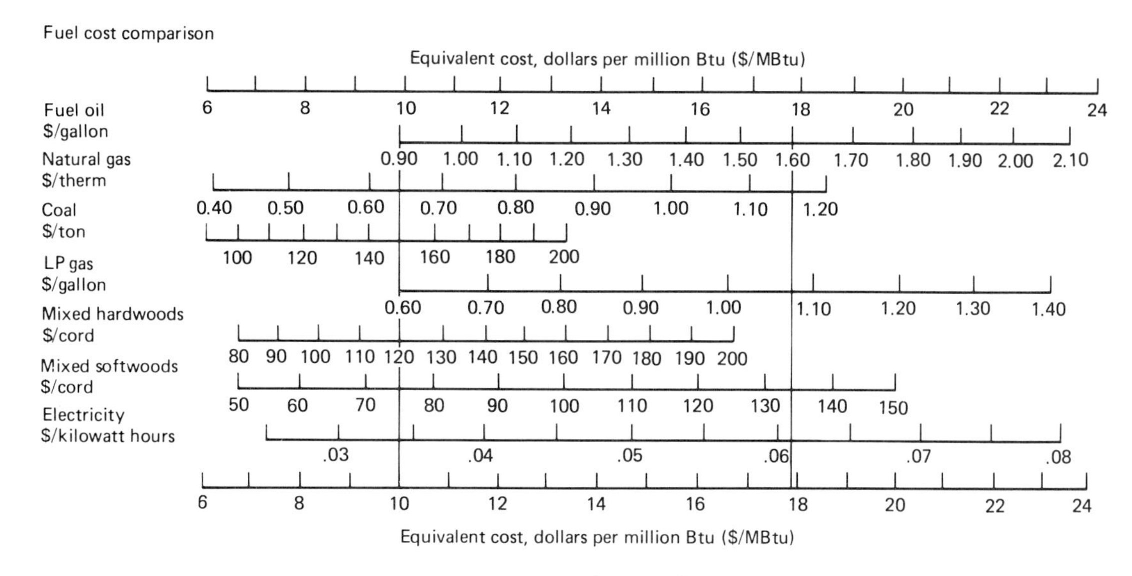


*Assumptions*

Natural Gas—Therm = 100,000 BTU = approx. 100 cu ft. 65% Efficiency. \$/MBtu = 15.38 × \$/Therm
Fuel Oil—138,000 Btu/gal. 65% Efficiency. \$/MBtu = 11.15 × \$/gal.
LP Gas—93,000 Btu/gal. 65% Efficiency. \$/MBtu = 16.54 × \$/gal.
Electricity—3413 Btu/kwh. 100% Efficiency. \$/MBtu = 293 × \$/kwh
Mixed Hardwoods—24 MBtu/cord. 50% Efficiency (Airtight stove). \$/MBtu = \$/cord ÷ 12
Mixed Softwoods—15 MBtu/cord. 50% Efficiency. \$/MBtu = \$/cord ÷ 7.5
Coal—12,500 Btu/lb. 60% Efficiency. \$/MBtu = \$/ton ÷ 15

**Figure 10-1.** Fuel cost comparison nomograph. (Reprinted with permission from the Northeast Regional Agricultural Engineering Service, Cornell University, Ithaca, N.Y.)

sunshine as determined from the accumulation of historical data. Table 10-2 expresses the percentage of the maximum possible amount of sunshine reaching the earth's surface in the absence of clouds, fog, smoke, or other restrictions for selected locations. These percentages do not represent the contribution of solar energy from diffuse radiation during cloudy days. This statistical percentage is entered in line E of Table 10-1a. The remaining lines in the worksheet can be determined as annotated.

Let us illustrate the use of Table 10-1*a* by continuing our discussion from Chapter 6 regarding a typical family of four in Billings, Montana. From the example of Table 6-3*b*, the daily BTU requirement (line G), the daily available solar radiation (line I), the collector system efficiency (line H), and the actual effective total collector area installed can be transferred to Table 10-1*b* as annotated. Calculating for estimated system output, we shall illustrate the use of Table 10-1*b* line by line for the month of January.

Line A = 31 days/month
Line B = 31 days/month × 55395 BTU/day = 1.72 MBTU/month
Line C = 31 days/month × 1478 BTU/ft$^2$-day = 45818 BTU/ft$^2$-month
Line D = Effective collector area × Line C
= 66.6 ft$^2$ × 45,818 BTU/ft$^2$-month = 3.05 MBTU/month

**Table 10-1a**
**Worksheet to Determine Solar Energy Contribution to DHW Energy Requirements**

*Latitude* ________________

*Collector Tilt Angle* ________________

[1]*Effective Collector Aperture Area* ________________ *ft²*

| *Line* | *Evaluation Factors* | | *Jan.* | *Feb.* | *Mar.* | *Apr.* | *May* | *Jun.* | *Jul.* | *Aug.* | *Sept.* | *Oct.* | *Nov.* | *Dec.* | *Total per Year* |
|---|---|---|---|---|---|---|---|---|---|---|---|---|---|---|---|
| A | Days in month | | 31 | 28 | 31 | 30 | 31 | 30 | 31 | 31 | 30 | 31 | 30 | 31 | 365 |
| B | BTU requirement | [1]Daily (BTU) | | | | | | | | | | | | | X |
| | | Monthly (MBTU) | | | | | | | | | | | | | |
| C | Available solar radiation (BTU/ft²) | [1]Daily | | | | | | | | | | | | | X |
| | | Monthly | | | | | | | | | | | | | X |
| D | Collector array output (MBTU's) (Collector area), * C | | | | | | | | | | | | | | X |
| E | Mean percentage of possible sun (Table 10-2) | | | | | | | | | | | | | | |
| F | Collector array output E * D (MBTU) | | | | | | | | | | | | | | |
| G | [1]Collector system efficiency, *Ns* | | | | | | | | | | | | | | |
| H | Estimated system output, G * F (MBTU) | | | | | | | | | | | | | | |
| I | Percentage (%) solar energy contribution, $H \div B$ | | | | | | | | | | | | | | |

[1]From Table 6-3.
* Multiplication.
÷ Division.

**Table 10-1b (Example)**
**Worksheet to Determine Solar Energy Contribution to DHW Energy Requirements**

*Latitude* 46°

*Collector Tilt Angle* 48°

[1]*Effective Collector Aperture Area* 66.6 *ft²*

| *Line* | *Evaluation Factors* | | *Jan.* | *Feb.* | *Mar.* | *Apr.* | *May* | *Jun.* | *Jul.* | *Aug.* | *Sept.* | *Oct.* | *Nov.* | *Dec.* | *Total per Year* |
|---|---|---|---|---|---|---|---|---|---|---|---|---|---|---|---|
| A | Days in month | | 31 | 28 | 31 | 30 | 31 | 30 | 31 | 31 | 30 | 31 | 30 | 31 | 365 |
| B | BTU requirement | [1]Daily (BTU) | 55395 | 55395 | 55395 | 52479 | 52479 | 49564 | 49564 | 49564 | 49564 | 42749 | 55395 | 55395 | |
| | | Monthly (MBTU) | 1.72 | 1.55 | 1.72 | 1.57 | 1.63 | 1.49 | 1.54 | 1.54 | 1.49 | 1.63 | 1.66 | 1.72 | 19.26 |
| C | Available solar radiation (BTU/ft²) | [1]Daily | 1478 | 1972 | 2228 | 2266 | 2234 | 2204 | 2200 | 2200 | 2118 | 1860 | 1448 | 1250 | |
| | | Monthly | 45818 | 55216 | 69068 | 67980 | 69254 | 66120 | 68200 | 68200 | 63540 | 57660 | 43440 | 38750 | |
| D | Collector array output (MBTU) (Collector area), * C | | 3.05 | 3.68 | 4.60 | 4.53 | 4.61 | 4.40 | 4.54 | 4.54 | 4.23 | 3.84 | 2.89 | 2.58 | |
| E | Mean percentage of possible sun (Table 10-2) | | 0.47 | 0.53 | 0.62 | 0.60 | 0.61 | 0.64 | 0.77 | 0.75 | 0.68 | 0.62 | 0.46 | 0.44 | 0.62 |
| F | Collector array output E * D (MBTU's) | | 1.43 | 1.95 | 2.85 | 2.72 | 2.81 | 2.82 | 3.50 | 3.41 | 2.88 | 2.38 | 1.33 | 1.14 | 29.22 |
| G | [1]Collector system efficiency (*Ns*) | | 0.6 | 0.5 | 0.5 | 0.5 | 0.5 | 0.5 | 0.4 | 0.4 | 0.5 | 0.5 | 0.5 | 0.6 | 0.5 |
| H | Estimated system output, G * F (MBTU) | | 0.86 | 0.98 | 1.43 | 1.36 | 1.41 | 1.41 | 1.40 | 1.36 | 1.44 | 1.19 | 0.67 | 0.68 | 14.19 |
| I | Percentage (%) solar contribution, $H \div B$ | | 50 | 63 | 83 | 87 | 87 | 95 | 91 | 88 | 97 | 73 | 40 | 40 | 74.5 |

[1]From Table 6-3.
* Multiplication.
÷ Division.

**Table 10-2**

SUNSHINE - AVERAGE PERCENTAGE OF POSSIBLE

| DATA THROUGH 1980 | YRS | JAN | FEB | MAR | APR | MAY | JUN | JUL | AUG | SEP | OCT | NOV | DEC | ANN |
|---|---|---|---|---|---|---|---|---|---|---|---|---|---|---|
| BIRMINGHAM, ALABAMA - AP | 34 | 42 | 50 | 55 | 63 | 66 | 65 | 59 | 63 | 61 | 66 | 55 | 46 | 58 |
| MONTGOMERY, ALABAMA | 30 | 47 | 54 | 58 | 65 | 65 | 67 | 63 | 66 | 62 | 67 | 56 | 50 | 59 |
| ANCHORAGE, ALASKA | 27 | 41 | 45 | 54 | 52 | 52 | 47 | 45 | 42 | 41 | 38 | 36 | 34 | 45 |
| JUNEAU, ALASKA | 33 | 32 | 32 | 37 | 39 | 39 | 34 | 31 | 32 | 26 | 19 | 23 | 20 | 32 |
| NOME, ALASKA | 32 | 38 | 52 | 52 | 52 | 47 | 39 | 36 | 29 | 32 | 33 | 29 | 35 | 40 |
| FLAGSTAFF, ARIZONA | 4 | 75 | 72 | 69 | 84 | 89 | 90 | 75 | 81 | 82 | 78 | 77 | 78 | 81 |
| PHOENIX, ARIZONA | 85 | 77 | 80 | 83 | 88 | 93 | 94 | 85 | 85 | 89 | 88 | 84 | 77 | 86 |
| TUCSON, ARIZONA | 33 | 80 | 83 | 86 | 92 | 93 | 93 | 78 | 82 | 87 | 89 | 86 | 80 | 86 |
| YUMA, ARIZONA | 30 | 83 | 87 | 91 | 94 | 96 | 97 | 90 | 92 | 94 | 92 | 87 | 82 | 91 |
| FORT SMITH, ARKANSAS | 35 | 48 | 55 | 55 | 60 | 63 | 69 | 72 | 71 | 65 | 66 | 56 | 52 | 62 |
| NO. LITTLE ROCK, AR | 32 | 46 | 54 | 57 | 62 | 68 | 73 | 71 | 73 | 68 | 69 | 56 | 48 | 64 |
| EUREKA, CALIFORNIA | 70 | 42 | 45 | 51 | 55 | 56 | 57 | 52 | 48 | 53 | 49 | 42 | 40 | 50 |
| FRESNO, CALIFORNIA | 31 | 49 | 65 | 79 | 85 | 89 | 94 | 96 | 96 | 94 | 88 | 67 | 48 | 82 |
| LOS ANGELES, CA - CITY | 32 | 69 | 72 | 73 | 70 | 66 | 65 | 82 | 83 | 79 | 73 | 74 | 71 | 73 |
| RED BLUFF, CALIFORNIA | 36 | 54 | 63 | 69 | 80 | 85 | 89 | 96 | 94 | 92 | 81 | 62 | 53 | 79 |
| SACRAMENTO, CALIFORNIA | 32 | 45 | 61 | 72 | 80 | 87 | 92 | 97 | 96 | 93 | 84 | 64 | 47 | 79 |
| SAN DIEGO, CALIFORNIA | 40 | 71 | 72 | 71 | 66 | 58 | 57 | 68 | 69 | 68 | 67 | 74 | 72 | 68 |
| SAN FRANCISCO, CA - CITY | 38 | 56 | 62 | 69 | 73 | 72 | 73 | 66 | 65 | 72 | 70 | 62 | 53 | 67 |
| DENVER, COLORADO | 31 | 72 | 71 | 71 | 67 | 65 | 71 | 71 | 73 | 75 | 73 | 65 | 68 | 70 |
| GRAND JUNCTION, COLORADO | 34 | 58 | 64 | 64 | 67 | 71 | 79 | 78 | 76 | 79 | 74 | 63 | 60 | 70 |
| PUEBLO, COLORADO | 40 | 75 | 74 | 75 | 74 | 73 | 79 | 78 | 78 | 80 | 79 | 74 | 73 | 76 |
| HARTFORD, CONNECTICUT | 26 | 58 | 59 | 57 | 57 | 58 | 60 | 63 | 62 | 59 | 57 | 46 | 50 | 58 |
| WASHINGTON, DC - NATIONAL AP | 32 | 48 | 53 | 56 | 58 | 59 | 65 | 63 | 63 | 62 | 60 | 52 | 48 | 58 |
| APALACHICOLA, FLORIDA | 45 | 59 | 63 | 64 | 74 | 78 | 71 | 63 | 63 | 64 | 74 | 67 | 57 | 67 |
| JACKSONVILLE, FLORIDA | 29 | 57 | 61 | 67 | 72 | 69 | 63 | 61 | 59 | 54 | 58 | 61 | 56 | 62 |
| KEY WEST, FLORIDA | 21 | 73 | 76 | 83 | 84 | 81 | 74 | 77 | 77 | 71 | 70 | 71 | 72 | 76 |
| MIAMI, FLORIDA | 4 | 61 | 61 | 78 | 80 | 66 | 76 | 78 | 74 | 73 | 70 | 63 | 60 | 70 |
| PENSACOLA, FLORIDA | 5 | 48 | 53 | 61 | 63 | 67 | 67 | 57 | 58 | 60 | 71 | 64 | 49 | 60 |
| TAMPA, FLORIDA | 34 | 65 | 66 | 71 | 75 | 75 | 67 | 62 | 60 | 60 | 64 | 65 | 62 | 66 |
| ATLANTA, GEORGIA | 45 | 47 | 54 | 57 | 66 | 68 | 67 | 63 | 65 | 63 | 68 | 60 | 51 | 61 |
| MACON, GEORGIA | 32 | 54 | 59 | 62 | 69 | 70 | 69 | 65 | 70 | 63 | 69 | 63 | 56 | 63 |
| SAVANNAH, GEORGIA | 30 | 54 | 59 | 62 | 70 | 67 | 64 | 63 | 63 | 57 | 64 | 62 | 55 | 62 |
| HILO, HAWAII | 30 | 46 | 43 | 38 | 34 | 36 | 42 | 42 | 42 | 44 | 39 | 34 | 36 | 40 |
| HONOLULU, HAWAII | 28 | 62 | 63 | 68 | 67 | 69 | 70 | 73 | 75 | 75 | 68 | 61 | 59 | 68 |
| KAHULUI, HAWAII | 18 | 66 | 66 | 66 | 65 | 71 | 74 | 74 | 74 | 77 | 72 | 68 | 66 | 70 |
| LIHUE, HAWAII | 30 | 51 | 53 | 53 | 51 | 57 | 60 | 62 | 65 | 67 | 57 | 49 | 47 | 56 |
| BOISE, IDAHO | 38 | 41 | 51 | 63 | 67 | 71 | 75 | 88 | 85 | 81 | 68 | 45 | 39 | 67 |
| POCATELLO, IDAHO | 31 | 37 | 51 | 61 | 64 | 67 | 73 | 83 | 80 | 79 | 70 | 47 | 38 | 65 |
| CAIRO, ILLINOIS | 38 | 44 | 51 | 56 | 63 | 66 | 73 | 74 | 76 | 69 | 69 | 53 | 45 | 63 |
| MOLINE, ILLINOIS | 37 | 48 | 49 | 50 | 53 | 57 | 63 | 69 | 67 | 64 | 60 | 44 | 41 | 57 |
| PEORIA, ILLINOIS | 37 | 46 | 50 | 51 | 55 | 59 | 66 | 68 | 66 | 65 | 62 | 45 | 41 | 57 |
| SPRINGFIELD, ILLINOIS | 32 | 46 | 51 | 50 | 56 | 64 | 69 | 73 | 72 | 70 | 65 | 50 | 43 | 61 |
| EVANSVILLE, INDIANA | 40 | 43 | 49 | 55 | 60 | 66 | 74 | 75 | 76 | 70 | 67 | 49 | 42 | 62 |
| FORT WAYNE, INDIANA | 34 | 44 | 50 | 52 | 58 | 67 | 71 | 74 | 73 | 66 | 62 | 41 | 38 | 60 |
| INDIANAPOLIS, INDIANA | 36 | 41 | 50 | 51 | 55 | 62 | 67 | 67 | 70 | 67 | 63 | 43 | 40 | 58 |
| DES MOINES, IOWA | 30 | 50 | 53 | 54 | 55 | 60 | 68 | 72 | 70 | 65 | 63 | 50 | 45 | 60 |
| SIOUX CITY, IOWA | 40 | 58 | 58 | 59 | 60 | 63 | 68 | 75 | 72 | 68 | 66 | 54 | 51 | 64 |
| CONCORDIA, KANSAS | 18 | 64 | 64 | 65 | 66 | 72 | 79 | 79 | 78 | 72 | 74 | 63 | 61 | 71 |
| DODGE CITY, KANSAS | 38 | 67 | 65 | 66 | 68 | 68 | 76 | 78 | 77 | 74 | 75 | 67 | 65 | 71 |
| TOPEKA, KANSAS | 31 | 54 | 54 | 54 | 56 | 60 | 65 | 69 | 70 | 65 | 65 | 55 | 52 | 61 |

*Source.* National Climatic Center, Asheville, North Carolina.

## SUNSHINE - AVERAGE PERCENTAGE OF POSSIBLE

| DATA THROUGH 1980 | YRS | JAN | FEB | MAR | APR | MAY | JUN | JUL | AUG | SEP | OCT | NOV | DEC | ANN |
|---|---|---|---|---|---|---|---|---|---|---|---|---|---|---|
| WICHITA, KANSAS | 27 | 58 | 59 | 61 | 63 | 65 | 70 | 75 | 74 | 67 | 68 | 59 | 59 | 65 |
| LOUISVILLE, KENTUCKY | 33 | 42 | 48 | 49 | 55 | 61 | 65 | 66 | 67 | 65 | 62 | 47 | 42 | 57 |
| NEW ORLEANS, LOUISIANA | 7 | 44 | 56 | 54 | 64 | 58 | 68 | 58 | 59 | 59 | 72 | 56 | 58 | 59 |
| SHREVEPORT, LOUISIANA | 28 | 47 | 55 | 56 | 57 | 64 | 71 | 73 | 71 | 67 | 71 | 60 | 53 | 63 |
| PORTLAND, MAINE | 40 | 56 | 59 | 56 | 56 | 56 | 60 | 64 | 64 | 62 | 58 | 47 | 53 | 58 |
| BALTIMORE, MARYLAND | 30 | 51 | 56 | 56 | 56 | 56 | 62 | 64 | 62 | 60 | 59 | 51 | 49 | 57 |
| BLUE HILL OBSERVATORY - MA | 94 | 46 | 50 | 48 | 50 | 52 | 55 | 57 | 58 | 56 | 55 | 47 | 46 | 52 |
| BOSTON, MASSACHUSETTS | 45 | 53 | 57 | 57 | 57 | 59 | 64 | 66 | 66 | 64 | 61 | 51 | 53 | 60 |
| ALPENA, MICHIGAN | 21 | 39 | 46 | 53 | 55 | 61 | 64 | 68 | 61 | 54 | 45 | 30 | 29 | 53 |
| DETROIT, MI - CITY AP | 32 | 32 | 43 | 49 | 52 | 59 | 65 | 70 | 65 | 61 | 56 | 35 | 32 | 54 |
| DETROIT, MI - METRO AP | 15 | 38 | 47 | 49 | 53 | 60 | 65 | 69 | 68 | 61 | 51 | 34 | 30 | 54 |
| GRAND RAPIDS, MICHIGAN | 17 | 31 | 42 | 45 | 52 | 56 | 62 | 64 | 62 | 56 | 45 | 27 | 24 | 49 |
| LANSING, MICHIGAN | 26 | 36 | 44 | 48 | 53 | 63 | 66 | 70 | 66 | 59 | 52 | 31 | 30 | 53 |
| MARQUETTE, MICHIGAN - CITY | 41 | 35 | 43 | 51 | 56 | 58 | 60 | 67 | 62 | 52 | 45 | 27 | 28 | 51 |
| SAULT STE. MARIE, MICHIGAN | 39 | 35 | 46 | 54 | 55 | 57 | 58 | 63 | 58 | 46 | 40 | 23 | 27 | 48 |
| DULUTH, MINNESOTA | 30 | 49 | 54 | 55 | 54 | 56 | 58 | 66 | 61 | 52 | 47 | 35 | 40 | 54 |
| MINNEAPOLIS-ST.PAUL, MINNESOTA | 42 | 51 | 58 | 55 | 56 | 59 | 63 | 71 | 67 | 61 | 57 | 39 | 40 | 58 |
| JACKSON, MISSISSIPPI | 16 | 45 | 55 | 58 | 62 | 60 | 70 | 62 | 63 | 58 | 68 | 54 | 49 | 59 |
| COLUMBIA, MISSOURI | 11 | 52 | 52 | 51 | 57 | 62 | 72 | 72 | 68 | 60 | 62 | 49 | 46 | 60 |
| KANSAS CITY, MISSOURI-INTL AP | 8 | 59 | 57 | 64 | 72 | 70 | 76 | 77 | 70 | 65 | 66 | 53 | 56 | 67 |
| KANSAS CITY, MISSOURI-DNTN AP | 38 | 53 | 56 | 58 | 60 | 63 | 68 | 76 | 73 | 68 | 67 | 57 | 50 | |
| ST. LOUIS, MISSOURI | 21 | 52 | 52 | 54 | 56 | 62 | 68 | 71 | 65 | 64 | 63 | 48 | 44 | 59 |
| SPRINGFIELD, MISSOURI | 35 | 50 | 54 | 57 | 61 | 64 | 68 | 72 | 73 | 69 | 67 | 55 | 49 | 63 |
| BILLINGS, MONTANA | 41 | 47 | 53 | 62 | 60 | 61 | 64 | 77 | 75 | 68 | 62 | 46 | 44 | 62 |
| GREAT FALLS, MONTANA | 38 | 48 | 56 | 65 | 61 | 63 | 65 | 80 | 76 | 68 | 61 | 46 | 44 | 63 |
| HAVRE, MONTANA | 20 | 48 | 59 | 70 | 68 | 72 | 75 | 85 | 82 | 74 | 64 | 49 | 43 | 68 |
| HELENA, MONTANA | 40 | 46 | 53 | 60 | 57 | 60 | 62 | 78 | 74 | 67 | 61 | 45 | 42 | 61 |
| MISSOULA, MONTANA | 36 | 29 | 39 | 49 | 52 | 56 | 58 | 79 | 74 | 66 | 52 | 33 | 26 | 54 |
| LINCOLN, NEBRASKA | 24 | 58 | 57 | 58 | 60 | 63 | 72 | 74 | 73 | 67 | 67 | 55 | 52 | 64 |
| NORTH PLATTE, NEBRASKA | 28 | 59 | 58 | 60 | 63 | 63 | 70 | 75 | 73 | 70 | 71 | 60 | 60 | 66 |
| OMAHA (NORTH), NEBRASKA | 44 | 54 | 54 | 55 | 58 | 61 | 68 | 74 | 71 | 68 | 67 | 52 | 48 | 62 |
| VALENTINE, NEBRASKA | 25 | 65 | 62 | 60 | 60 | 64 | 70 | 77 | 77 | 71 | 70 | 61 | 61 | 67 |
| ELY, NEVADA | 41 | 66 | 66 | 70 | 67 | 71 | 80 | 80 | 81 | 83 | 75 | 66 | 65 | 73 |
| LAS VEGAS, NEVADA | 31 | 76 | 81 | 83 | 86 | 88 | 92 | 87 | 88 | 92 | 86 | 81 | 78 | 85 |
| RENO, NEVADA | 37 | 66 | 68 | 76 | 81 | 82 | 85 | 93 | 93 | 92 | 84 | 71 | 64 | 81 |
| WINNEMUCCA, NEVADA | 31 | 48 | 53 | 57 | 63 | 70 | 75 | 86 | 84 | 83 | 74 | 55 | 50 | 68 |
| CONCORD, NEW HAMPSHIRE | 39 | 51 | 55 | 52 | 53 | 55 | 58 | 63 | 60 | 56 | 54 | 42 | 47 | 55 |
| MT. WASHINGTON, NH | 42 | 32 | 34 | 33 | 35 | 37 | 32 | 31 | 31 | 36 | 39 | 29 | 30 | 33 |
| ATLANTIC CITY, NJ - NAFEC | 20 | 49 | 52 | 53 | 55 | 54 | 58 | 59 | 63 | 59 | 56 | 48 | 44 | 55 |
| TRENTON, NEW JERSEY | 48 | 51 | 56 | 55 | 58 | 59 | 62 | 64 | 62 | 61 | 59 | 52 | 49 | 58 |
| ALBUQUERQUE, NEW MEXICO | 41 | 72 | 73 | 73 | 77 | 79 | 83 | 76 | 76 | 80 | 80 | 77 | 72 | 77 |
| ROSWELL, NEW MEXICO | 6 | 61 | 68 | 77 | 78 | 81 | 83 | 76 | 73 | 74 | 81 | 74 | 69 | 75 |
| ALBANY, NEW YORK | 42 | 45 | 51 | 52 | 53 | 55 | 59 | 63 | 60 | 56 | 51 | 36 | 39 | 53 |
| BINGHAMTON, NEW YORK | 29 | 37 | 42 | 44 | 51 | 56 | 62 | 65 | 60 | 56 | 48 | 30 | 28 | 50 |
| BUFFALO, NEW YORK | 37 | 33 | 39 | 46 | 52 | 58 | 66 | 68 | 66 | 60 | 51 | 29 | 28 | 52 |
| NEW YORK, NY - CENTRAL PARK | 100 | 50 | 55 | 56 | 59 | 61 | 64 | 65 | 64 | 63 | 61 | 52 | 49 | 59 |
| ROCHESTER, NEW YORK | 40 | 36 | 42 | 49 | 55 | 59 | 67 | 69 | 67 | 60 | 49 | 31 | 31 | 54 |
| SYRACUSE, NEW YORK | 31 | 35 | 40 | 46 | 51 | 55 | 59 | 64 | 60 | 54 | 44 | 25 | 25 | 49 |
| ASHEVILLE, NORTH CAROLINA | 16 | 55 | 63 | 64 | 68 | 62 | 65 | 59 | 55 | 54 | 63 | 60 | 58 | 61 |
| CAPE HATTERAS, NORTH CAROLINA | 18 | 49 | 54 | 60 | 65 | 61 | 61 | 62 | 62 | 59 | 59 | 56 | 47 | 59 |

## Table 10-2 (*continued*)

| DATA THROUGH 1980 | YRS | JAN | FEB | MAR | APR | MAY | JUN | JUL | AUG | SEP | OCT | NOV | DEC | ANN |
|---|---|---|---|---|---|---|---|---|---|---|---|---|---|---|
| | SUNSHINE - AVERAGE PERCENTAGE OF POSSIBLE | | | | | | | | | | | | | |
| CHARLOTTE, NORTH CAROLINA | 30 | 55 | 61 | 63 | 70 | 69 | 70 | 69 | 70 | 66 | 69 | 62 | 58 | 66 |
| GREENSBORO, NORTH CAROLINA | 52 | 52 | 58 | 60 | 64 | 65 | 66 | 63 | 63 | 63 | 66 | 60 | 54 | 62 |
| RALEIGH, NORTH CAROLINA | 26 | 54 | 60 | 62 | 64 | 59 | 61 | 61 | 61 | 59 | 62 | 61 | 56 | 60 |
| WILMINGTON, NORTH CAROLINA | 29 | 57 | 60 | 63 | 70 | 66 | 65 | 63 | 63 | 61 | 65 | 65 | 59 | 63 |
| BISMARCK, NORTH DAKOTA | 41 | 55 | 54 | 60 | 58 | 64 | 65 | 76 | 73 | 66 | 59 | 45 | 47 | 62 |
| FARGO, NORTH DAKOTA | 38 | 51 | 57 | 58 | 58 | 59 | 60 | 71 | 68 | 60 | 56 | 40 | 43 | 58 |
| WILLISTON, NORTH DAKOTA | 18 | 50 | 55 | 61 | 58 | 62 | 66 | 76 | 74 | 64 | 59 | 43 | 48 | 61 |
| CINCINNATI, OHIO | 65 | 41 | 45 | 50 | 55 | 60 | 67 | 68 | 66 | 66 | 58 | 44 | 38 | 56 |
| CLEVELAND, OHIO | 37 | 31 | 37 | 43 | 52 | 58 | 65 | 67 | 63 | 60 | 54 | 31 | 27 | 51 |
| COLUMBUS, OHIO | 29 | 36 | 42 | 43 | 51 | 57 | 60 | 61 | 61 | 62 | 57 | 38 | 31 | 52 |
| DAYTON, OHIO | 37 | 41 | 46 | 49 | 53 | 60 | 67 | 67 | 69 | 67 | 61 | 42 | 37 | 56 |
| TOLEDO, OHIO | 25 | 43 | 47 | 49 | 53 | 61 | 64 | 67 | 63 | 60 | 56 | 38 | 34 | 55 |
| OKLAHOMA CITY, OKLAHOMA | 26 | 58 | 60 | 64 | 64 | 65 | 73 | 77 | 78 | 70 | 69 | 60 | 59 | 67 |
| TULSA, OKLAHOMA | 38 | 51 | 55 | 57 | 57 | 59 | 67 | 72 | 72 | 65 | 65 | 58 | 54 | 62 |
| PORTLAND, OREGON | 31 | 26 | 37 | 45 | 51 | 56 | 54 | 69 | 64 | 60 | 43 | 30 | 21 | 49 |
| GUAM, PACIFIC | 18 | 56 | 60 | 66 | 68 | 64 | 61 | 49 | 45 | 42 | 44 | 49 | 52 | 55 |
| JOHNSTON ISLAND, PACIFIC | 18 | 69 | 72 | 73 | 68 | 73 | 77 | 79 | 75 | 73 | 66 | 59 | 63 | 71 |
| KOROR ISLAND, PACIFIC | 20 | 56 | 58 | 65 | 64 | 57 | 49 | 51 | 48 | 55 | 51 | 53 | 50 | 55 |
| MAJURO, MARSHALL IS. PACIFIC | 20 | 62 | 62 | 65 | 53 | 56 | 53 | 55 | 61 | 59 | 54 | 54 | 53 | 57 |
| PAGO PAGO, AMERICAN SAMOA | 13 | 45 | 46 | 49 | 37 | 36 | 36 | 48 | 54 | 60 | 46 | 41 | 38 | 45 |
| PONAPE ISLAND, PACIFIC | 22 | 46 | 46 | 51 | 43 | 42 | 44 | 47 | 48 | 48 | 42 | 45 | 40 | 45 |
| TRUK, CAROLINE IS. PACIFIC | 20 | 58 | 60 | 61 | 56 | 52 | 50 | 53 | 56 | 52 | 48 | 52 | 50 | 54 |
| WAKE ISLAND, PACIFIC | 11 | 68 | 72 | 76 | 75 | 75 | 77 | 74 | 70 | 67 | 70 | 64 | 66 | 71 |
| YAP ISLAND, PACIFIC | 21 | 61 | 64 | 67 | 69 | 62 | 56 | 50 | 50 | 51 | 48 | 57 | 56 | 57 |
| HARRISBURG, PENNSYLVANIA | 42 | 48 | 54 | 57 | 59 | 59 | 64 | 68 | 67 | 62 | 58 | 47 | 45 | 59 |
| PHILADELPHIA, PENNSYLVANIA | 38 | 50 | 54 | 56 | 57 | 57 | 62 | 62 | 62 | 60 | 59 | 52 | 50 | 57 |
| PITTSBURGH, PA - INTL AP | 28 | 34 | 39 | 44 | 48 | 51 | 57 | 57 | 56 | 58 | 53 | 39 | 30 | 48 |
| AVOCA, PENNSYLVANIA | 25 | 44 | 48 | 49 | 54 | 56 | 60 | 61 | 60 | 55 | 52 | 35 | 35 | 52 |
| SAN JUAN, PUERTO RICO | 25 | 66 | 69 | 74 | 67 | 61 | 61 | 67 | 66 | 60 | 59 | 57 | 57 | 64 |
| PROVIDENCE, RHODE ISLAND | 27 | 57 | 58 | 56 | 56 | 57 | 59 | 61 | 58 | 60 | 60 | 50 | 53 | 57 |
| CHARLESTON, SOUTH CAROLINA | 22 | 57 | 61 | 67 | 71 | 71 | 68 | 68 | 67 | 62 | 67 | 64 | 59 | 66 |
| COLUMBIA, SOUTH CAROLINA | 27 | 57 | 62 | 64 | 69 | 67 | 66 | 66 | 67 | 64 | 67 | 65 | 61 | 65 |
| GREENVILLE-SPARTANBURG, SC | 18 | 53 | 60 | 63 | 66 | 59 | 60 | 60 | 62 | 60 | 67 | 61 | 56 | 61 |
| HURON, SOUTH DAKOTA | 41 | 57 | 60 | 58 | 60 | 65 | 69 | 77 | 74 | 68 | 63 | 52 | 49 | 64 |
| RAPID CITY, SOUTH DAKOTA | 38 | 55 | 59 | 61 | 59 | 57 | 62 | 72 | 73 | 68 | 66 | 55 | 54 | 62 |
| CHATTANOOGA, TENNESSEE | 50 | 41 | 48 | 51 | 60 | 65 | 65 | 61 | 62 | 63 | 63 | 53 | 43 | 57 |
| KNOXVILLE, TENNESSEE | 38 | 39 | 46 | 53 | 61 | 62 | 63 | 61 | 62 | 59 | 62 | 50 | 40 | 57 |
| MEMPHIS, TENNESSEE | 30 | 49 | 54 | 57 | 65 | 70 | 74 | 74 | 76 | 69 | 72 | 58 | 51 | 65 |
| NASHVILLE, TENNESSEE | 38 | 40 | 47 | 52 | 59 | 62 | 67 | 64 | 65 | 63 | 64 | 51 | 42 | 57 |
| ABILENE, TEXAS | 32 | 62 | 65 | 70 | 70 | 70 | 78 | 79 | 77 | 68 | 72 | 69 | 66 | 71 |
| AMARILLO, TEXAS | 39 | 68 | 69 | 71 | 73 | 72 | 77 | 78 | 77 | 74 | 75 | 72 | 68 | 73 |
| AUSTIN, TEXAS | 39 | 48 | 52 | 55 | 53 | 57 | 70 | 76 | 75 | 67 | 66 | 57 | 51 | 61 |
| BROWNSVILLE, TEXAS | 38 | 43 | 49 | 51 | 57 | 66 | 74 | 81 | 76 | 68 | 66 | 52 | 44 | 61 |
| CORPUS CHRISTI, TEXAS | 38 | 46 | 52 | 56 | 57 | 62 | 74 | 82 | 78 | 68 | 69 | 58 | 47 | 64 |
| EL PASO, TEXAS | 38 | 77 | 82 | 85 | 88 | 89 | 89 | 80 | 81 | 82 | 84 | 83 | 79 | 83 |
| GALVESTON, TEXAS | 90 | 48 | 51 | 55 | 61 | 68 | 76 | 72 | 71 | 68 | 72 | 60 | 49 | 63 |
| HOUSTON, TEXAS | 11 | 40 | 52 | 47 | 52 | 59 | 67 | 67 | 63 | 59 | 65 | 55 | 60 | 57 |
| LUBBOCK, TEXAS | 8 | 63 | 70 | 78 | 77 | 76 | 81 | 78 | 80 | 70 | 79 | 70 | 71 | 75 |
| PORT ARTHUR, TEXAS | 26 | 42 | 52 | 52 | 52 | 64 | 69 | 65 | 63 | 62 | 67 | 57 | 47 | 58 |
| SAN ANTONIO, TEXAS | 38 | 48 | 53 | 58 | 55 | 56 | 68 | 75 | 74 | 67 | 65 | 56 | 51 | 61 |

SUNSHINE - AVERAGE PERCENTAGE OF POSSIBLE

| DATA THROUGH 1980 | YRS | JAN | FEB | MAR | APR | MAY | JUN | JUL | AUG | SEP | OCT | NOV | DEC | ANN |
|---|---|---|---|---|---|---|---|---|---|---|---|---|---|---|
| MILFORD, UTAH | 7 | 56 | 61 | 61 | 67 | 74 | 86 | 78 | 84 | 85 | 79 | 67 | 69 | 73 |
| SALT LAKE CITY, UTAH | 42 | 47 | 55 | 64 | 67 | 73 | 79 | 84 | 83 | 84 | 73 | 55 | 45 | 70 |
| BURLINGTON, VERMONT | 37 | 41 | 48 | 51 | 50 | 56 | 59 | 64 | 60 | 54 | 48 | 30 | 33 | 50 |
| LYNCHBURG, VIRGINIA | 36 | 51 | 56 | 58 | 62 | 63 | 66 | 62 | 62 | 61 | 61 | 55 | 53 | 60 |
| NORFOLK, VIRGINIA | 21 | 56 | 59 | 63 | 66 | 65 | 68 | 65 | 65 | 64 | 60 | 59 | 57 | 63 |
| RICHMOND, VIRGINIA | 30 | 51 | 57 | 59 | 64 | 64 | 68 | 66 | 65 | 64 | 60 | 56 | 52 | 61 |
| QUILLAYUTE, WASHINGTON | 14 | 22 | 30 | 31 | 35 | 37 | 36 | 44 | 42 | 46 | 35 | 22 | 17 | 35 |
| SEATTLE, WA - URBAN SITE | 31 | 28 | 34 | 42 | 47 | 52 | 49 | 63 | 56 | 53 | 37 | 28 | 23 | 45 |
| SEATTLE, WA - INTL AP | 14 | 24 | 38 | 49 | 52 | 57 | 55 | 66 | 60 | 56 | 44 | 28 | 18 | 48 |
| SPOKANE, WASHINGTON | 32 | 26 | 39 | 53 | 60 | 62 | 65 | 80 | 77 | 70 | 52 | 28 | 21 | 56 |
| WALLA WALLA, WASHINGTON | 63 | 23 | 34 | 50 | 61 | 66 | 72 | 85 | 82 | 73 | 59 | 30 | 19 | 58 |
| PARKERSBURG, WEST VIRGINIA | 82 | 32 | 37 | 43 | 50 | 56 | 59 | 62 | 59 | 59 | 54 | 37 | 30 | 49 |
| GREEN BAY, WISCONSIN | 31 | 46 | 52 | 52 | 51 | 59 | 63 | 65 | 63 | 55 | 49 | 37 | 37 | 54 |
| MADISON, WISCONSIN | 34 | 49 | 53 | 53 | 52 | 59 | 65 | 69 | 67 | 62 | 56 | 40 | 40 | 56 |
| MILWAUKEE, WISCONSIN | 40 | 45 | 47 | 50 | 54 | 59 | 64 | 70 | 66 | 60 | 56 | 41 | 38 | 56 |
| CHEYENNE, WYOMING | 44 | 62 | 65 | 65 | 61 | 59 | 65 | 68 | 67 | 69 | 69 | 60 | 59 | 64 |
| LANDER, WYOMING | 34 | 65 | 68 | 72 | 67 | 65 | 73 | 77 | 76 | 74 | 71 | 61 | 63 | 70 |
| SHERIDAN, WYOMING | 40 | 55 | 57 | 61 | 58 | 58 | 63 | 75 | 73 | 67 | 63 | 52 | 53 | 62 |

Line E = Percentage of possible sun in Billings, Montana (Table 10-2)
= 0.47

Line F = Line E × Line D
= 0.47 × 3.05 MBTU/month = 1.43 MBTU/month

Line G = Collector system efficiency (Table 6-3)
= 0.6

Line H = Line G × Line F
= 0.6 × 1.43 MBTU/month = 0.86 MBTU/month

Line I = Line H divided by Line B
= 0.86 MBTU/month divided by 1.72 MBTU/month
= 0.50

This procedure can be followed to determine an estimated system output for each month. This three collector system in Billings, Montana, would yield an estimated 14.19 MBTU per year. Because the total demand is 19.26 MBTU, the solar domestic hot water system would provide approximately 74.5 percent of the total hot water needed.

## COST FACTORS

### Conventional Energy Costs

Projecting the cost of energy from conventional fuels with any degree of certainty is an impossible task. Many sources predict that fuel costs will inflate yearly at about 9 to 13 percent. Today's utility bills are already squeezing the typical family budget, and tomorrow's utility bills at continuing escalated rates may become a financial burden upon the homeowner. The cost of conventional DHW equipment is minimal compared with the cost of solar DHW equipment. A typical electric hot water system may only cost from $150 to $300 in comparison with a solar DHW equipment cost of $2500. It is the increasing cost of conventional fuels that makes the switch to solar a practical investment.

Typical costs to heat water electrically are represented in Table 10-3. Local electric rates per kilowatt-hour can be determined either from the monthly electric bills by dividing the cost of quick recovery water heating by the total number of kilowatt-hours used, or by direct inquiry from the utility company. The fuel surcharge should be included in the cost as applicable.

The cost to heat water electrically for our typical family of four in Billings, Montana, is determined by the same method from which Table 10-3 was derived. From Table 6-3, it was previously determined that this family has a 19.26 MBTU yearly requirement for hot water. Assuming an electric rate of $0.07 per kwh we have a yearly cost of $395.02.

$$19.26 \times 10^6 \text{ BTU} \times \frac{1 \text{ kWH}}{3413 \text{ BTU}} \times \frac{\$0.07}{\text{kWH}} = \$395.02$$

This same procedure can be used to determine the yearly cost of all types of fuels using the energy conversion factors presented in Figure 10-1.

## Solar Energy Costs

### Equipment and Labor Costs

Equipment and labor are the two major cost factors of a solar DHW system. The cost of system equipment appears to average approximately $2500 and the cost of labor for installation can cost anywhere from $800 to $1500. It is important to note that the price of equipment is rapidly inflating with other materials in the building industry. Wages for skilled mechanical/electrical technicians are also rapidly increasing at 8 to 10 percent per year. The commercially available flat plate collector is now well developed and dramatic technological advances in its improvement will be few. A solar DHW system will never be cheaper to install than it is right now.

### Maintenance Costs

Maintenance costs depend upon the durability of system components and the careful choice of the heat transfer medium. Maintenance on system components should be minimal over a 20-year life system. In general, heat transfer fluids can cause the highest maintenance cost, especially if they must be changed every two years. Glycol/water solutions are initially inexpensive, normally being one-third the cost of most synthetic hydrocarbons and one-fifth the cost of silicones. However, when maintenance costs are added to the price of this fluid plus the cost of new fluid every two or three years, one will find the total cost of silicones or hydrocarbons the better buy both monetarily and in terms of operational effectiveness. Maintenance costs can detract from the solar DHW investment. By using premium grade components and materials, and by observing daily temperatures and pressures either personally or with electronic monitoring, problems can be kept at a minimum.

### Operational Costs

Depending on system type, the only electrically operated components in the solar system may be the differential controller and the circulator/pump. A thermosiphon system has a distinct advantage over other systems at this point in that there are no operational costs. The differential controller uses a minimal amount of energy in operation and does not dramatically alter a cost analysis if omitted. If an electric circulator/pump is used, it should be noted that the higher the horsepower required, the higher the wattage needed, and therefore the higher the electricity costs. Examples of pump operational costs for various horsepower requirements are illustrated in Table 10-4. A closed loop system with its low horsepower circulator will cost less to operate than an open loop system with its higher horsepower pump. (Remember, costs in open loop systems can be lowered by operating a series of circulators as was explained previously in Chapter 8.)

For example, if the electric rate is $0.07 per kilowatt-hour and a 1/20 hp

**Table 10-3**
**Typical Domestic Hot Water Electrical Expenses**

| *Water Heated per Day (gal)* | *Yearly Requirement* | | *Yearly Cost to Heat Water from 40° F to 135° F* | | | | | | | | |
|---|---|---|---|---|---|---|---|---|---|---|---|
| | *MBTU* | *KWH* | *4¢/KWH* | *5¢/KWH* | *6¢/KWH* | *7¢/KWH* | *8¢/KWH* | *9¢/KWH* | *10¢/KWH* | *11¢/KWH* | *12¢/KWH* |
| 60 | 17.3 | 5068.9 | $202.76 | $253.45 | $304.13 | $354.82 | $405.51 | $456.20 | $506.89 | $557.58 | $608.27 |
| 70 | 20.2 | 5918.6 | $236.74 | $295.93 | $355.12 | $414.30 | $473.49 | $532.67 | $591.86 | $651.05 | $710.23 |
| 80 | 23.1 | 6768.2 | $270.73 | $338.41 | $406.09 | $473.77 | $541.46 | $609.14 | $676.82 | $744.50 | $812.18 |
| 90 | 26.0 | 7617.9 | $304.72 | $380.90 | $457.07 | $533.25 | $609.43 | $685.61 | $761.79 | $837.97 | $914.15 |
| 100 | 28.9 | 8467.6 | $338.70 | $423.38 | $508.06 | $592.73 | $677.41 | $762.08 | $846.76 | $931.44 | $1016.11 |
| 110 | 31.8 | 9317.3 | $372.69 | $465.87 | $559.04 | $652.21 | $745.38 | $838.56 | $931.73 | $1024.90 | $1118.08 |
| 120 | 34.7 | 10,167.0 | $406.68 | $508.35 | $610.02 | $711.69 | $813.26 | $915.03 | $1016.70 | $1118.37 | $1220.04 |

**Table 10-4**
**Typical Pump Operational Costs**

| *Horse-power of pump or Cir-culator* | *Kilo-watts Required* | *Annual KWH (Based on 8 hrs per Day)* | *Annual Costs at Various Electric Rates* | | | | | | | | |
|---|---|---|---|---|---|---|---|---|---|---|---|
| | | | *4¢/KWH* | *5¢/KWH* | *6¢/KWH* | *7¢/KWH* | *8¢/KWH* | *9¢/KWH* | *10¢/KWH* | *11¢/KWH* | *12¢/KWH* |
| 1/20 | 0.098 | 286.2 | $11.45 | $14.31 | $17.17 | $20.03 | $22.90 | $25.76 | $28.62 | $31.48 | $34.34 |
| 1/12 | 0.185 | 540.2 | $21.61 | $27.01 | $32.41 | $37.81 | $43.22 | $48.62 | $54.02 | $59.42 | $64.82 |
| 1/4 | 0.420 | 1226.4 | $49.06 | $61.32 | $73.58 | $85.85 | $98.11 | $110.38 | $122.64 | $134.90 | $147.17 |
| 1/3 | 0.530 | 1547.6 | $61.90 | $77.38 | $92.86 | $108.33 | $123.81 | $139.28 | $154.76 | $170.24 | $185.71 |
| 1/2 | 0.790 | 2306.8 | $92.27 | $115.34 | $138.41 | $161.48 | $184.54 | $207.61 | $230.68 | $253.75 | $276.82 |

circulator is operated an average 8 hours per day year round, then the electric operational cost would be approximately $1.67 per month.

## Solar Energy Tax Credits

To help lower the cost of solar equipment and installation and as a way of making solar a more attractive energy alternative, the federal government and many states have enacted various income tax credit and property tax credit incentives. There are two types of taxes that are affected by the installation of a solar DHW system. These include: (1) income taxes (federal and some state) and (2) property taxes. Federal income taxes presently allow for 40 percent tax credit off the entire solar DHW system expenditure to a maximum credit of $4000. This credit can be obtained upon filing Internal Revenue Service form 5695 with the yearly personal income tax IRS form 1040. This credit is based on the cost of items installed in a principle residence before January 1, 1986. State income taxes are also credited as a percentage return of personal income taxes on a solar investment depending on each state legislation as illustrated in Table 10-5.

**Table 10-5**
**State Tax Breaks for Residential Solar Systems**

| *State* | *Property Tax Exemption* | *Income Tax Incentive* | *Sales Tax Exemption* |
|---|---|---|---|
| Alabama | No | Up to $1,000 credit | No |
| Alaska | No | Up to $200 credit | Not applicable |
| Arizona | Exemption | Up to $1,000 credit | Exemption |
| Arkansas | No | 100% deduction | No |
| California | No | Up to $3,000 credit per application | No |
| Colorado | Exemption | Up to $3,000 credit | No |
| Connecticut | Local option | Not applicable | Exemption |
| Delaware | No | $200 credit for DHW systems | Not applicable |
| Florida | Exemption | Not applicable | Exemption |
| Georgia | Local option | No | Refund |
| Hawaii | Exemption | 10% credit | No |
| Idaho | No | 100% deduction | No |
| Illinois | Exemption | No | No |
| Indiana | Exemption | Up to $300 credit | No |
| Iowa | Exemption | No | No |
| Kansas | Exemption; refund based on efficiency of system | Up to $1,500 credit | No |
| Kentucky | No | No | No |
| Louisiana | Exemption | No | No |
| Maine | Exemption | Up to $100 credit | Refund |
| Maryland | Exemption statewide plus credit at local option | No | No |

**Table 10-5 (*continued*)**

| *State* | *Property Tax Exemption* | *Income Tax Incentive* | *Sales Tax Exemption* |
|---|---|---|---|
| Massachusetts | Exemption | Up to $1,000 credit | Exemption |
| Michigan | Exemption | Up to $1,700 credit | Exemption |
| Minnesota | Exemption | Up to $2,000 credit | No |
| Mississippi | No | No | Exemption for colleges, junior colleges and universities |
| Missouri | No | No | No |
| Montana | Exemption | Up to $125 credit | Not applicable |
| Nebraska | Yes | Up to $3,000 credit | Refund |
| Nevada | Limited exemption | Not applicable | No |
| New Hampshire | Local option | Not applicable | Not applicable |
| New Jersey | Exemption | No | Exemption |
| New Mexico | No | Up to $4,000 credit | No |
| New York | Exemption | Up to $2750 credit | No |
| North Carolina | Exemption | Up to $1,000 credit | No |
| North Dakota | Exemption | 5% credit for 2 years | No |
| Ohio | Exemption | Up to $1,000 credit | Exemption |
| Oklahoma | No | Up to $2,800 credit | Refund |
| Oregon | Exemption | Up to $1,000 credit | Not applicable |
| Pennsylvania | No | No | No |
| Rhode Island | Exemption | Up to $1,000 credit | Refund |
| South Carolina | No | Up to 1,000 deduction | No |
| South Dakota | Exemption | Not applicable | No |
| Tennessee | Exemption | Not applicable | No |
| Texas | Exemption | Not applicable | Exemption |
| Utah | No | Up to $1,000 credit | No |
| Vermont | Local option | Up to $1,000 credit | No |
| Virginia | Local option | No | No |
| Washington | Exemption | Not applicable | No |
| West Virginia | No | No | No |
| Wisconsin | Exemption | No* | No |
| Wyoming | No | Not applicable | No |

*Wisconsin offers a direct rebate for part of solar expenditures; the rebate is unrelated to taxes.

*Source:* Courtesy of the Conservation and Renewable Energy Inquiry and Referral Service; formerly the Solar Heating and Cooling Information Center.

Appendix C (State Tax Credit Incentives) explains the state tax incentives for each state in further detail. Savings on income taxes can also be realized through the interest paid on borrowed money as a tax deduction. In some states increased property taxes may result by adding a solar DHW system if the assessed value of the building is increased. In other states this additional structure is exempt from property taxes. Additional tax credits and property assessment exclusions for solar DHW systems are either already available or are subject to exclusion in legislation at both federal and state levels. The availability of these tax credits and property-assessment exclusions should be investigated to ensure an accurate determination of actual system costs.

There appears to be much confusion about the difference between tax credits and tax deductions. It is important to understand this difference. First it must be pointed out that a tax ***credit*** is *not* a tax ***deduction***. *Deductions* are subtracted ***from income***. *Credits* are subtracted ***from taxes***. For example, if you are entitled to a $1000 tax credit, and you owe $100 for income taxes, then you would subtract the total credit from the tax owed, and find you don't owe any taxes. Let's illustrate this with another example. Assume you earn $20,000 per year and assume the personal earned liability is 20 percent of the income. The total tax liability would therefore be $4000. If you installed a solar DHW system for $2500, a 40 percent tax credit would reduce the $4000 tax liability by $1000. You would then have a tax liability of only $3000. If this $1000 credit was incorrectly taken as a tax deduction, the earned income would be reduced from $20,000 to $19,000 which would result in a tax liability of $3800. To reiterate, a solar tax credit results in a tax credit not a tax deduction.

## TIME VALUE OF MONEY FACTORS

### Present and Future Worth Factors

To provide a more accurate account of the actual savings and payback period provided by using solar energy to heat domestic water, the "time value of money" must be considered. This concept is illustrated by considering a loan of $2500 to be paid back at the end of a 5-year period with an interest rate of 10 percent a year. The amount owed at the end of the first year is the original sum of $2500 plus the $250 cost for the use of capital, for a total of $2750. At the end of the second year the amount owed is $2750 plus the cost for the use of capital for a total of $3025. This process of compounding continues as illustrated in Table 10-6 until the end of the 5-year loan, at which time the original $2500 borrowed has actually cost $4026.28 for the principal and undistributed interest.

The total interest paid of $1526.25 is the rate of return on the money loaned. The lender can say that the "future worth" of the $2500 loaned at 10 percent over 5 years is $4026.25. The computation of Table 10-6 can be more expeditiously determined using Equation 10-1.

**Table 10-6**
**Cost of $2500 at 10% Compounded Interest**

| *Compounded Amount Due* | *Loan Term in Years* |
|---|---|
| $2750.00 | 1 |
| $3025.00 | 2 |
| $3327.50 | 3 |
| $3660.25 | 4 |
| $4026.25 | 5 |

***Equation 10-1***

$$S = P\,(1 + i)^n$$

*where* $S$ = a sum of money at a specified future date
$P$ = a present sum of money
$i$ = interest rate earned at the end of each period
$n$ = the number of interest periods

The "time value of money" can be displayed graphically as Figure 10-2.

At the end of the first period of time, the time value of $P$ is $P + Pi$ or $P(1 + i)$; at the second interval of time, the time value of $P$ is $P(1 + i) + P(1 + i)\,(i)$ or $P(1 + i)^2$. The sum $S$ at the end of the $n$th period will result in Equation 10-1. The factor $(1 + i)^n$ is called the ***single payment compound amount factor*** (SPCAF). Equation 10-1 can be rewritten as Equation 10-2.

***Equation 10-2***

$$S = P(i - n \text{ SPCAF})$$

For example, if \$2500, $P$, is borrowed at 10 percent ($i$) over 5 years, $n$, the future worth, $S$ of the initial \$2500 can be found using Equation 10-2 and the "single payment compound amount factor" (SPCAF), which can be obtained from Table 10-8 under the applicable interest rate, as follows:

$$S = P\,(i - n \text{ SPCAF})$$

*where* $S$ = future worth of money
$P$ = \$2500
$(i - n \text{ SPCAF}) = (0.10 - 5 \text{ SPCAF}) = 1.6105$
and $S = \$2500\,(1.6105) = \$4026.25$

Because of inflation, future money is not as valuable as money at the present and must be discounted by the factor $1/(1 + i)^n$, which is called the ***single-payment present worth factor*** (SPPWF). Simply stated, the present worth of money is simply the inverse of Equation 10-1 and can be written as Equation 10-3.

***Equation 10-3***

$$P = \frac{S}{(1 + i)^n}$$

In terms of mnemonic symbols we have Equation 10-4:

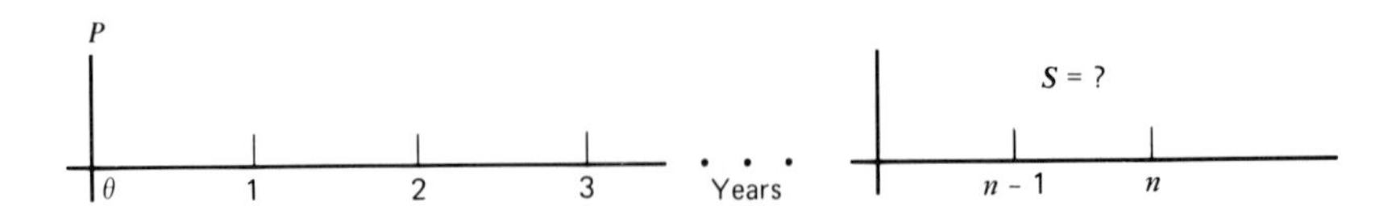


**Figure 10-2.** Time scales.

*Equation 10-4*

$$P = S\ (i - n\ \text{SPPWF})$$

For example, the time value of a future sum of $4026.25 occurring 5 years from the initial investment can be found from Equation 10-4 and the single payment present worth factor (SPPWF), which can be obtained from Table 10-8 under the applicable interest rate as follows:

$$P = S\ (i - n\ \text{SPPWF})$$

*where* $P$ = present worth of money

$S$ = \$4026.25

$(i - n\ \text{SPPWF}) = (0.10 - 5\ \text{SPPWF}) = 0.62092$

and $P = \$4026.25\ (0.62092) = \$2500$

## Capital Recovery Factor

It is always convenient to discuss economics in terms of cash. But what if a person does not have the cash to purchase a solar DHW system? We need to determine what the total cost of the system will be if the money is borrowed and yearly or monthly payments are made. The future series of end-of-period payments that will just recover a sum $P$ over $n$ periods with compound interest is illustrated in Figure 10-3. End-of-period payments can be determined using Equation 10-5. The factor by which a present capital sum $P$ is multiplied to find the future series $R$ that will exactly recover it with interest is called the ***capital recovery factor*** (CRF).

*Equation 10-5*

$$R = P\ (i - n\ \text{CRF})$$

To illustrate the use of this factor, suppose $2500 is borrowed for solar DHW system equipment at 10 percent interest compounded annually for a 5-year term. The series of repayments can be found by using Equation 10-5 and the capital recovery factor (CRF), which can be obtained from Table 10-8 under the applicable interest rate, as follows:

$$R = P\ (i - n\ \text{CRF})$$

*where* $R$ = repayment made at end of each year

$P$ = \$2500

$(i - n\ \text{CRF}) = (0.10 - 5\ \text{CRF}) = 0.26380$

and $R = \$2500\ (0.26380) = \$659.50$

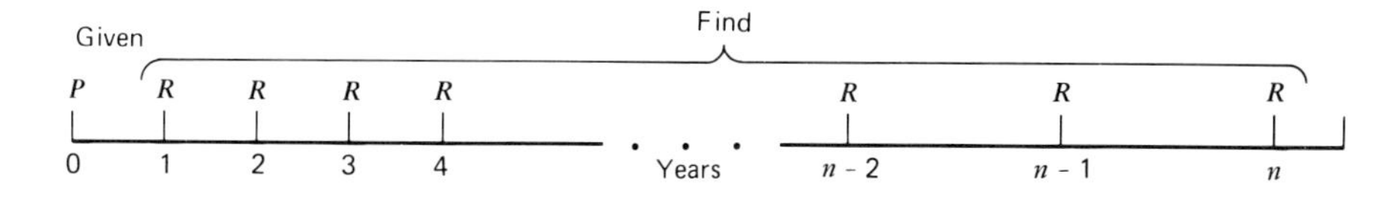


**Figure 10-3.** Capital recovery scale.

This would equate to a repayment of $54.96 per month. Table 10-7 illustrates the cost of capital of $2500 at 10 percent interest, and with five end-of-year uniform payments for the recovery of capital.

Table 10-7 shows that the money on deposit at the beginning of each period (Column 1) earns interest during that period (Column 2), and the payment at the end of the period (Column 4) repays this interest plus some of the principal (Column 6). For example, the unpaid principal at the beginning of year 3 is $1640.05, the interest earned that year at 10 percent is $164.01, and the payment at the end of that year of $659.50, consists of $164.01 in interest and $495.49 in principal.

The actual cost of the solar DHW system in terms of present worth money at 10 percent interest compounded annually is the sum of the initial amount of money borrowed plus the present worth of the interest due at the end of each repayment period. In this case, using Equation 10-4 and Table 10-8, the present worth of money invested would be:

| | | |
|---|---|---|
| Original Amount | = | $2500.00 |
| 250.00 (0.90909) | = | $ 227.27 |
| 209.05 (0.82645) | = | $ 172.77 |
| 164.01 (0.75131) | = | $ 123.22 |
| 114.46 (0.68301) | = | $ 78.18 |
| 59.95 (0.62092) | = | $ 37.22 |
| Present worth cost | = | $3138.66 |

Therefore, in a loan repayment situation of $2500 at 10 percent compounded interest with end-of period repayments, the present worth of money invested in a solar DHW system would be $3138.66. A solar DHW system is an investment, and if money is borrowed to install such a system, the payments

**Table 10-7**
**Visualizing the Capital Recovery Factor**

| *Year* | *(1)* *Money Owed at Start of Year* | *(2)* *Interest Owed at End of Year* | *(3)* *Principal and Interest Owed at End of Year* | *(4)* *Series of Repayment* | *(5)* *Money Owed at End of Year After Repayment* | *(6)* *Recovery Capital* |
|---|---|---|---|---|---|---|
| 1 | $2500.00 | $250.00 | $2750.00 | $659.50 | $2090.50 | $ 409.50 |
| 2 | $2090.50 | $209.05 | $2299.55 | $659.50 | $1640.05 | $ 450.45 |
| 3 | $1640.05 | $164.01 | $1804.06 | $659.50 | $1144.56 | $ 495.49 |
| 4 | $1144.56 | $114.46 | $1259.02 | $659.50 | $ 599.52 | $ 545.05 |
| 5 | $ 599.52 | $ 59.95 | $ 659.50 | $659.50 | .00 | $ 599.52 |
| | | | | | | Total $2500.00 |

## Table 10-8
## Discrete Rate-of-Return Factors

**5%**

| n | SPCAF | SPPWF | CRF |
|---|---|---|---|
| 1 | 1.0500 | .95238 | 1.0500 |
| 2 | 1.1025 | .90703 | .53780 |
| 3 | 1.1576 | .86384 | .36721 |
| 4 | 1.2155 | .82270 | .28201 |
| 5 | 1.2763 | .78353 | .23097 |
| 6 | 1.3401 | .74622 | .19702 |
| 7 | 1.4071 | .71068 | .17282 |
| 8 | 1.4775 | .67684 | .15472 |
| 9 | 1.5513 | .64461 | .14069 |
| 10 | 1.6289 | .61391 | .12950 |
| 11 | 1.7103 | .58468 | .12039 |
| 12 | 1.7959 | .55684 | .11283 |
| 13 | 1.8856 | .53032 | .10646 |
| 14 | 1.9799 | .50507 | .10102 |
| 15 | 2.0789 | .48102 | .09634 |
| 16 | 2.1829 | .45811 | .09227 |
| 17 | 2.2920 | .43630 | .08870 |
| 18 | 2.4066 | .41552 | .08555 |
| 19 | 2.5269 | .39573 | .08275 |
| 20 | 2.6533 | .37689 | .08024 |
| 21 | 2.7860 | .35894 | .07800 |
| 22 | 2.9253 | .34185 | .07597 |
| 23 | 3.0715 | .32557 | .07414 |
| 24 | 3.2251 | .31007 | .07247 |
| 25 | 3.3864 | .29530 | .07095 |
| 26 | 3.5557 | .28124 | .06956 |
| 27 | 3.7335 | .26785 | .06829 |
| 28 | 3.9201 | .25509 | .06712 |
| 29 | 4.1161 | .24295 | .06605 |
| 30 | 4.3219 | .23138 | .06505 |

**7%**

| n | SPCAF | SPPWF | CRF |
|---|---|---|---|
| 1 | 1.0700 | .93458 | 1.0700 |
| 2 | 1.1449 | .87344 | .55309 |
| 3 | 1.2250 | .81630 | .38105 |
| 4 | 1.3108 | .76290 | .29523 |
| 5 | 1.4026 | .71299 | .24389 |
| 6 | 1.5007 | .66634 | .20980 |
| 7 | 1.6058 | .62275 | .18555 |
| 8 | 1.7182 | .58201 | .16747 |
| 9 | 1.8385 | .54393 | .15349 |
| 10 | 1.9672 | .50835 | .14238 |
| 11 | 2.1049 | .47509 | .13336 |
| 12 | 2.2522 | .44401 | .12590 |
| 13 | 2.4098 | .41496 | .11965 |
| 14 | 2.5785 | .38782 | .11434 |
| 15 | 2.7590 | .36245 | .10979 |
| 16 | 2.9522 | .33873 | .10586 |
| 17 | 3.1588 | .31657 | .10243 |
| 18 | 3.3799 | .29586 | .09941 |
| 19 | 3.6165 | .27651 | .09675 |
| 20 | 3.8697 | .25842 | .09439 |
| 21 | 4.1406 | .24151 | .09229 |
| 22 | 4.4304 | .22571 | .09041 |
| 23 | 4.7405 | .21095 | .08871 |
| 24 | 5.0724 | .19715 | .08719 |
| 25 | 5.4274 | .18425 | .08581 |
| 26 | 5.8074 | .17220 | .08456 |
| 27 | 6.2139 | .16093 | .08343 |
| 28 | 6.6488 | .15040 | .08239 |
| 29 | 7.1143 | .14056 | .08145 |
| 30 | 7.6123 | .13137 | .08059 |

**9%**

| n | SPCAF | SPPWF | CRF |
|---|---|---|---|
| 1 | 1.0900 | .91743 | 1.0900 |
| 2 | 1.1881 | .84168 | .56847 |
| 3 | 1.2950 | .77218 | .39505 |
| 4 | 1.4116 | .70843 | .30867 |
| 5 | 1.5386 | .64993 | .25709 |
| 6 | 1.6771 | .59627 | .22292 |
| 7 | 1.8280 | .54703 | .19869 |
| 8 | 1.9926 | .50187 | .18067 |
| 9 | 2.1719 | .46043 | .16680 |
| 10 | 2.3674 | .42241 | .15582 |
| 11 | 2.5804 | .38753 | .14695 |
| 12 | 2.8127 | .35553 | .13965 |
| 13 | 3.0658 | .32618 | .13357 |
| 14 | 3.3417 | .29925 | .12843 |
| 15 | 3.6425 | .27454 | .12406 |
| 16 | 3.9703 | .25187 | .12030 |
| 17 | 4.3276 | .23107 | .11705 |
| 18 | 4.7171 | .21199 | .11421 |
| 19 | 5.1417 | .19449 | .11173 |
| 20 | 5.6044 | .17843 | .10955 |
| 21 | 6.1088 | .16370 | .10762 |
| 22 | 6.6586 | .15018 | .10590 |
| 23 | 7.2579 | .13778 | .10438 |
| 24 | 7.9111 | .12640 | .10302 |
| 25 | 8.6231 | .11597 | .10181 |
| 26 | 9.3992 | .10639 | .10072 |
| 27 | 10.245 | .09761 | .09974 |
| 28 | 11.167 | .08955 | .09885 |
| 29 | 12.172 | .08216 | .09806 |
| 30 | 13.268 | .07537 | .09734 |

**12%**

| n | SPCAF | SPPWF | CRF |
|---|---|---|---|
| 1 | 1.1200 | .89286 | 1.1200 |
| 2 | 1.2544 | .79719 | .59170 |
| 3 | 1.4049 | .71178 | .41635 |
| 4 | 1.5735 | .63552 | .32923 |
| 5 | 1.7623 | .56743 | .27741 |
| 6 | 1.9738 | .50663 | .24323 |
| 7 | 2.2107 | .45235 | .21912 |
| 8 | 2.4760 | .40388 | .20130 |
| 9 | 2.7731 | .36061 | .18768 |
| 10 | 3.1058 | .32197 | .17698 |
| 11 | 3.4786 | .28748 | .16842 |
| 12 | 3.8960 | .25668 | .16144 |
| 13 | 4.3635 | .22917 | .15568 |
| 14 | 4.8871 | .20462 | .15087 |
| 15 | 5.4736 | .18270 | .14682 |
| 16 | 6.1304 | .16312 | .14339 |
| 17 | 6.8660 | .14564 | .14046 |
| 18 | 7.6900 | .13004 | .13794 |
| 19 | 8.6128 | .11611 | .13576 |
| 20 | 9.6463 | .10367 | .13388 |
| 21 | 10.804 | .09256 | .13224 |
| 22 | 12.100 | .08264 | .13081 |
| 23 | 13.552 | .07379 | .12956 |
| 24 | 15.179 | .06588 | .12846 |
| 25 | 17.000 | .05882 | .12750 |
| 26 | 19.040 | .05252 | .12665 |
| 27 | 21.325 | .04689 | .12590 |
| 28 | 23.884 | .04187 | .12524 |
| 29 | 26.750 | .03738 | .12466 |
| 30 | 29.960 | .03338 | .12414 |

**17%**

| n | SPCAF | SPPWF | CRF |
|---|---|---|---|
| 1 | 1.1700 | .85470 | 1.1700 |
| 2 | 1.3689 | .73051 | .63083 |
| 3 | 1.6016 | .62437 | .45257 |
| 4 | 1.8739 | .53365 | .36453 |
| 5 | 2.1924 | .45611 | .31256 |
| 6 | 2.5652 | .38984 | .27861 |
| 7 | 3.0012 | .33320 | .25495 |
| 8 | 3.5115 | .28478 | .23769 |
| 9 | 4.1084 | .24340 | .22469 |
| 10 | 4.8068 | .20804 | .21466 |
| 11 | 5.6240 | .17781 | .20676 |
| 12 | 6.5801 | .15197 | .20047 |
| 13 | 7.6987 | .12989 | .19538 |
| 14 | 9.0075 | .11102 | .19123 |
| 15 | 10.539 | .09489 | .18782 |
| 16 | 12.330 | .08110 | .18500 |
| 17 | 14.426 | .06932 | .18266 |
| 18 | 16.879 | .05925 | .18071 |
| 19 | 19.748 | .05064 | .17907 |
| 20 | 23.106 | .04328 | .17769 |
| 21 | 27.034 | .03699 | .17653 |
| 22 | 31.629 | .03162 | .17555 |
| 23 | 37.006 | .02702 | .17472 |
| 24 | 43.297 | .02310 | .17402 |
| 25 | 50.658 | .01974 | .17342 |
| 26 | 59.270 | .01687 | .17292 |
| 27 | 69.345 | .01442 | .17249 |
| 28 | 81.134 | .01233 | .17212 |
| 29 | 94.927 | .01053 | .17181 |
| 30 | 111.06 | .00900 | .17154 |

**6%**

| n | SPCAF | SPPWF | CRF |
|---|---|---|---|
| 1 | 1.0600 | .94340 | 1.0600 |
| 2 | 1.1236 | .89000 | .54544 |
| 3 | 1.1910 | .83962 | .37411 |
| 4 | 1.2625 | .79209 | .28859 |
| 5 | 1.3382 | .74726 | .23740 |
| 6 | 1.4185 | .70496 | .20336 |
| 7 | 1.5036 | .66506 | .17914 |
| 8 | 1.5938 | .62741 | .16104 |
| 9 | 1.6895 | .59190 | .14702 |
| 10 | 1.7908 | .55839 | .13587 |
| 11 | 1.8983 | .52679 | .12679 |
| 12 | 2.0122 | .49697 | .11928 |
| 13 | 2.1329 | .46884 | .11296 |
| 14 | 2.2609 | .44230 | .10758 |
| 15 | 2.3966 | .41727 | .10296 |
| 16 | 2.5404 | .39365 | .09895 |
| 17 | 2.6928 | .37136 | .09545 |
| 18 | 2.8543 | .35034 | .09236 |
| 19 | 3.0256 | .33051 | .08962 |
| 20 | 3.2071 | .31180 | .08719 |
| 21 | 3.3996 | .29416 | .08501 |
| 22 | 3.6035 | .27751 | .08304 |
| 23 | 3.8197 | .26180 | .08128 |
| 24 | 4.0489 | .24698 | .07968 |
| 25 | 4.2919 | .23300 | .07823 |
| 26 | 4.5494 | .21981 | .07690 |
| 27 | 4.8223 | .20737 | .07570 |
| 28 | 5.1117 | .19563 | .07459 |
| 29 | 5.4184 | .18456 | .07358 |
| 30 | 5.7435 | .17411 | .07265 |

**8%**

| n | SPCAF | SPPWF | CRF |
|---|---|---|---|
| 1 | 1.0800 | .92593 | 1.0800 |
| 2 | 1.1664 | .85734 | .56077 |
| 3 | 1.2597 | .79383 | .38803 |
| 4 | 1.3605 | .73503 | .30192 |
| 5 | 1.4693 | .68058 | .25046 |
| 6 | 1.5869 | .63017 | .21632 |
| 7 | 1.7138 | .58349 | .19207 |
| 8 | 1.8509 | .54027 | .17401 |
| 9 | 1.9990 | .50025 | .16008 |
| 10 | 2.1589 | .46319 | .14903 |
| 11 | 2.3316 | .42888 | .14008 |
| 12 | 2.5182 | .39711 | .13270 |
| 13 | 2.7196 | .36770 | .12652 |
| 14 | 2.9372 | .34046 | .12130 |
| 15 | 3.1722 | .31524 | .11683 |
| 16 | 3.4259 | .29189 | .11298 |
| 17 | 3.7000 | .27027 | .10963 |
| 18 | 3.9960 | .25025 | .10670 |
| 19 | 4.3157 | .23171 | .10413 |
| 20 | 4.6610 | .21455 | .10185 |
| 21 | 5.0338 | .19866 | .09983 |
| 22 | 5.4365 | .18394 | .09803 |
| 23 | 5.8715 | .17032 | .09642 |
| 24 | 6.3412 | .15770 | .09498 |
| 25 | 6.8485 | .14602 | .09368 |
| 26 | 7.3964 | .13520 | .09251 |
| 27 | 7.9881 | .12519 | .09145 |
| 28 | 8.6271 | .11591 | .09049 |
| 29 | 9.3173 | .10733 | .08962 |
| 30 | 10.063 | .09938 | .08883 |

**10%**

| n | SPCAF | SPPWF | CRF |
|---|---|---|---|
| 1 | 1.1000 | .90909 | 1.1000 |
| 2 | 1.2100 | .82645 | .57619 |
| 3 | 1.3310 | .75131 | .40211 |
| 4 | 1.4641 | .68301 | .31547 |
| 5 | 1.6105 | .62092 | .26380 |
| 6 | 1.7716 | .56447 | .22961 |
| 7 | 1.9487 | .51316 | .20541 |
| 8 | 2.1436 | .46651 | .18744 |
| 9 | 2.3579 | .42410 | .17364 |
| 10 | 2.5937 | .38554 | .16275 |
| 11 | 2.8531 | .35049 | .15396 |
| 12 | 3.1384 | .31863 | .14676 |
| 13 | 3.4523 | .28966 | .14078 |
| 14 | 3.7975 | .26333 | .13575 |
| 15 | 4.1772 | .23939 | .13147 |
| 16 | 4.5950 | .21763 | .12782 |
| 17 | 5.0545 | .19784 | .12466 |
| 18 | 5.5599 | .17986 | .12193 |
| 19 | 6.1159 | .16351 | .11955 |
| 20 | 6.7275 | .14864 | .11746 |
| 21 | 7.4003 | .13513 | .11562 |
| 22 | 8.1403 | .12285 | .11401 |
| 23 | 8.9543 | .11168 | .11257 |
| 24 | 9.8497 | .10153 | .11130 |
| 25 | 10.835 | .09230 | .11017 |
| 26 | 11.918 | .08391 | .10916 |
| 27 | 13.110 | .07628 | .10826 |
| 28 | 14.421 | .06934 | .10745 |
| 29 | 15.863 | .06304 | .10673 |
| 30 | 17.449 | .05731 | .10608 |

**15%**

| n | SPCAF | SPPWF | CRF |
|---|---|---|---|
| 1 | 1.1500 | .86957 | 1.1500 |
| 2 | 1.3225 | .75614 | .61512 |
| 3 | 1.5209 | .65752 | .43798 |
| 4 | 1.7490 | .57175 | .35027 |
| 5 | 2.0114 | .49718 | .29832 |
| 6 | 2.3131 | .43233 | .26424 |
| 7 | 2.6600 | .37594 | .24036 |
| 8 | 3.0590 | .32690 | .22285 |
| 9 | 3.5179 | .28426 | .20957 |
| 10 | 4.0456 | .24718 | .19925 |
| 11 | 4.6524 | .21494 | .19107 |
| 12 | 5.3503 | .18691 | .18448 |
| 13 | 6.1528 | .16253 | .17911 |
| 14 | 7.0757 | .14133 | .17469 |
| 15 | 8.1371 | .12289 | .17102 |
| 16 | 9.3576 | .10686 | .16795 |
| 17 | 10.761 | .09293 | .16537 |
| 18 | 12.375 | .08081 | .16319 |
| 19 | 14.232 | .07027 | .16134 |
| 20 | 16.367 | .06110 | .15976 |
| 21 | 18.822 | .05313 | .15842 |
| 22 | 21.645 | .04620 | .15727 |
| 23 | 24.891 | .04017 | .15628 |
| 24 | 28.625 | .03493 | .15543 |
| 25 | 32.919 | .03038 | .15470 |
| 26 | 37.857 | .02642 | .15407 |
| 27 | 43.535 | .02297 | .15353 |
| 28 | 50.066 | .01997 | .15306 |
| 29 | 57.575 | .01737 | .15265 |
| 30 | 66.212 | .01510 | .15230 |

**20%**

| n | SPCAF | SPPWF | CRF |
|---|---|---|---|
| 1 | 1.2000 | .83333 | 1.2000 |
| 2 | 1.4400 | .69444 | .65455 |
| 3 | 1.7280 | .57870 | .47473 |
| 4 | 2.0736 | .48225 | .38629 |
| 5 | 2.4883 | .40188 | .33438 |
| 6 | 2.9860 | .33490 | .30071 |
| 7 | 3.5832 | .27908 | .27742 |
| 8 | 4.2998 | .23257 | .26061 |
| 9 | 5.1598 | .19381 | .24808 |
| 10 | 6.1917 | .16151 | .23852 |
| 11 | 7.4301 | .13459 | .23110 |
| 12 | 8.9161 | .11216 | .22526 |
| 13 | 10.699 | .09346 | .22062 |
| 14 | 12.839 | .07789 | .21689 |
| 15 | 15.407 | .06491 | .21388 |
| 16 | 18.488 | .05409 | .21144 |
| 17 | 22.186 | .04507 | .20944 |
| 18 | 26.623 | .03756 | .20781 |
| 19 | 31.948 | .03130 | .20646 |
| 20 | 38.338 | .02608 | .20536 |
| 21 | 46.005 | .02174 | .20444 |
| 22 | 55.206 | .01811 | .20369 |
| 23 | 66.247 | .01510 | .20307 |
| 24 | 79.497 | .01258 | .20255 |
| 25 | 95.396 | .01048 | .20212 |
| 26 | 114.48 | .00874 | .20176 |
| 27 | 137.37 | .00728 | .20147 |
| 28 | 164.84 | .00607 | .20122 |
| 29 | 197.81 | .00506 | .20102 |
| 30 | 237.38 | .00421 | .20085 |

may be worked out with a bank so that repayments are approximately the same as the conventional monthly utility bill for hot water.

## COMPARATIVE ANALYSIS—CONVENTIONAL VERSUS SOLAR DHW

We must now discuss payback of the initial invested capital and determine the cost per unit energy of a conventional fuel versus solar. A comparative analysis of a typical solar DHW system should impart the fact that solar DHW is an economically sound and viable investment. ***Payback*** is defined as the time at which initial cost and annual operating and maintenance expenses of a ***solar DHW system*** equal the total savings from not using ***conventional energy sources*** both at compounded interest rates of inflation for the same amount of energy. We shall illustrate this convention of payback and the tangible savings resulting from the use of solar energy to heat water by comparing electric hot water costs with solar hot water costs. This same economic analysis can be conducted using any of the other energy sources at their current costs.

The solar contribution of a solar DHW system for the typical family of four in Billings, Montana, was determined in Table 10-1*b* to be 14.19 MBTU per year. At $0.07 per kilowatt-hour it would cost $291.03 per year to heat the same amount of water with conventional electric heating. Using a 1/20 hp circulator the solar operational cost can be determined from Table 10-4 to be $20.03 per year. As an additional measure we shall assume a $20 per year maintenance charge for solar. The cost and savings comparison is depicted in Table 10-9.

The cost comparison depicted in Table 10-9 is perhaps a little lenient toward electric water heating costs. For instance, this table depicts a conservative energy inflation rate of only 5 percent, whereas fossil fuels and electricity are continuing to inflate at much higher rates. Most conventional hot water systems do not last over the 20-year life and have to be replaced. Salvage value of the solar DHW components is also not considered. By omitting these parameters, a conservative view of payback should be attained. The example presented in Table 10-9 is graphically illustrated in Figure 10-4.

The payback period from savings realized prior to federal tax credits can be seen to be 7½ years. Considering a federal tax credit of 40 percent (i.e., 40 percent × $2500 = $1000), the payback period would be reduced to less than 4½ years. State tax credits could reduce this payback period even further. The total actual savings after federal tax credits are the savings realized from solar energy ($8299.70) minus the estimated cost of the solar energy equipment ($1500.00) plus the estimated cost of the conventional energy equipment ($300.00) for a cumulative total savings of $7099.70 over 20 years. A dealer installation fee of a nominal $1000 rather than a "do-it-yourself" installation would still yield a cumulative total savings of $6099.70 over 20 years. If further worth of savings is considered throughout a 20-year

**Table 10-9**
**Example of Energy Costs, Savings Realized, and Future Worth of Savings**

| | *Conventional DHW Fuel Costs at 5% per year Inflation* | | *Solar Operational and Maintenance Costs at 5% per Year Inflation* | | *Savings Realized From Solar* | | *Future Worth of Savings Through 20 Years at 6% Compounded Interest* | |
|---|---|---|---|---|---|---|---|---|
| *Years* | *Yearly* | *Cummulative* | *Yearly* | *Cummulative* | *Yearly* | *Cummulative* | *Yearly* | *Cummulative* |
| 1 | $291.03 | $ 291.03 | $ 40.03 | $ 40.03 | $251.00 | $ 251.00 | $804.98 | $ 804.98 |
| 2 | 305.58 | 596.61 | 42.03 | 82.06 | 263.55 | 514.55 | 797.40 | 1602.38 |
| 3 | 320.86 | 917.47 | 44.13 | 126.19 | 276.73 | 791.28 | 789.87 | 2392.25 |
| 4 | 336.90 | 1254.37 | 46.34 | 172.53 | 290.56 | 1081.84 | 782.42 | 3174.67 |
| 5 | 353.75 | 1608.12 | 48.66 | 221.19 | 305.09 | 1386.93 | 775.05 | 3949.72 |
| 6 | 371.44 | 1979.56 | 51.09 | 272.28 | 320.35 | 1707.28 | 767.75 | 4717.47 |
| 7 | 390.01 | 2369.57 | 53.64 | 325.92 | 336.37 | 2043.65 | 760.50 | 5477.97 |
| 8 | 409.51 | 2779.08 | 56.32 | 382.24 | 353.19 | 2396.84 | 753.32 | 6231.29 |
| 9 | 429.99 | 3209.07 | 59.14 | 441.38 | 370.85 | 2767.69 | 746.22 | 6977.51 |
| 10 | 451.49 | 3660.56 | 62.10 | 503.48 | 389.39 | 3157.08 | 739.20 | 7716.71 |
| 11 | 474.06 | 4134.62 | 65.20 | 568.68 | 408.86 | 3565.94 | 732.19 | 8448.90 |
| 12 | 497.76 | 4632.38 | 68.46 | 637.14 | 429.30 | 3995.24 | 725.30 | 9174.20 |
| 13 | 522.65 | 5155.03 | 71.88 | 709.02 | 450.77 | 4446.01 | 718.44 | 9892.64 |
| 14 | 548.78 | 5703.81 | 75.47 | 784.49 | 473.31 | 4919.32 | 711.67 | 10604.31 |
| 15 | 576.22 | 6280.03 | 79.24 | 863.73 | 496.98 | 5416.30 | 704.97 | 11309.28 |
| 16 | 605.03 | 6885.06 | 83.20 | 946.93 | 521.83 | 5938.13 | 698.31 | 12007.59 |
| 17 | 635.28 | 7520.34 | 87.36 | 1034.29 | 547.92 | 6486.05 | 691.75 | 12699.34 |
| 18 | 667.04 | 8187.38 | 91.73 | 1126.02 | 575.31 | 7061.36 | 685.20 | 13384.54 |
| 19 | 700.39 | 8887.77 | 96.32 | 1222.34 | 604.07 | 7665.43 | 678.73 | 14063.27 |
| 20 | 734.41 | 9623.18 | 101.14 | 1323.48 | 634.27 | 8299.70 | 672.33 | 14735.60 |
| | Total | $9623.18 | Total | $1323.48 | Total | $8299.70 | Total | $14735.60 |

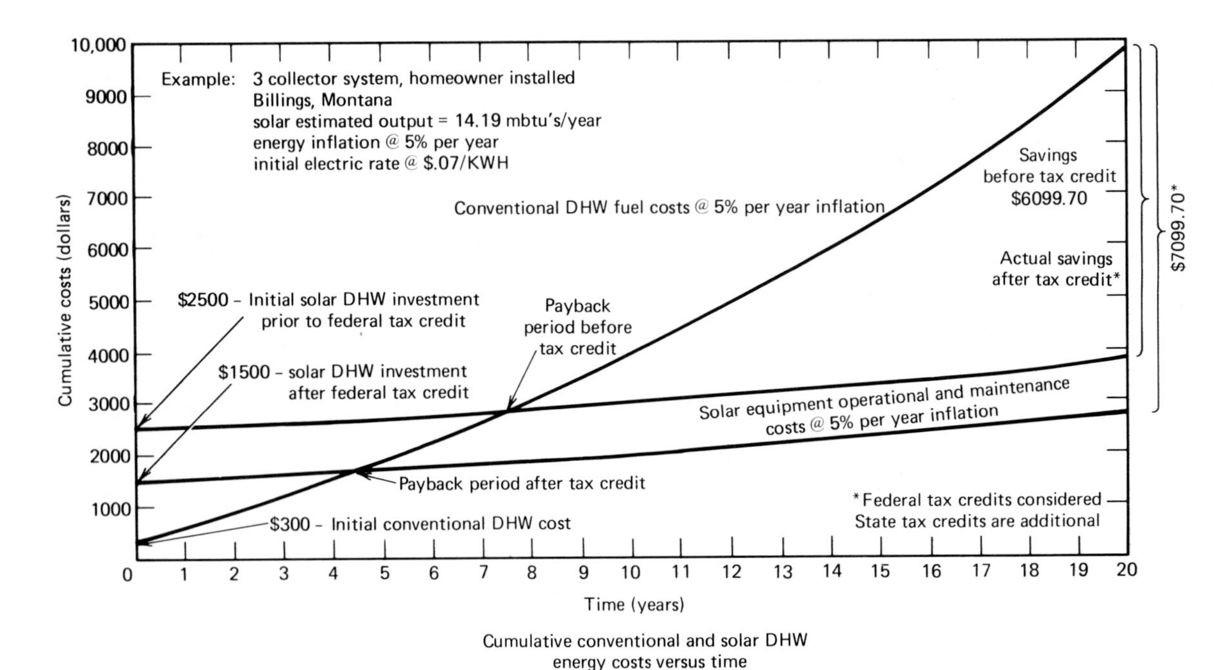


**Figure 10-4.** Solar domestic hot water energy savings over 20 years.

period at a 6 percent compounded interest rate, the resultant savings can be determined using Equation 10-2 as follows:

$$S_{20} = P(6\% - 20 \text{ SPCAF}) = 251.00\ (3.2071) = \$\ \ 804.98$$
$$S_{19} = P(6\% - 19 \text{ SPCAF}) = 263.55\ (3.0256) = \$\ \ 797.40$$
$$\vdots$$
$$S_{1} = P(6\% - 1 \text{ SPCAF}) = 634.27\ (1.0600) = \$\ \ 672.33$$
$$\text{Total Future worth savings } \$14735.60$$

At this point, one might ask if it would be better to simply leave $2500 in a savings account rather than invest the money in a solar DHW system. The following comparison should be observed:

***Bank Investment***

| | |
|---|---|
| Future worth of $2500 savings for 20 years @ 6 percent compounded interest | $8,017.75 |
| Less expense to heat domestic water | 9,623.18 |
| Less estimated equipment cost | 300.00 |
| Actual money lost from banking $2500 over 20 years | (−) $1,905.43 |

**Table 10-10**
**Typical Example of Electricity and Solar Fuel Costs[a] (dollars per BTU)**

| *Years* | *Solar Operation and Maintenance Costs (Table 10-9)* | *Electricity Cost to Heat Water (Table 10-9)* |
|---|---|---|
| 1 | $ 40.03 | $ 291.03 |
| 2 | 42.03 | 305.58 |
| 3 | 44.13 | 320.86 |
| 4 | 46.34 | 336.90 |
| 5 | 48.66 | 353.75 |
| 6 | 51.09 | 371.44 |
| 7 | 53.64 | 390.01 |
| 8 | 56.32 | 409.51 |
| 9 | 59.14 | 429.99 |
| 10 | 62.10 | 451.49 |
| 11 | 65.20 | 474.06 |
| 12 | 68.46 | 497.76 |
| 13 | 71.88 | 522.65 |
| 14 | 75.47 | 548.78 |
| 15 | 79.24 | 576.22 |
| 16 | 83.20 | 605.03 |
| 17 | 87.36 | 635.28 |
| 18 | 91.73 | 667.04 |
| 19 | 96.32 | 700.39 |
| 20 | 101.14 | 735.41 |
| Subtotal | $1323.48 | $9623.18 |
| Cost of Equipment[b] | 2500.00 | 300.00 |
| Total Cost | $3823.48 | $9923.18 |
| Less federal tax credit at 40 %[b] | −1000.00 | 0 |
| Less state tax credit (i.e., Montana) (Appendix C)[b] | −87.50 | 0 |
| 20 yr total costs | $2735.98 | $9923.18 |
| 20 yr total MBTU | 283.8 | 283.8 |
| Cost per MBTU | $9.64 | $34.97 |

[a]Assume: (1) Data from Table 10-9.
(2) Solar system is self-installed.

[b]Credit based on percentage of solar DHW system cost.

***Solar DHW Investment***

| | |
|---|---|
| Future worth of energy savings for 20 years @ 6 percent compounded interest | $14,735.60 |
| Less expense for solar operation and maintenance | 1,323.48 |
| Less expense for solar DHW equipment after federal tax credit | 1,500.00 |
| Actual money saved from banking energy savings over 20 years | (+) $11,912.12 |

The banking investment in comparison with the solar DHW investment clearly illustrates that the solar investment in heating water provides a positive cash savings. The solar investment also provides a better use of cash flow throughout the 20-year period. The solar DHW system cost is "self-liquidating." Once the system cost has been repaid, only electricity costs for pump operation and maintenance costs remain.

Finally, disregarding potential savings, we can relate a true cost comparison for all fuels on a cost per BTU basis as illustrated in Table 10-10.

The cost per BTU figures in Table 10-10 represents our typical example in Billings, Montana, as discussed throughout this Chapter.

Each solar DHW system represents an individual case, and an economical evaluation of each individual case indicates that solar energy is an excellent "first" application in using the sun's energy rather than using conventional fuels to heat domestic water.

## REVIEW QUESTIONS

1. A family of four uses 80 gal of 140° F water per day. Inlet water temperature averages 50° F per day. Neglecting equipment costs and assuming the cost of electricity is 8 cents per kwh and the cost of fuel oil is $1.60 per gallon, which of the two fuels would provide the most economical hot water? What is the comparison of these two energy sources in dollars per MBTU?

2. If a solar collector array is located in Portland, Maine, and its output has been determined to be 20 MBTU per year with a collector system efficiency ($n_s$) of 0.6, what is the estimated annual system output?

3. What is the yearly cost to electrically heat 100 gal of water from 40° to 135° F at 9 cents per kwh? At 12 cents per kwh? 105 gallons at 10 cents per kwh?

4. What would the actual installed cost of a solar DHW system be after federal taxes if the before-tax costs were $3200? If the system is installed in California, what is the final actual cost of the solar system?

5. If $1000 is invested at 12 percent compounded interest over 5 years, what is the future worth of the initial $1000?

6. If $3000 is borrowed at 10 percent compounded interest over 3 years, what are the monthly repayments? Tabulate this recovery of capital for each year.

7. What is the present worth of the money invested in review question 6?

8. If a 1/12 horsepower pump is needed to pump transfer fluid 8 hours per day from the collector to storage loop and requires 0.25 kilowatts, what is the annual power consumption costs at an electric rate of 8 cents per kwh?

9. Savings from a solar DHW system for a 5-year period were found to be $250, $280, $325, $410, and $440 for each year, respectively. If these yearly savings were invested at the end of each year at 8 percent compounded interest, what would be the future worth of the total savings for the 5-year period?

10. Determine the cost comparison between solar and conventional DHW heating as depicted in Table 10-9 with a compounded inflation rate at 12 percent per year. What is the total savings realized over 20 years by using solar energy?

APPENDIX A

# Construction of Useful Tools

## PLOTTING SOLAR SHADE

It was shown in Chapter 6, that a Mercator projection should be plotted to determine the potential shading problems for any particular location of a solar collector array. A surveyor's transit or abney level is excellent for plotting solar shade. If such instruments are not available, however, a homemade shade plotter can readily be substituted. The construction of such a simple device is shown below.

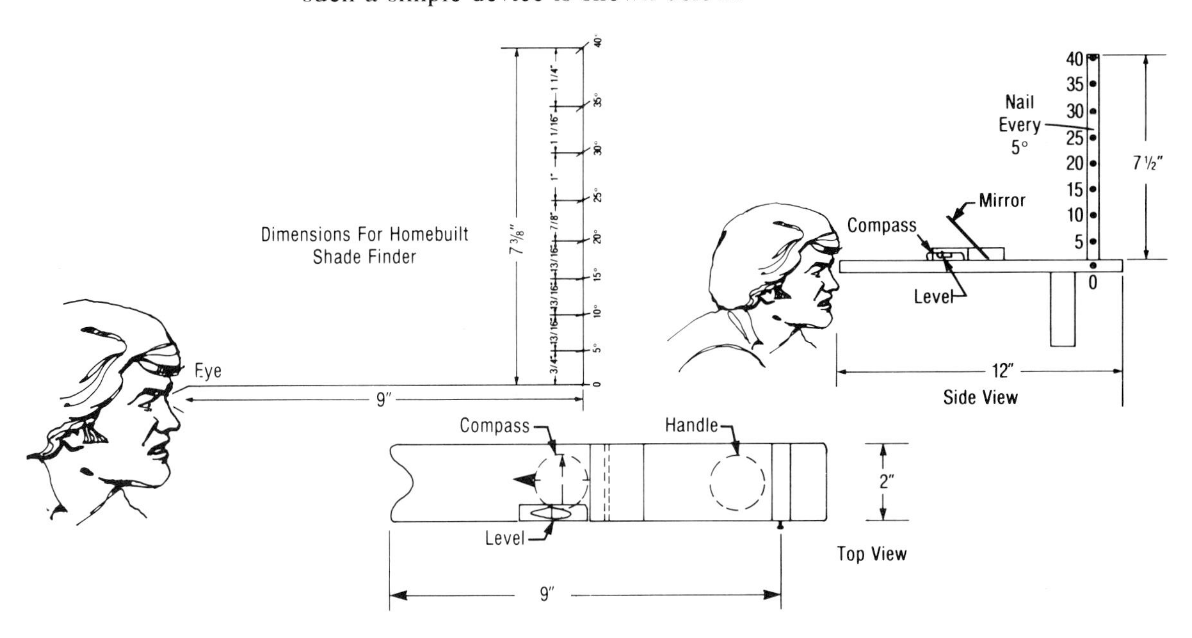


To use this instrument, hold it level with your eye at the bridge of your nose. Then sight along the zero-angle-indicator nail to the horizon. Without moving the instrument, glance up at the obstruction and note where its top falls on the vertical stick. Again, without moving the instrument, glance in the mirror and read the compass. The compass heading (corrected for local magnetic deviation) and the height of the obstruction can then be combined and plotted on the Mercator projection of Figure 6-9 in Chapter 6 for the applicable latitude. By moving across the horizon, all shading problems can be plotted quickly. If desired, a small line level can be added to this shade plotter. It should be mounted parallel with the horizontal horizon sighting stick. It will provide an artificial horizon.

## ROOF LADDER CONSTRUCTION

Chapter 7 stressed the importance of safety precautions when roof mounting a collector array. A simple roof ladder as depicted below can be constructed to facilitate installation.

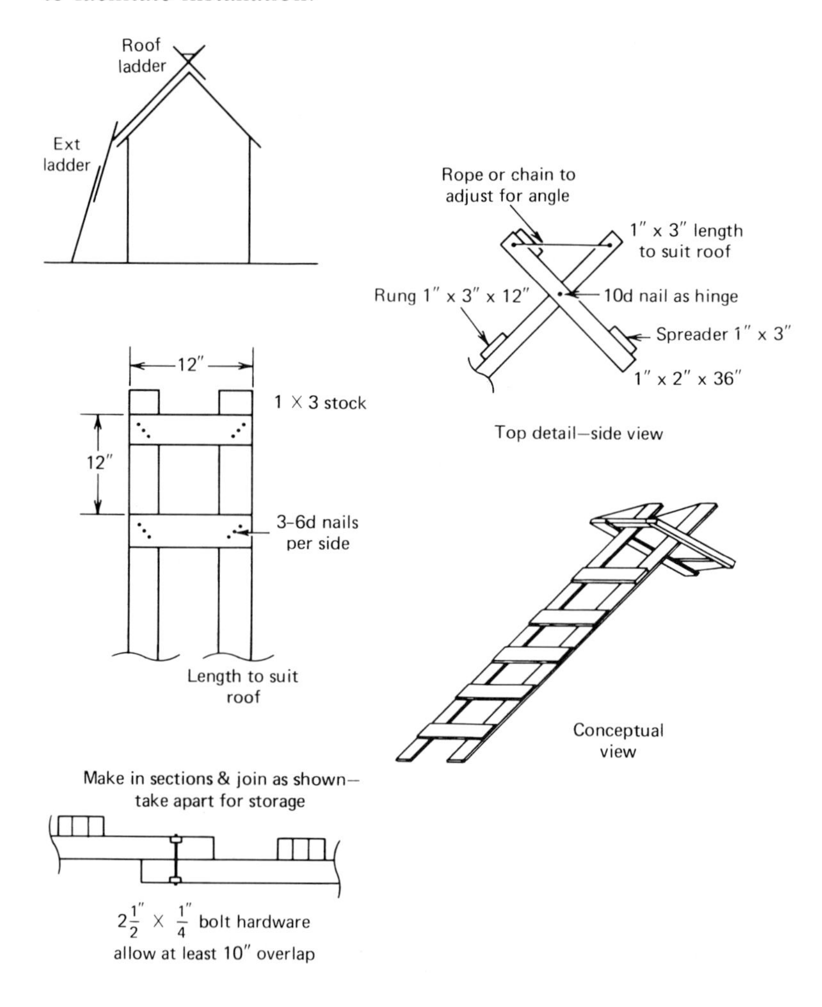

APPENDIX B

# Mathematical Tables, Relationships, and Conversions

## MATHEMATICAL TABLES

### Prefixes

| | | | | |
|---|---|---|---|---|
| atto | a | one-quintillionth | 0.000 000 000 000 000 001 | $10^{-18}$ |
| femto | f | one-quadrillionth | 0.000 000 000 000 001 | $10^{-15}$ |
| pico | p | one-trillionth | 0.000 000 000 001 | $10^{-12}$ |
| nano | n | one-billionth | 0.000 000 001 | $10^{-9}$ |
| micro | | one-millionth | 0.000 001 | $10^{-6}$ |
| milli | m | one-thousandth | 0.001 | $10^{-3}$ |
| centi | c | one-hundredth | 0.01 | $10^{-2}$ |
| deci | d | one-tenth | 0.1 | $10^{-1}$ |
| uni | | one | 1.0 | $10^{0}$ |
| deka | da | ten | 10.0 | $10^{1}$ |
| hecto | h | one hundred | 100.0 | $10^{2}$ |
| kilo | k | one thousand | 1000.0 | $10^{3}$ |
| mega | M | one million | 1 000 000.0 | $10^{6}$ |
| giga | G | one billion | 1 000 000 000.0 | $10^{9}$ |
| tera | T | one trillion | 1 000 000 000 000.0 | $10^{12}$ |

### Equivalent Measurements

**Length**

1 inch = 25.4 millimeter (mm)
1 inch = 2.54 centimeters (cm)
1 mm = 0.03937 inches
1 mm = 0.00328 feet
1 foot = 304.8 mm
1 meter = 3.281 feet
1 yard = 0.914 meter
1 kilometer (km) = 0.621 miles
1 mile = 1.609 km
1 micron = 0.000001 meter

**Area**

| | |
|---|---|
| 1 square inch (in.$^2$) | = 6.4516 square centimeters (cm$^2$) |
| 1 square foot (ft$^2$) | = 0.0929 square meters (m$^2$) |
| 1 square mile (mi$^2$) | = 2.590 square meters (m$^2$) |
| 1 square centimeter (cm$^2$) | = 0.155 square inches (in$^2$) |
| 1 square meter (m$^2$) | = 10.764 square feet (ft$^2$) |
| 1 square kilometer (km$^2$) | = 0.386 square miles (miles$^2$) |

**Volume**

| | |
|---|---|
| 1 gallon (U.S.) | = 3.785 liters |
| 1 liter | = 0.264 gallons (U.S.) |
| 1 cubic centimeter (cm$^3$) | = 0.061 cubic inches (in.$^3$) |
| 1 barrel of oil | = 42 gallons (U.S.) |
| 1 gallon of water at 60° F | = 8.33 pounds |

**Mass**

| | |
|---|---|
| 1 kilogram (kg) | = 1000 grams = 2.205 pounds |
| 1 metric ton = 1000kg | = 0.984 tons |
| 1 cubic inch of water (60° F) | = 0.07355 cubic inches of mercury (32° F) |
| 1 cubic inch of mercury (32° F) | = 0.4905 pounds |
| 1 cubic inch of mercury (32° F) | = 13.596 cubic inches of water (60° F) |

**Density**

1 lb/ft$^3$ = 16.0184 kg/m$^3$
1 kg/m$^3$ = 0.06243 lb/ft$^3$

**Pressure**

| | |
|---|---|
| 1 standard atmosphere | = 14.696 lb/in$^2$ |
| | = 760 mm mercury (Hg) |
| 1 bar | = 0.987 atmospheres |
| 1 pound per square inch (psi) | = 2.307 ft of water |
| | = 0.06805 atmospheres |

### Energy

| | |
|---|---|
| 1 BTU | = $1.055 \times 10^3$ joules (J) |
| | = 778.169 ft-lb |
| | = 252 calories |
| 1 Calorie = 4.187 J | = 0.003968 BTU |
| 1 Watt = 1 joule/sec | = 0.00134 horsepower |
| 1 Kilowatt-hour (kwh) | = $3.6 \times 10^6$ J |
| 1 Horsepower | = 550 ft-lb/sec |
| | = 2546.4 BTU/hr |
| 1 langley/hr = 11.63 watts/$m^2$ | = 3.687 BTU/$ft^2$ |

### Fuel to Energy

| | |
|---|---|
| 1 kwh | = 3,413 BTU |
| 1 gallon of oil | = 138,000 BTU |
| 1 Therm | = 100,000 BTU |
| | = 100 $ft^3$ natural gas |
| 1 gallon LP gas | = 93,000 BTU |
| 1 cord mixed hardwood | = 24 MBTU |
| 1 cord mixed softwood | = 15 MBTU |
| 1 pound of coal | = 12,500 BTU. |

### Physical Constants

| | |
|---|---|
| Acceleration of gravity (g) | = 32.174 ft/$sec^2$ |
| | = 980.665 cm/$sec^2$ |
| Pi ($\pi$) | = 3.1415926536 |
| Base of Natural Logarithms (e) | = 2.7182818285 |
| Absolute zero = $-273.15°$ centigrade | = $-459.67°$ Fahrenheit |
| Degrees Centigrade | = 5/9 (F $-32$) |
| Degrees Fahrenheit | = 9/5 C $+32$ |

## MATHEMATICAL RELATIONSHIPS

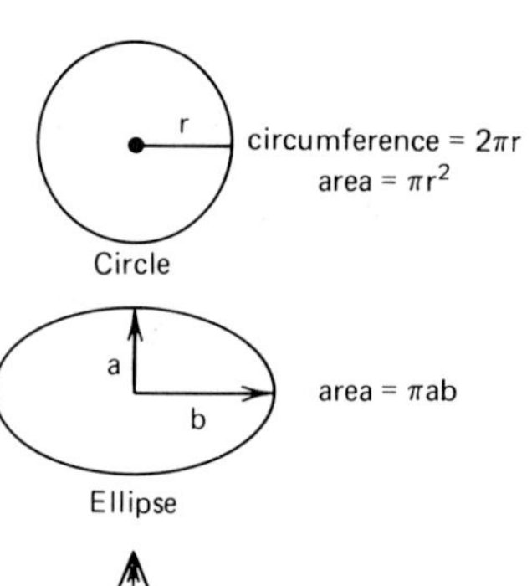


circumference = $2\pi r$
area = $\pi r^2$

area = $\pi ab$

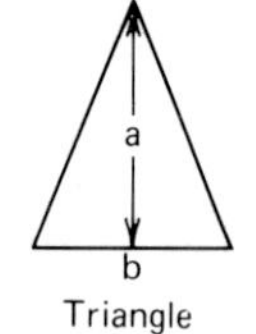


area = $\frac{1}{2}ab$

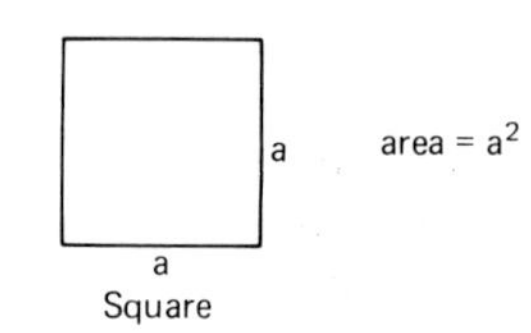


area = $a^2$

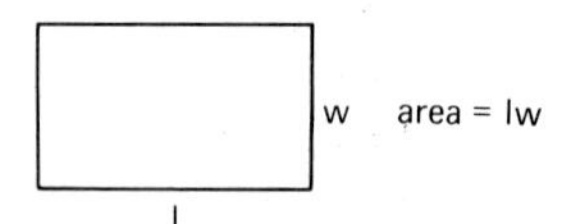


area = lw

Areas of common plane figures.

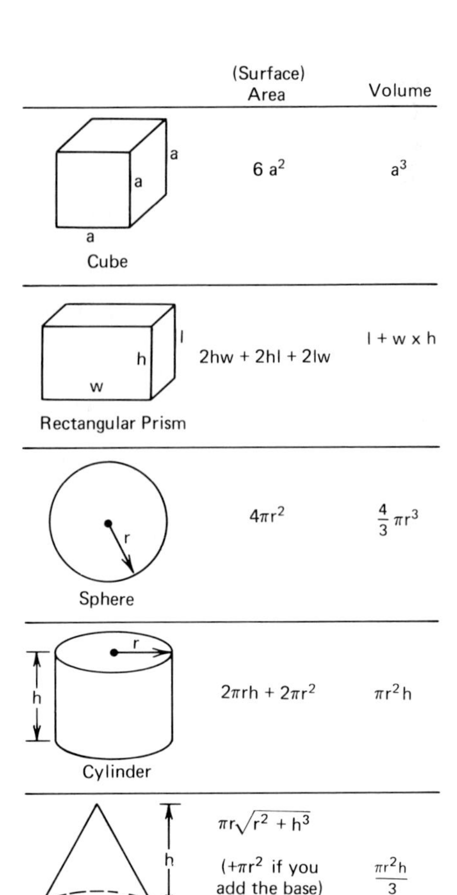

| | (Surface) Area | Volume |
|---|---|---|
| Cube | $6a^2$ | $a^3$ |
| Rectangular Prism | $2hw + 2hl + 2lw$ | $l + w \times h$ |
| Sphere | $4\pi r^2$ | $\frac{4}{3}\pi r^3$ |
| Cylinder | $2\pi rh + 2\pi r^2$ | $\pi r^2 h$ |
| Cone | $\pi r\sqrt{r^2 + h^3}$ (+$\pi r^2$ if you add the base) | $\frac{\pi r^2 h}{3}$ |

Areas and volumes of common shapes.

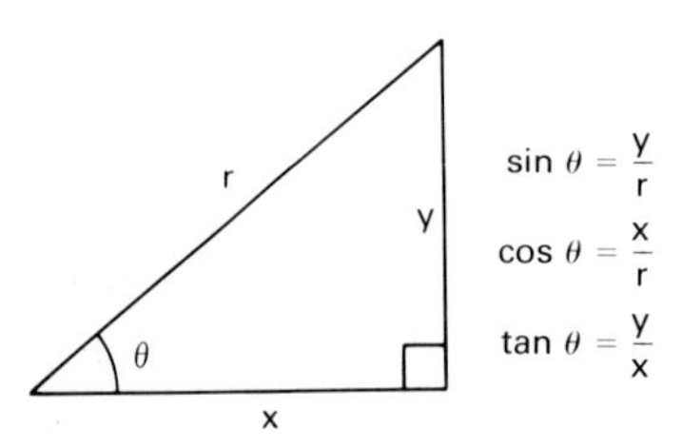


$\sin\theta = \frac{y}{r}$

$\cos\theta = \frac{x}{r}$

$\tan\theta = \frac{y}{x}$

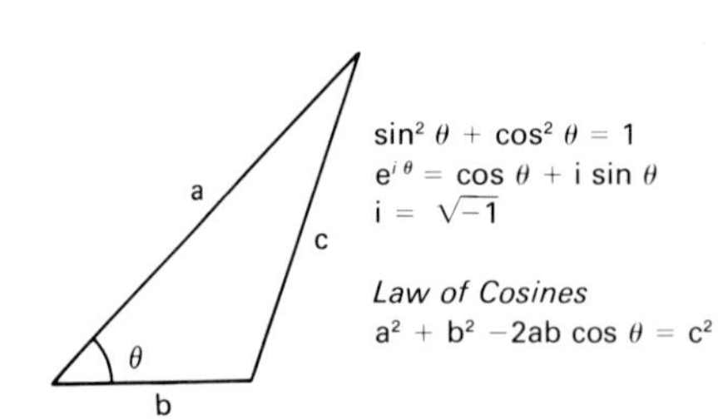


$\sin^2\theta + \cos^2\theta = 1$
$e^{i\theta} = \cos\theta + i\sin\theta$
$i = \sqrt{-1}$

*Law of Cosines*
$a^2 + b^2 - 2ab\cos\theta = c^2$

Trigonometric Relationships

*Law of Exponents*

$a^x \times a^x = a^{x+y}$  $\frac{1}{a^x} = a^{-x}$

$(ab)^x = a^x \times b^x$  $\frac{a^x}{a^y} = a^{x-y}$

$(a^x)^y = a^{xy}$  $a^0 = 1$

*Laws of Logarithms*

$\mathrm{Ln}(y^x) = x\mathrm{Ln}\, y$

$\mathrm{Ln}(ab) = \mathrm{Ln}\, a + \mathrm{Ln}\, b$

$\mathrm{Ln}\left(\frac{a}{b}\right) = \mathrm{Ln}\, a - \mathrm{Ln}\, b$

*Fundamental Identities and Reciprocal Relations (A represents any angle)*

$\sin A = \frac{1}{\csc A}$, $\cos A = \frac{1}{\sec A}$, $\tan A = \frac{1}{\cot A}$

$\csc A = \frac{1}{\sin A}$, $\sec A = \frac{1}{\cos A}$, $\cot A = \frac{1}{\tan A}$

## MATHEMATICAL CONVERSION (ALPHABETICAL ORDER)

| *Multiply* | *By* | *To Obtain* |
|---|---|---|
| Atmospheres | 76.0 | Centimeters-mercury |
| Atmosperes | 29.92 | Inches-mercury |
| Atmospheres | 14.70 | Pounds/in.$^2$ |
| Atmospheres | 1.058 | Tons/ ft$^2$ |
| Barrels-oil | 42 | Gallons-oil |
| British Thermal Units | 777.6 | Foot-pounds |
| British Thermal Units | $3.927 \times 10^{-4}$ | Horsepower-hours |
| British Thermal Units | $2.928 \times 10^{-4}$ | Killowatt-hours |
| British Thermal Units | 252 | Calories |
| BTU/ft$^2$ | 0.271 | Langleys |
| BTU/ft$^2$-hr | 3.154 | watts/m$^2$ |
| BTU/ft$^2$-hr | 0.075 | Calories/cm-sec |
| BTU/ft$^2$-hr | 0.00452 | Langleys/min |
| BTU/ft$^2$-hr-°F | 0.488 | Calories/hr-cm$^2$-°C |
| BTU/hr | 0.216 | Foot-pounds |
| BTU/hr | 0.00039 | Horsepower |
| BTU/hr | 0.29283 | Watts |
| Calories | 0.00397 | BTU |
| Calories/cm-sec | 13.272 | BTU/hr-ft$^2$ |
| Calories/hr-cm$^2$-°C | 2.048 | BTU/hr-ft$^2$-°F |
| Centimeters | 0.3937 | Inches |
| Centimeters | 0.01 | Meters |
| Centimeters | 10 | Millimeters |
| Centimeters-mercury | 0.0132 | Atmospheres |
| Centimeters-mercury | 0.4460 | Feet-water (4° C) |
| Centimeters-mercury | 136.0 | Kilograms/m$^2$ |
| Centimeters-mercury | 27.85 | Pounds/ft$^2$ |
| Centimeters-mercury | 0.1934 | Pounds/in.$^2$ |
| Centimeters/second | 0.0328 | Feet/second |
| Centimeters/second | 0.0224 | Miles/hour |
| Cubic centimeters | $3.531 \times 10^{-5}$ | Cubic feet |
| Cubic centimeters | 0.0610 | Cubic inches |
| Cubic centimeters | $1 \times 10^{-6}$ | Cubic meters |
| Cubic centimeters | $1.3079 \times 10^{-6}$ | Cubic yards |
| Cubic centimeters | $2.642 \times 10^{-4}$ | Gallons |
| Cubic centimeters | 0.0010 | Liters |
| Cubic feet | 1728 | Cubic inches |
| Cubic feet | 0.0283 | Cubic meters |
| Cubic feet | 7.4805 | Gallons |
| Cubic feet | 28.32 | Liters |
| Cubic feet/min | 0.1247 | Gallons/second |
| Cubic feet/min | 0.4719 | Liters/second |
| Cubic feet/second | 448.831 | Gallons/min |
| Cubic ft of water @ 60° F | 62.37 | Pounds |

| *Multiply* | *By* | *To Obtain* |
|---|---|---|
| Cubic inches | 16.39 | Cubic centimeters |
| Cubic inches | 0.0005787 | Cubic feet |
| Cubic inches | $1.6387 \times 10^{-5}$ | Cubic meters |
| Cubic inches | $2.1433 \times 10^{-5}$ | Cubic years |
| Cubic inches | 0.004329 | Gallons |
| Cubic inches | 0.0164 | Liters |
| Cubic meters | $1 \times 10^{6}$ | Cubic centimeters |
| Cubic meters | 35.31 | Cubic feet |
| Cubic meters | 61023 | Cubic inches |
| Cubic meters | 1.308 | Cubic yards |
| Cubic meters | 264.2 | Gallons |
| Cubic meters | 1000 | Liters |
| Cubic yards | 27 | Cubic feet |
| Cubic yards | 46.656 | Cubic inches |
| Cubic yards | 0.7645 | Cubic meters |
| Cubic yards | 202.0 | Gallons |
| Cubic yards | 764.5 | Liters |
| Degrees (angle) | 60 | Minutes |
| Degrees (angle) | 0.0174 | Radians |
| Degrees (angle) | 3600 | Seconds |
| Degree/second | 0.1667 | Revolutions/min |
| Degree/second | 0.0028 | Revolutions/second |
| Feet | 30.48 | Centimeters |
| Feet | 12 | Inches |
| Feet | 0.3048 | Meters |
| Feet | 0.3333 | Yards |
| Feet-water(4° C) | 0.8826 | Inches-mercury |
| Feet-water | 0.434 | Pounds/inch$^2$ |
| Feet-water | 62.43 | Pounds/ft$^2$ |
| Feet/min | 0.5080 | Centimeters/second |
| Feet/min | 0.0114 | Miles/hour |
| Feet/second | 30.48 | Centimeters/second |
| Feet/second | 0.6818 | Miles/hour |
| Foot-pounds | 0.0013 | British Thermal Units |
| Foot-pounds/min | $3.030 \times 10^{-5}$ | Horsepower |
| Foot-pounds/min | $2.2597 \times 10^{-5}$ | Kilowatts |
| Gallons | 3785 | Cubic centimeters |
| Gallons | 0.1337 | Cubic feet |
| Gallons | 231 | Cubic inches |
| Gallons | 0.0038 | Cubic meters |
| Gallons | 3.785 | Liters |
| Gallons, Imperial | 1.2009 | U.S. Gallons |
| Gallons, U.S. | 0.8327 | Imperial gallons |
| Gallons-(water @ 60° F) | 8.33 | Pounds-water |

| *Multiply* | *By* | *To Obtain* |
|---|---|---|
| Gallons/min | 0.00223 | Cubic feet/second |
| Grams | 980.7 | Dynes |
| Grams | 0.0022 | Pounds |
| Grams/$cm^3$ | 0.0361 | Pounds/$in.^3$ |
| Horsepower | 2544 | BTU/hour |
| Horsepower | 550 | Foot-pounds/second |
| Horsepower | 0.7457 | Kilowatts |
| Horsepower-hours | 0.7457 | Kilowatt-hours |
| Inches | 2.540 | Centimeters |
| Inches-mercury | 0.033 | Atmospheres |
| Inches-mercury | 0.491 | Pounds/$inch^2$ |
| Inches-mercury | 70.73 | Pounds/$ft^2$ |
| Inches-water | 0.0735 | Inches-mercury |
| Kilograms | 980665 | Dynes |
| Kilograms | 2.205 | Pounds |
| Kilometers | 3281 | Feet |
| Kilometers | 1000 | Meters |
| Kilometers | 0.6214 | Miles |
| Kilometers | 1094 | Yards |
| Kilometers/hour | 54.68 | Feet/min |
| Kilometers/hour | 0.5396 | Knots |
| Kilowatts | 56.8 | BTU/min |
| Kilowatts | 737.6 | Foot-pounds/second |
| Kilowatts | 1.341 | Horsepower |
| Kilowatt-hours | 3413 | BTU's |
| Kilowatt-hours | $2.655 \times 10^6$ | Foot-pounds |
| Kilowatt-hours | 1.341 | Horsepower-hours |
| Langleys | 3.687 | BTU/$ft^2$ |
| Langleys/min | 221.2 | BTU/$ft^2$-hr |
| Liters | 0.0353 | Cubic feet |
| Liters | 61.02 | Cubic inches |
| Liters | 0.0010 | Cubic meters |
| Liters | 0.2642 | Gallons |
| Meters | 3.281 | Feet |
| Meters | 39.37 | Inches |
| Meters | 0.001 | Kilometers |
| Meters | 1.094 | Yards |
| Meters/min | 0.06 | Kilometers/hr |
| Meters/min | 0.0373 | Miles/hr |
| Meters/second | 196.8 | Feet/min |
| Meters/second | 3.281 | Feet/second |
| Meters/second | 0.03728 | Miles/min |
| Microns | $1 \times 10^6$ | Meters |
| Miles | 5280 | Feet |

| *Multiply* | *By* | *To Obtain* |
|---|---|---|
| Miles | 1.609 | Kilometers |
| Miles/hr | 88 | Feet/min |
| Miles/hr | 1.467 | Feet/second |
| Miles/hr | 1.609 | Kilometers/hr |
| Miles/hr | 0.8690 | Knots |
| Miles/min | 2682 | Centimeters/second |
| Miles/min | 88 | Feet/second |
| Miles/min | 1.609 | Kilometers/min |
| Milligrams | 0.001 | Grams |
| Milliliters | 0.001 | Liters |
| Millimeters | 0.1 | Centimeters |
| Millimeters | 0.0394 | Inches |
| Ounces | 0.0625 | Pounds |
| Pounds | 16 | Ounces |
| Pounds | 454 | Grams |
| Pounds-water | 0.1198 | Gallons |
| Pounds/$in^3$ | 1728 | Pounds/$ft.^3$ |
| Pounds/ft. | 1488 | Kilograms/m |
| Pounds/in. | 178.6 | Grams/cm |
| Pounds/$ft.^2$ | 4.882 | Kilograms/$m^2$ |
| Pounds/$in.^2$ | 0.0680 | Atmospheres |
| Pounds/$in.^2$ | 2.036 | Inches-mercury |
| Quadrants (angle) | 1.571 | Radians |
| Quarts (liq.) | 57.75 | Cubic inches |
| Radians | 57.30 | Degrees |
| Radians | 3438 | Minutes |
| Radians | 0.637 | Quadrants |
| Radians/second | 9.549 | Revolutions/min |
| Revolutions/second | 360 | Degrees/second |
| Revolutions/second | 6.283 | Radians/second |
| Revolutions/second | 60 | Revolutions/min |
| Seconds (angle) | $4.8481 \times 10^{-6}$ | Radians |
| Square centimeters | 0.0011 | Square feet |
| Square centimeters | 0.1550 | Square inches |
| Square centimeters | 0.0001 | Square meters |
| Square centimeters | 100 | Square millimeters |
| Square feet | 929.0 | Square centimeters |
| Square feet | 144 | Square inches |
| Square feet | 0.0929 | Square meters |
| Square feet | $3.5870 \times 10^{-8}$ | Square miles |
| Square feet | 0.1111 | Square yards |
| Square inches | 6.452 | Square centimeters |
| Square inches | 0.0069 | Square feet |
| Square kilometers | $1.0764 \times 10^{7}$ | Square feet |

| *Multiply* | *By* | *To Obtain* |
|---|---|---|
| Square kilometers | $1 \times 10^6$ | Square meters |
| Square kilometers | 0.3861 | Square miles |
| Square kilometers | $1.1960 \times 10^6$ | Square yards |
| Square meters | 10.76 | Square feet |
| Square meters | 1.1960 | Square yards |
| Square miles | 2.590 | Square kilometers |
| Square miles | $3.0976 \times 10^6$ | Square yards |
| Square millimeters | 0.01 | Square centimeters |
| Square millimeters | 0.0016 | Square inches |
| Square yards | 9 | Square feet |
| Square yards | 0.8361 | Square meters |
| Square yards | $3.2283 \times 10^{-7}$ | Square miles |
| Tons (metric) | 2205 | Pounds |
| Tons (short) | 2000 | Pounds |
| Tons of Air Conditioning | 12,000 | BTU/hr |
| Watts | 0.7377 | Foot-pounds/second |
| Watts | 0.0013 | Horsepower |
| Watts | 0.001 | Kilowatts |
| Watt-hours | 3.413 | British Thermal Units |
| Watt-hours | 2655 | Foot-pounds |
| Watt-hours | 0.00134 | Horsepower-hours |
| Watt-hours | 0.001 | Kilowatt-hours |
| Watts/$m^2$ | 0.317 | BTU/$ft^2$-hr |
| Yards | 91.44 | Centimeters |
| Yards | 3 | Feet |
| Yards | 36 | Inches |
| Yards | 0.9144 | Meters |

APPENDIX C

# State Tax Credit Incentives

## ALABAMA
### Tax Incentives

This law provides an individual income tax deduction for active and passive solar energy systems up to a maximum of $1000 at the following rates:

| *Year* | *Active* | *Passive* |
|---|---|---|
| 1981 | 25% of expenditures | 12½% of expenditures |
| 1982 | 20% of expenditures | 10% of expenditures |
| 1983 | 15% of expenditures | 7½% of expenditures |
| 1984 | 10% of expenditures | 5% of expenditures |
| 1985 | 5% of expenditures | 2½% of expenditures |

Systems must be approved by the Alabama Solar Energy Center and the Department of Revenue.

## ALASKA
### Tax incentives

Alaska allows a 10% residential fuel conservation credit of up to $200 per individual or married couple for money spent on the following; (1) insulation, (2) insulating windows; (3) labor related to items 1 and 2; (4) alternate energy systems which are not dependent on fossil fuel, including solar, wind, tidal, and geothermal. Expires 12/31/82 (Chapter 94, Laws of 1977).

This law allows a business tax credit of 35% of the cost of an alternative energy system (not dependent on oil or gas) or an energy conservation improvement. The maximum credit is $5,000. An alternative energy system must provide at least 10% of the average thermal, electrical or mechanical energy or 30% of the hot water needs of the building (Chapter 83, Laws of 1980).

Contact State Department of Revenue
Income Tax Division
Pouch SA
State Office Building
Juneau, AK 99811
(907) 465-2326

## ARIZONA

### Tax Incentives

Solar energy devices are exempt from property taxes through 12/31/89 (Chapter 146, Laws of 1979).

Solar energy devices are exempt from Transaction Privilege and Use Taxes through 12/31/89 (Chapter 146, Laws of 1979).

Individuals may claim an income tax credit equal to 35% of the cost of a solar energy device through 1983. Thereafter, the percentage decreases 5% per year until the credit expires on 12/31/89. The maximum credit is $1,000. In lieu of claiming a credit, taxpayers may amortize qualifying devices over a period of 36 months. Home builders may claim a credit on new speculative homes in lieu of the purchaser.

Commercial and industrial solar installations qualify for the same tax credits and accelerated amortization privileges as residential applications. In this case, amortization is in lieu of ordinary depreciation.

Solar devices qualify for all tax incentives if they furnish heating or cooling, electrical or mechanical power or interior daylighting. Both passive and active systems are included. All systems must be warranted for at least one year and must meet standards of the Solar Energy Commission (Chapter 39, Laws of 1980).

A residential income tax credit equal to 25% of the cost of insulation and other energy conserving devices is available through 1989. Maximum credit is $100 (Chapter 146, Laws of 1979).

Contact State Department of Revenue
Box 29002
Phoenix, AZ 85038
(602) 255-3381

## ARKANSAS

### Tax Incentives

An energy conservation income tax deduction is available to any individual, fiduciary or corporation. Eligible items include insulation and energy conservation adjustments made to buildings constructed before 1/1/79. Also eligible are these additions made to new or existing buildings: devices using solar, bioconversion, geothermal, or hydroelectric energy (for mechanical or electrical power) for space heating or cooling or water heating; devices using wind energy; and devices creating energy from woodburning in stoves, furnaces, or fireplaces with controllable drafts and dampers. Deductions exceeding income may be carried forward until exhausted. Eligible expenditures must be made by 12/31/84. The Commissioner of Revenue may promulgate regulations for the administration of the deduction (Act 535, 1977; Act 742, 1979; Act 59, 1980).

Contact State Department of Revenue
Income Tax Section
7th and Wolfe Streets
Little Rock, AR 72201
(501) 371-2193

## CALIFORNIA

### Tax Incentives

California provides personal income tax credit of 55% of the cost of a solar energy system, up to a maximum of $3000. If a system is installed in other than a single-family dwelling and the cost exceeds $6000, the credit equals 25% of the cost, or $3000, whichever is greater. In both single-family dwellings and other buildings, the cost of energy-conserving devices installed in conjunction with the solar energy system may also be included in the total cost used to calculate the tax credit. If a federal credit is claimed, the state credit is reduced by the amount of the federal credit. The same provisions apply to corporate taxpayers. Taxpayers who partially own and partially lease a solar system for a public utility are also eligible for the credit. The system must meet the criteria of the California Energy Commission. Eligible expenses for credit include attorney's fees, compensation, and recording fees associated with obtaining a solar easement. Credit expires 12/31/80 (Chapter 168, Laws of 1976; Chapter 1082, Laws of 1977; Chapter 1154, Laws of 1978; Chapter 816, Laws of 1979).

Contact Franchise Tax Board
Attn: Correspondence
Sacramento, CA 95807
(916) 355-0370

## COLORADO

### Tax Incentives

Alternative energy devices will be assessed for real estate tax at 5% of their value. Eligible devices must use solar or geothermal energy, renewable biomass, or wind resources. Passive solar designs are included, but devices for the direct combustion of wood are ineligible. The exemption expires 12/31/89 (Chapter 344, Laws of 1975; Chapter 363, Laws of 1979).

Contact Local tax assessor

Creates an income tax credit equal to 20% of the cost of energy conserving measures eligible for the federal residential energy conservation tax credit. Among eligible items are storm windows and doors, insulation and clock thermostats. Maximum credit is $400.

All types of taxpayers may claim an income tax credit for installing renewable energy systems. Individual taxpayers may claim a credit equal to 30% of the cost of solar, wind and geothermal equipment that qualifies for the federal residential credit. Maximum credit is $3,000 and excess credit may be carried forward. The credit expires 12/31/85. Business taxpayers may claim a 30% credit on solar and wind energy equipment through the 1986 tax year (HB 1264, 1980).

Contact Colorado Department of Revenue
Capitol Annex Building
1375 Sherman Street
Denver, CO 80203
(303) 839-5600

## CONNECTICUT

### Tax Incentives

Municipalities are authorized to exempt windmills, waterwheels and solar heating, cooling and electrical systems from real estate tax. Installation must take place between 4/20/77 and 10/1/91. The exemption will be effective for 15 years after installation. Systems must meet standards of the Office of Policy and Management (Public Acts 76-409, 77-490, 79-479, 80-406).

Corporations whose gross annual revenues do not exceed $100,000,000 and who are engaged in the design, manufacture, sale or installation of alternative energy (solar, wind, water, biomass) systems are exempt from corporation business tax through 7/1/85 (Public Act 80-406).

A sales tax exemption is created for alternative energy (solar, wind, water, biomass) systems, including materials and installation (Public Act 77-457, 80-406).

Contact Commissioner of Revenue Service
92 Farmington Avenue
Hartford, CT 06115
(203) 566-7120

## DELAWARE

### Tax Incentive

This law provides an income tax credit of $200 for solar energy devices designed to produce domestic hot water. Systems must meet HUD Intermediate Minimum Property Standards Supplement for Solar Heating and Domestic Hot Water Systems. Systems must be warranted according to criteria set out in the law (Chapter 512, Laws of 1978).

Contact Division of Revenue
State Office Building
820 French Street
Wilmington, DE 19801
(302) 571-3360

## FLORIDA

### Tax Incentives

Effective 8/5/79, solar energy systems are exempt from sales and use tax until 6/30/84 (Chapter 339, Laws of 1979).

Contact Florida Department of Revenue
Sales Tax Division
Carlton Building
Tallahassee, FL 32304
(940) 488-6800

A real estate tax exemption applies to solar, wind, and geothermal energy systems installed between 1/1/80 and 12/31/90. Heat pumps and waste heat recovery systems installed in buildings constructed before 1/1/85 are also covered (Chapter 183, Laws of 1980).

**GEORGIA**

**Tax Incentives**

Real estate owners may claim a refund of sales tax paid for the purchase of solar equipment. Expires 7/1/86 (Act 1030, 1976; Act 1309, 1978).

Contact State Department of Revenue, Sales Tax Division
309 Trinity-Washington Building
Atlanta, GA 30334
(404) 656-4065

Any county or municipality may exempt solar heating and cooling equipment and machinery used to manufacture solar equipment from property taxes. Expires 7/1/86 (Georgia Constitution, Article VII, Section 1, Paragraph IV).

Contact Local city council or county board of supervisors

**HAWAII**

**Tax Incentives**

A 10% income tax credit is provided to individuals and corporations who purchase solar energy devices that are placed in service by 12/31/81. The law also provides property tax exemptions for solar energy systems through 12/31/81. This exemption also applies to any non-nuclear and non-fossil fuel system and to any improvement that increases the efficiency of systems which use fossil fuel (Act 189, 1976).

Corporate and individual taxpayers may claim an income tax credit for the cost of installing insulating material on a hot water heating tank and exposed hot water pipes. The maximum credit is $30. Expires 12/31/84 (Act 19, 1978).

Contact State Tax Department
P. O. Box 259
Honolulu, HI 96809
(808) 548-3270

**IDAHO**

**Tax Incentives**

This law allows an income tax deduction for a solar heating/cooling or solar electrical system installed in the taxpayer's residence. The deduction equals 40% of the cost in the first year and 20% of the cost in each of the next 3 years; the maximum deduction in any year is $5000. This deduction also applies to systems fueled by wind, geothermal energy, wood, or wood products. Built-in fireplaces qualify if they have control doors, regulated draft, and heat exchangers that deliver heated air to substantial portions of the residence (Chapter 212, Laws of 1976).

Contact Idaho State Tax Commission
P.O. Box 36
Boise, ID 83722
(208) 334-3560

## ILLINOIS

### Tax Incentives

A property owner who installs a solar or wind energy system may claim an alternative valuation for property taxes. The property is assessed twice: with the solar or wind energy system and also as though it were equipped with a conventional system. The lesser of the two assessments is used to compute the tax due. Owners must file a claim with the local Board of Assessors. (Public Act 79-943, 1975; Public Act 80-430, 1977; Public Act 80-1218, 1978).

## INDIANA

### Tax Incentives

The law permits the property owner who installs a solar heating and cooling system to have property assessment reduced by the difference between the assessment of the property with the system and the assessment of the property without the system. The owner must apply to the county auditor (Public Law 15, 1974; Public Law 68, 1977).

Contact Local assessor or board of assessors

Public Law 20, 1980 provides an income tax credit for individuals, corporations, and partnerships using solar or wind to heat space, water, or for generating electricity. It applies to all types of buildings at the following rates:

| *Residential Buildings* | *Nonresidential Buildings* |
|---|---|
| 25% of eligible expenditures to a maximum credit of $3000 | 25% of eligible expenditures to a maximum credit of $10,000 |

All systems must meet standards promulgated by the State Department of Commerce. Taxpayers must obtain certification before obtaining credit.

Contact State Department of Revenue
(317) 232-2101

## IOWA

### Tax Incentives

Installation of a solar energy system will not increase the assessed, actual, or taxable values of property for 1979-1985 (Section 441.21, Code of 1979).

Contact Local assessor or board of assessors

## KANSAS

### Tax Incentives

Individual taxpayers may claim an income tax credit equal to 30% of the cost of a residential solar (active or passive) or wind system to a maximum of $1500. Excess credit may be carried forward. Expires 7/1/83.

A solar or wind installation on business or investment property is eligible for a 30% credit to a maximum of $4500 of that year's tax bill, whichever is less. The cost of an installation on business or investment property may also be amortized over 60 months. Expires 7/1/83 (Chapter 434. Laws of 1976; Chapter 346, Laws of 1977; Chapter 409, Laws of 1978; Chapter 310, Laws of 1980).

If a solar system supplies 70% of the energy for heating and cooling, the property owner may be reimbursed 35% of his property tax for up to 5 consecutive years. Installation must be completed before 12/31/80 (Chapter 345, Laws of 1977; Chapter 419, Laws of 1978; Chapter 310, Laws of 1980).

Solar energy systems are exempt from real estate tax through 12/31/85 (Chapter 310, Laws of 1980).

Provides an income tax deduction of 50% (maximum $500) of the cost of insulating residential buildings owned by the taxpayer. The deduction can be applied separately to each residential building owned and insulated by the taxpayer. To qualify, a building must have been constructed before 7/1/77. The insulating materials must meet minimum standards for energy conservation in new buildings prescribed by the Federal Housing Administration (Chapter 410, Laws of 1978).

Contact State Department of Revenue
P.O. Box 692
Topeka, KS 66601
(913) 296-3909

## LOUISIANA

### Tax Incentives

Solar energy equipment installed in owner-occupied residential buildings or in swimming pools are exempt from property tax (Act 591, 1978).

Contact Local parish tax assessor

## MAINE

### Tax Incentives

Solar space or water heating systems are exempt from property tax for 5 years after installation. Eligible taxpayers must apply to the local Board of Assessors. Purchasers of solar energy systems may also receive a sales tax rebate from the Office of Energy Resources (Chapter 542, Laws of 1977).

Contact (for property tax) Local assessor or board of assessors.
Contact (for sales tax) Office of Energy Resources
55 Capitol Street
Augusta, ME 04330
(207) 289-3811

This law creates an income tax credit for solar, wind, and wood energy systems which provide space or water heating or electrical or mechanical power. Fireplaces and woodstoves not operating as central heating systems are ineligible. Both active and passive solar systems qualify. The credit equals the lessor of $100 or 20% of eligible expenditures. Retroactive to 1/1/79 (Chapter 557, Laws of 1979).

Contact Bureau of Taxation
Department of Finance and Administration
State Office Building
Augusta, ME 04333
(207) 289-2076

## MARYLAND

### Tax Incentives

A solar energy unit will be assessed at no more than a conventional system needed to serve the building (Chapter 509, Laws of 1975; Chapter 509, Laws of 1978).

Contact Local assessor or board of assessors

Baltimore city and any other city or county may offer property tax credits for the installation of solar space heating or cooling or water heating systems. Credits may be applied over a three year period (Chapter 740, Laws of 1976; Chapter 286, Laws of 1980).

Contact Local city or county department of revenue

## MASSACHUSETTS

### Tax Incentives

Solar energy systems are exempt from tax for 20 years from the date of installation (Chapter 734, Laws of 1975; Chapter 388, Laws of 1978).

Contact Local assessor or board of assessors

A personal income tax credit of 35% of the cost of renewable energy equipment is created. The equipment must be installed in the taxpayer's principal residence in the state. Maximum credit is $1000. Eligible equipment must use solar energy for space heating or cooling or water heating or must use wind energy for any nonbusiness residential purpose. If a federal income tax credit or grant is received by the taxpayer, the state credit will be reduced. The credit expires 12/31/83.

Sales of equipment for residential solar energy systems, wind power systems, or heat pumps are exempt from sales tax. Wood-fueled central heating systems installed in a person's principal residence in the state and costing more than $900 are exempt from sales tax through 12/31/83. Eligible furnaces must be approved by the State Fire Marshall or Building Code Commissioner (Chapter 796, Laws of 1979).

Corporations may deduct the cost of a solar or wind energy system from income. The system will also be exempt from tangible property tax (Chapter 487, Laws of 1977).

Contact State Department of Corporations & Taxation
100 Cambridge Street
Boston, MA 02204
(617) 727-4201 (income tax)
(617) 727-4601 (sales tax)

## MICHIGAN

### Tax Incentives

This law exempts solar, wind, or water energy conversion devices from real and personal property tax. An application must be filed with local tax assessor, who will submit it to the state tax commission for certification. Authority to exempt expires 7/1/85, but exemptions made by that time stay in force (Public Act 135, 1976).

Contact Local Government Services
Treasury Building
Lansing, MI 48922
(517) 373-3232

Proceeds from sales of solar, wind, or water energy conversion devices used for heating, cooling, or electrical generation in new or existing residential or commercial buildings are excluded from business activities tax. Expires 1/1/85 (Public Act 132, 1976).

Tangible property used for solar, wind, or water energy devices is excluded from excise tax if it is used to heat, cool, or electrify a new or existing commercial or residential building. Expires 1/1/85 (Public Act 133, 1976).

Income tax credit may be claimed for a residential solar, wind, or water energy device that is used for heating, cooling or electricity. This includes devices designed to use the difference between water temperatures in a body of water. Energy conservation measures installed in connection with such devices are also eligible; these include insulation, water-flow reduction devices, and some wood furnaces. Swimming pool heaters are eligible only if 25% or more of their heating capacity is used for residential purposes. The credit is refundable. The law instructs the Department of Commerce to establish system eligibility standards within 180 days of the law's passage. To be eligible, expenditures must be made by 12/31/83. The rate of credit changes annually. For 1980, the rate for single-family dwellings is 25% of the first $2000 spent, plus 15% of the next $8000 spent. In 1980, the rate for other buildings is 25% of the first $2000, plus 15% of next $13,000 (Public Act 605, 1978; Public Act 41, 1979).

Contact State Department of Treasury
State Tax Commission
State Capitol Building
Lansing, MI 48922
(517) 373-2910

## MINNESOTA

### Tax Incentives

The market value of solar, wind, or agriculturally derived methane gas systems used for heating, cooling, or electricity in a building or structure is excluded from property tax. The installations must be done prior to 1/1/84 (Chapter 786, Laws of 1978).

Contact Local assessor or board of assessors

This law provides an individual income tax credit of 20% of the first $10,000 spent on renewable energy source equipment installed on a Minnesota building of six dwelling units or less. Eligible expenditures include: those eligible as federal renewable energy source property (solar, wind, and geothermal); earth-sheltered dwellings; equipment producing ethanol, methanol or methane for fuel, but not for resale; passive solar energy systems. Excess credit can be carried forward through 1984. Federal regulations of the U.S. Internal Revenue Service shall be used to administer relevant portions of the credit. Expenditures must be made between 1/1/79 and 12/31/82 (Chapter 303, Laws of 1979).

Contact Department of Revenue<br>Centennial Office Building<br>658 Cedar Street<br>St. Paul, MN 55145<br>(612) 296-3781

## MISSISSIPPI

### Tax Incentives

Labor, property or services used in the construction of solar energy heating, lighting, or electric generating facilities used by universities, colleges, or junior colleges are exempted from sales tax. Expires 1/1/83 (§27-65-105 of the Mississippi Code).

Contact Mississippi Tax Commission<br>Sales Tax Division<br>P. O. Box 960<br>Jackson, MS 39205<br>(601) 354-6274

## MONTANA

### Tax Incentives

Energy systems using non-fossil fuel energy (such as solar, wind, solid wastes, decomposition of organic wastes, solid wood wastes, and small scale hydroelectric) installed in an income taxpayer's dwelling before 12/31/82 are eligible for an income tax credit of 5% of the first $1000 and 2½% of the next $3000 (Chapter 548, Laws of 1975; Chapter 574, Laws of 1977; Chapter 652, Laws of 1979).

This law provides individual or corporate income tax deductions for energy conservation improvements, including storm windows and insulation. It applies to all types of buildings at the following rates:

| Residential buildings | Non-residential buildings |
|---|---|
| 100% of 1st $1000 | 100% of 1st $2000 |
| 50% of 2nd $1000 | 50% of 2nd $2000 |
| 20% of 3rd $1000 | 20% of 3rd $2000 |
| 10% of 4th $1000 | 10% of 4th $2000 |

(Chapter 576, Laws of 1977).

Contact State Department of Revenue
Income Tax Section
Sam Mitchell Building
Helena, MT 59601
(406) 449-2837

This law provides a 10-year real estate tax exemption for capital investments in non-fossil forms of energy generation as defined in the state income tax law (Montana Code Annotated 15-32-102). The maximum exemptions are $20,000 for a single-family dwelling and $10,000 for other buildings (Chapter 639, Laws of 1979).

Contact State Department of Revenue
Property Assessment Division
Sam Mitchell Building
Helena, MT 59601
(406) 449-2808

## NEBRASKA

### Tax Incentives

This bill amends the state constitution to allow the legislature to exempt energy conservation improvement from real estate tax.

A sales tax rebate is available to purchasers of alternative energy systems, including active and passive solar, wind, biomass and geothermal systems. Expires 12/31/83 (Legislative Bill 954, 1980).

Contact Department of Revenue
301 Centennial Mall, S.
Lincoln, NE 68509
(402) 471-2971

An income tax credit is available for renewable energy resources for individuals or corporations of 20% of the cost to a maximum of $3000 and for agricultural, commercial or industrial purposes of 20% of the cost to a maximum of $4000. A one year carry-forward provision is included for active, passive, biomass, and wind energy. Solar, wind, and geothermal systems are exempted from property taxes for 5 years following installation. Systems must be installed by December 1, 1985. State Energy Office shall promulgate rules. The tax commission shall issue rules and guidance for the county assessors.

## NEVADA

### Tax Incentives

This law establishes a property tax allowance on solar, wind, geothermal, water-powered, or solid waste energy systems in residential buildings. The property tax allowance equals the difference in tax on the property with the energy system and the tax on the property without the energy system. The allowance may not exceed the tax accrued or $2000, whichever is less. Claims are to be filled with the county assessor (Chapter 345, Laws of 1977).

Contact Local county assessor

## NEW HAMPSHIRE

### Tax Incentives

Cities and towns are enabled to grant property tax exemptions to property owners with solar heating, cooling, or hot water systems and will decide the amount of the exemption and the manner of determination. An application for the exemption must be filled with the local assessor (Chapter 391, Laws of 1975; Chapter 5202, Laws of 1977).

Contact Local assessor or board of assessors

## NEW JERSEY

### Tax Incentives

Solar heating and cooling systems, including those which derive their energy from indirect sources, such as wind and sea thermal gradients, are exempt from increased property tax assessment. Local Construction Code Official must certify that the system meets the State Department of Energy's standards (Chapter 256, Laws of 1977).

Solar energy devices designed to provide heating, cooling, electrical or mechanical power are exempt from sales tax. These systems must meet the standards established by the State Department of Energy (Chapter 465, Law of 1977).

Contact Application Guidelines for Property and Sales Tax Exemptions for Solar Systems in New Jersey
New Jersey Department of Energy
Office of Alternative Technology
101 Commerce Street
Newark, NJ 07102
(201) 648-6293

## NEW MEXICO

### Tax Incentives

This law provides for an income tax credit of 25% of the cost of a solar energy system or a maximum of $4000. It is available for solar energy systems which heat or cool the taxpayer's residence and for swimming pool heating systems. The criteria of the Solar Heating and Cooling Demonstration Act of 1974 (42 USC 5506) must be met. Credit in excess of the taxes due will be refunded (Chapter 12, Laws of 1975; Chapter 170, Laws of 1978; Chapter 353, Laws of 1979).

Individuals may claim an income tax credit for a solar energy system used in an irrigation pumping system. The system design must be approved by the Energy Resources Board prior to installation, and it must result in a 75% reduction in the use of fossil fuel. This law is not applicable if federal credit were claimed or if credit were claimed for this equipment under other provisions of the state law. Credit in excess of the taxes due will be refunded (Chapter 114, Laws of 1977).

Contact State Department of Taxation & Revenue
Income Tax Division
P.O. Box 630
Santa Fe, NM 87503
(505) 827-3221

## NEW YORK

### Tax Incentives

This law provides a property tax exemption for solar or wind energy systems. Passive systems qualify to a limited extent. The system must conform to guidelines of the state energy office and must be installed before 7/1/88. The exemption is good for 15 years after it is granted (Chapter 322, Laws of 1977; Chapter 220, Laws of 1979).

Contact Local assessor or board of assessors

An income tax credit is available for installation of active/passive solar systems and wind for principal residences as follows:

55% of the cost to purchase and install to a maximum of $2750, with a combined State and Federal tax credit of $6750.

Credit is available after Jan. 1 1981 and ends before December 31, 1986.

## NORTH CAROLINA

### Tax Incentives

This law provides for a corporate and individual income tax credit of 25% of the cost of a solar heating, cooling, or hot water system. There is a maximum credit of $1000 per unit or building. Although this credit may be taken only once, the amount of credit may be spread over 3 years. The system may be in any type of building, and it must meet the performance criteria of the U.S. Secretary of the Treasury or the North Carolina Secretary of Revenue (Chapter 792, Laws of 1977; Chapter 892, Laws of 1979).

Contact State Department of Revenue
Income Tax Division
P.O. Box 25000
Raleigh, NC 27640
(919) 733-3991

Buildings with solar heating or cooling systems shall be assessed as though they had a conventional system. Expires 12/1/85 (Chapter 965, Laws of 1977).

Contact Local assessor or board of assessors

## NORTH DAKOTA

### Tax Incentives

This law provides for an income tax credit (corporate or individual) for solar or wind energy devices. The credit is 5% per year for 2 years. The system must provide heating, cooling, mechanical, or electrical power (Chapter 537, Laws of 1977).

Contact State Tax Commission
Income Tax Division
Capitol Building
Bismarck, ND 58505
(701) 224-3450

Solar heating or cooling systems in any building are exempt from property tax for 5 years after installation (Chapter 508, Laws of 1975).

Contant Local assessor or board of assessors

## OHIO

### Tax Incentives

Solar, wind, and hydrothermal energy systems installed through 12/31/85 are exempt from real estate tax. A corporate franchise tax credit of 10% of the cost of a solar or wind energy system is created. The law creates a sales tax exemption for materials sold to a construction contractor for use in a solar, wind or hydrothermal energy system. Sales of these systems are exempt from sales tax; components and charges for installations are also covered by the exemption. This exemption is valid through 12/31/85.

A personal income tax credit of 10% of the cost of a solar, wind, or hydrothermal energy system is created. The system must be installed in a building that is owned by the taxpayer and located in Ohio. The credit includes a two-year carry forward provision and a maximum credit of $1000. To qualify for any of these tax advantages, the taxpayer's system must provide space heating or cooling, hot water, industrial process heat, or mechanical or electrical energy and must meet guidelines established by the Department of Energy. Passive designs are included (Amended Substitute House Bill 154, 1979).

Contact for income and franchise tax information) Department of Taxation
Income Tax Division
1030 Freeway Drive
Columbus, OH 43229
(614) 466-7910

Contact (for guidelines) | Ohio Department of Energy
30 E. Broad Street, 34th Floor
Columbus, OH 43215
(614) 466-7915

Contact (for sales tax information) | Department of Taxation
Sales Tax Division
30 W. Broad Street
Columbus, OH 43215
(614) 466-7350

## OKLAHOMA

### Tax Incentives

An income tax credit is created for the installation of solar or wind energy systems on all types of buildings. For residential applications, the credit in 1980 equals 80% of the federal percentage available to individuals. Thereafter the proportion of the federal percentage decreases 10% per year until expiration in 1989. There is a carry-forward provision for systems installed at the taxpayer's principal residence.

For installing solar or wind energy systems on non-residential buildings, the taxpayer may claim a 15% credit through 1988 (Chapter 209, Laws of 1980).

Contact State Tax Commission
Income Tax Division
2501 Lincoln Boulevard
Oklahoma City, OK 73194
(405) 521-3125

## OREGON

### Tax Incentives

This law exempts, in addition to solar systems, geothermal, wind, water and methane gas energy systems for real estate tax until 1/1/98. Property owned or leased by persons producing, transporting, or distributing energy is not included (Chapter 196, Laws of 1977; Chapter 670, Laws of 1979).

Contact Local assessor or board of assessors.

An income tax credit is provided for owners, tenants, and lessors of property used as a principal or secondary residence. The credit equals 25% of the cost of alternative energy devices using solar, water, wind, or geothermal resources for 10% of the energy requirements of the dwelling unit or 50% of the water heating requirements of the unit. Maximum credit is $1000 per dwelling unit. A carry forward provision is included. Systems must be certified by the Department of Energy before a credit may be claimed (Chapter 196, Laws of 1977; Chapter 670, Laws of 1979).

This law provides an income tax credit for weatherization materials installed in the taxpayer's principal residence, including a mobile or floating home or a unit of a multifamily dwelling. A list of qualifying items is available from the Department of Energy or the Department of Revenue. Items must be installed before 1/1/85. The credit equals 25% of the cost of eligible items to a maximum of $125. A 5-year carry forward provision is included (Chapter 534, Laws of 1979).

This law provides a corporation excise tax credit for commercial lending institutions making loans at 6½% interest or less for the installation of certified alternative energy devices. Maximum loans are $10,000 for home improvement loans. The tax credit is equal to the difference between 6½% and the average interest rate for home improvement loans made in a previous calendar year (Chapter 483, Laws of 1979).

This law creates a corporate income tax credit of 35% of the cost of energy conservation facilities. The credit is claimed at a rate of 10 for the first two years and 5% for each of the next three years. A carry-forward provision is included.

Energy conservation facilities means facilities used in trade or business and employing or processing renewable energy sources to: (1) replace a substantial part of an existing use of electricity, petroleum, or natural gas; (2) provide initial use of energy where such resources would have been used; (3) generate electricity to replace an existing source of electricity or provide a new source of electricity; or (4) perform a process that obtains energy resources from material that would otherwise be solid waste. Renewable energy resources include, but are not limited to straw, forest slash, wood waste or other forms of forest waste, industrial or municipal waste, solar energy, wind power, water power or geothermal energy (Chapter 512, Laws of 1979).

Contact State Department of Revenue
State Office Building
Salem, OR 97310
(503) 378-3366

## RHODE ISLAND

### Tax Incentives

Solar, wind and cogeneration systems are exempt from real estate tax through 7/11/80 (Chapter 202, Laws of 1977; Chapter 35, Laws of 1980).

Cities and towns are empowered to exempt solar and wind energy systems from any local property taxes (Chapter 36, Laws of 1980).

Contact Local Assessor or
Board of Assessors

A rebate is available for sales tax paid on solar, wind and other renewable energy systems and their components. Systems must provide heating, cooling, hot water or electricity. Expires 6/30/85 (Chapter 156, Laws of 1980).

An income tax credit is created for the installation of solar or wind energy equipment. Both active and passive solar systems are included. The credit for the taxpayer's principal residence equals 10% of the system's cost to a maximum of $1000. A carry-forward provision is included. If a system is installed in business or investment property, the credit equals 10% to a maximum of $1500. This credit may not reduce tax liability to less than $100, but excess credit can be carried forward. All credits expire on 6/30/85 (Chapter 36, Laws of 1980).

An income tax credit is created for installing energy conservation materials, waste heat recovery systems and cogeneration systems in industrial and commercial buildings. Credit equals 10% of the cost of qualifying expenditures to a maximum of $5000. Excess credit may be carried forward. Expires 6/30/85 (Chapter 149, Laws of 1980).

An income tax credit is created for the installation of energy conservation materials in residential rental property. The owner of the structure may claim a credit equal to 20% of qualifying expenditures to a maximum of $500 per structure and $500 as a total credit. No more than $1000 in credit may be claimed in any tax year, but excess credit may be carried forward. Expires 6/30/85 (Chapter 148, Laws of 1980).

Contact Division of Taxation
Department of Administration
289 Promenade Street
Providence, RI 02908
(401) 277-3050

## SOUTH CAROLINA

### Tax Incentives

This law creates an income tax deduction for energy conservation materials and renewable energy systems. The deduction equals 25% of eligible expenditures to a maximum of $1000. Individuals, fiduciaries and corporations may claim the deduction. Conservation materials include insulation, storm windows and other materials installed after 7/1/80 in existing buildings. Renewable energy systems include active and passive solar, biomass conversion, geothermal, hydroelectric, wind and woodburning systems, excluding fireplaces. Purchase must take place before 7/1/83. Deductions exceeding income may be carried forward. All eligible materials and systems must meet state and federal standards in effect at the time of purchase (Act 519, 1980).

Contact Tax Commission
John C. Calhoun Office Building
Box 125
Columbia, SC 29214
(803) 758-3391

## SOUTH DAKOTA

### Tax Incentives

This law provides a property tax assessment credit for renewable energy systems (solar, wind, geothermal, biomass) and ethyl alcohol plants. For residential property, the amount of the credit equals the assessed value of the structure with the system, minus the assessed value of the structure without the system for three years after installation. For the following three years, the credit equals 75%, 50%, and 25% of the base credit. For commercial structures the credit is 50% of the difference in the assessments for the first three years after installation. For the following three years the credit is 37.5%, 25%, and 12.5% of the base credit. Taxpayers should contact the county auditor for credit applications (Chapter 74, Laws of 1978; Chapter 84, Laws of 1980).

Contact Local county auditor or
Department of Revenue
Capitol Lake Plaza
Pierre, SD 57501
(605) 773-3311

## TENNESSEE

### Tax Incentives

Solar or wind energy systems for heating, cooling, or electrical power shall be exempt from property taxation. Law expires 1/1/88 (Chapter 837, Laws of 1978).

Contact Local assessor or board of assessors

## TEXAS

### Tax Incentives

This law exempts solar and wind energy devices from real estate tax assessments. The devices must be used for thermal, mechanical or electrical energy. The Comptroller of Public Accounts shall develop guidelines to assist tax assessors in carrying out this law. Effective 1/1/80 (Chapter 107, Laws of 1979).

This law provides a franchise tax exemption for corporations exclusively engaged in manufacturing, selling, or installing solar energy devices for heating, cooling, or electrical power (Chapter 584, Laws of 1977).

Solar energy systems used for heating, cooling, or electrical power are exempt from sales tax. Corporations may deduct from taxable capital the amortized cost of a solar energy device over a period of 60 months or more. (Chapter 719, Laws of 1975).

Contact Comptroller of Public Accounts
LBJ State Office Building
Austin, TX 78774
(512) 475-2206
(800) 252-5555

## UTAH

### Tax Incentives

This law creates an individual income tax credit for the installation of an active or passive solar, wind or hydroelectric energy system at the taxpayer's residence. The credit equals 10% of the system's cost to a maximum of $1,000. Systems must be installed between 7/1/77 and 6/20/85. A four-year carry-forward provision is included. Businesses may claim a 10% credit against income or franchise tax for installing these systems in commercial properties owned by them to a maximum of $3,000 per unit. Home builders may elect to pass the credit on to the home purchaser. The maximum credit for home builders is $1,000 per unit. The State Energy Office must certify systems before a credit may be claimed. The State Tax Commission may promulgate rules to administer the credits (Chapter 66, Laws of 1980).

Contact State Tax Commission
200 State Office
Salt Lake City, UT 84114
(801) 533-5831

## VERMONT

### Tax Incentives

Towns may enact a property tax exemption for alternate energy systems. Systems exempted are grist mills, windmills, solar energy systems, and devices to convert organic matter to methane. All components are exempt, including land on which the facility is situated, up to one-half acre (Act 226, 1976).

Contact Local assessor or board of assessors

Wood-fired central heating and solar or wind systems for heating, cooling, or electrical power are eligible for income tax credit if they are installed in the taxpayer's dwelling before 7/1/83. The credit is equal to 25% of the cost of the system or $1000, whichever is less. Businesses may deduct 25% of the cost of the system or $3000, whichever is less (Act 210, 1978).

Contact State Tax Department
Income Tax Division
State Street
Montpelier, VT 05602
(802) 828-2517

## VIRGINIA

### Tax Incentives

Any county, city, or town may exempt solar energy equipment used for heating, cooling, or other applications from property tax. The State Board of Housing must certify the system. The exemption is good for not less than 5 years (Chapter 561, Laws of 1977).

This law creates a separate class of tangible personal property for local taxation. The class includes energy conversion equipment purchased by a manufacturer for the purpose of changing the energy source of a plant from oil or gas to coal, wood, or alternative energy resources. Co-generation equipment is also included. Tax years covered are those beginning after 7/1/79. This class of property may be taxed at a rate different from the rate on other tangible personal property, but not higher than the rate on machinery and tools. To be eligible, equipment must have been purchased after 12/31/74 (Chapter 351, Laws of 1979).

Contact Local taxing authority

## WASHINGTON

### Tax Incentives

Buildings equipped with unconventional (including active and passive solar) heating, cooling, domestic water heating or electrical systems shall be assessed for real estate tax as though they were equipped with conventional systems. Expires 12/31/87 (Chapter 155, Laws of 1980).

Contact Local county assessor

## WISCONSIN

### Tax Incentives

Active solar and wind energy systems are exempt from real estate tax through 12/31/95. All systems must be certified by the Department of Industry, Labor and Human Relations (Chapter 349, Laws of 1979).

# Glossary

**Abscissa** The horizontal coordinate of a point in a two dimensional plane, parallel to the $x$-axis.

**Absolute pressure** Pressure measured relative to a perfect vacuum as a base.

**Absorber** The surface of a collector (normally black) that absorbs the solar radiation and converts it to heat energy.

**Absorbtance ($\alpha$)** The ratio of solar energy absorbed by a surface to the solar energy striking it. Energy not absorbed is transmitted or reflected.

**Acceleration** The time rate of change of velocity.

**Active solar system** A solar heating or cooling system that requires external mechanical power to move the collected heat.

**Air mass** The length of the path through the earth's atmosphere traversed by direct solar radiation, expressed as a multiple of the path length with the sun directly overhead.

**Air type collector** A collector that uses air as the heat transfer fluid.

**Altitude** The angle of the sun's position in the sky with respect to the earth's horizontal.

**Ambient temperature** Temperature of the surroundings (i.e., for collectors, outdoor temperature).

**Angle of incidence** The angle measured between an incoming beam of radiation and a line drawn perpendicular to the surface that it strikes.

**Anodic** More electrically positive.

**Aperture, solar** The effective radiant energy collection area of a solar collector.

**Aqueous** Relating to water.

**ASHRAE** The American Society of Heating, Refrigerating, and Air Conditioning Engineers, Inc.

**Avoirdupois weight** Series of units of weight based on the pound of 16 ounces.

**Azimuth** The angular distance between true south and the point on the horizon directly below the sun.

**Barometric pressure** The level of the atmospheric pressure above a perfect vacuum.

**Black body emitter** An ideal body which absorbs all radiation falling upon it and emits nothing.

**British Thermal Unit (BTU)** The quantity of heat energy needed to raise the temperature of one pound of water one degree Fahrenheit.

**Calorie** The quantity of heat energy needed to raise the temperature of one gram of water one degree Centigrade.

**Cal rod** An electric element in an electric hot water tank normally 4500 W, 240 VAC. Most electric tanks have two electric elements, one top and one bottom. Only the top cal rod is needed for solar.

**Cathodic** More electrically negative.

**Centrifugal pump** A type of pump which has blades that rotate and whirl the fluid around so that it acquires sufficient momentum to discharge from the pump body.

**Closed loop** Any loop in a system that is not exposed to the atmosphere.

**Coefficient** A number that serves as a measure of some property or characteristic.

**Collector** Any of a wide variety of solar devices used to collect radiant energy and convert it to heat.

**Collector efficiency** The ratio of the heat energy extracted from a collector to the solar energy striking the cover, expressed in percent.

**Collector tilt** The angle between the horizontal plane and the solar collector plane.

**Concentrating collector** A device that concentrates the sun's rays on an absorber surface which is significantly smaller than the overall collector area.

**Condensate** The liquid phase that develops on a heat exchanger surface as a hot gaseous fluid vapor is cooled below its dew point.

**Conductance** See Thermal conductance.

**Conduction** The transfer of heat energy through a material by the motion of adjacent atoms and molecules.

**Conductivity** The ease with which heat will flow through a material as determined by its physical characteristics.

**Convection, forced** Heat transfer through moving currents of air or liquid induced mechanically by a pump or blower in order to increase mass flow rates and velocities to yield a maximum heat transfer.

**Convection, natural** Heat transfer through moving currents of air or liquid as a result of thermal gradients and resulting density differences creating the necessary mass flow to promote heat transfer.

**Corrosion** Deterioration of metal by the chemical action of a fluid or components of a fluid.

**Declination** The angle between the plane of the earth's orbit and the equatorial plane.

**Demand load** Domestic water heating needs to be supplied by solar or conventional energy.

**Density** Weight per unit or mass per unit volume.

**Design life** The period of time for which a solar DHW system is expected to perform its intended function without requiring major maintenance or replacement.

**DHW** Abbreviation for domestic hot water.

**Diffuse radiation** Solar radiation that has been scattered by clouds and particles in the atmosphere and casts no shadow. Flat plate collectors can absorb it but concentrating collectors cannot.

**Direct-beam radiation** Solar energy received at the earth's surface uninterrupted by particles in the atmosphere and casting shadows on a clear day.

**Dyne** The basic unit of force in the centimeter-gram-second system of measurement. Defined as the force required to give a mass of one gram an acceleration of one centimeter per second.

**Ecliptic plane** The plane of the earth's orbit extended to meet the celestial sphere.

**Effectiveness** See Heat exchanger effectiveness.

**Efficiency** The ratio of the useful energy supplied by a system (output) to the energy supplied to the system (input).

**Elastomer** Any of various elastic substances resembling rubber.

**Electrolyte** A fluid solution capable of carrying an electrical current.

**Electron** An elementary particle consisting of a charge of negative electricity.

**Electrostatic** Attraction or repulsion of electric charges.

**Emmissivity** A measure of the thermal energy reradiated from a solar collector surface as a fraction of the energy which would be radiated by a totally black body surface at the same temperature.

**Energy** Defined as the ability to do work. A conserved quantity which is neither created nor destroyed. It can, however, be converted from one form to another or interconverted with matter according to Einstein's equation, $E = mc^2$ where $m$ is mass and $c$ is the speed of light.

**Equinox** The point of intersection of the ecliptic and celestial equator when the declination is zero.

**External manifold** A distribution pipe that runs outside the collector housing and connects to the inside header of each absorber plate.

**Fill-drain assembly** Comprised of two boiler drains and a check valve; installed for the filling and draining of transfer fluids in a closed loop freeze resistant system.

**Flash point** The temperature at which fluid vapors will flashover if an ignition source is present.

**Flat plate collector** A solar collector that converts sunlight to heat on a plane surface without the aid of reflecting surfaces to concentrate the sun's rays.

**Fluid** Any substance, gas or liquid, used to capture heat in the collector and transport the energy from the point of collection to storage or direct use.

**Flux** For collectors, the intensity of heat flow; for soldering, a substance used to clean the pipe surfaces and to promote a good mechanical bond upon soldering.

**Fossil fuels** Combustible substances of organic origin established in past geologic ages consisting of hydrocarbons formed from the decay of vegetation under heat and pressure (coal, oil, natural gas).

**Freon** Used for any of various nonflammable gaseous and liquid fluorinated hydrocarbons as refrigerants.

**FRP** Fiberglass reinforced polyester.

**Fusion** The union of atomic nuclei to form heavier nuclei resulting in the release of enormous quantities of energy when certain light elements unite.

**Galvanic corrosion** Material degradation caused by an electrochemical reaction between two or more different metals in a system, which are not properly isolated from one another.

**Galvanic series** Metals ranked from electrically positive to electrically negative to provide a relative measure of "corrodibility" of each metal when used in a multimetal system.

**Galvanized** To coat with zinc for protection from corrosion.

**Gauge pressure** That pressure measured above atmospheric pressure.

**Generic** Relating to or characteristic of an entire class.

**Glazing** A transparent/translucent sheet of glass/fiberglass that reduces heat loss from a solar collector and traps the thermal energy.

**Head** For pumping considerations, the vertical rise to the highest point of a piping system.

**Header** The pipe that runs across the top or bottom of an absorber plate, gathering or distributing the heat transfer fluid to or from the risers that run across the absorber surface (Also called manifolds if connected internal to each collector housing.)

**Heat** The sum total of all molecular energy of a body; a vector quantity.

**Heat exchanger** A device that transfers heat from one substance to another without mixing the two.

**Heat exchanger effectiveness** The ratio of the actual rate of heat transfer to the theoretical maximum rate of heat transfer.

**Heat pump** A mechanical device that transfers heat from one medium (heat source) to another (heat sink), thereby cooling the first and warming the second.

**Heat transfer medium** Air or liquid that is heated and used to transmit energy from its point of collection to its point of storage and/or end use.

**Hydrodynamics** The study of the motion of liquids and the forces acting upon them.

**Hydrometer** An instrument used to determine the specific gravity of a material

**Hydronic** A heating system using circulating hot water.

**Hydrostatic** Relating to liquids at rest or to the pressure they exert or transmit.

**Hysteresis setting** A time delay adjustment inherent in differential controller circuitry which prevents the constant cycling of a blower of pump following an "on" or "off" event.

**IMPS** An acronym for the U.S. Department of Housing and Urban Development's "Intermediate Minimum Property Standards Supplement: Solar Heating and Domestic Hot Water Systems."

**Inertia** See Thermal inertia.

**Infrared radiation** Electromagnetic radiation from the sun that has wavelengths slightly longer than visible light, not visible to the naked eye.

**Inhibitors** Additives to storage water or transfer fluids to prevent algae and/or corrosion.

**Insolation** The total amount of solar energy received at the earth's surface at any location and time (BTU/ft$^2$-hr).

**Insulation** A material with a high thermal resistance ($R$) to heat flow.

**Integral** Composed of in constituent parts essential to completeness.

**Intercept** The distance from the origin to a point where a graph crosses a coordinate axis.

**Internal manifold** A distribution pipe that connects the headers of the collectors internally and in turn becomes the header itself.

**Isogonic chart** A chart depicting magnetic compass deviations from true south.

**Kilowatt-Hour (KWH)** The amount of energy equivalent to 1 kilowatt of power being used for 1 hour (3,413 BTU).

**Kinetic energy** Energy of motion.

**Langley** A unit of measure of insolation named for the American astronomer Samuel P. Langley (3.687 BTU/$ft^2$).

**Latitude** Referring to a point on the earth as determined by an angle formed by a line intersecting the center of the earth to a particular point on the earth's surface and the plane cutting the earth at the equator.

**Liquid-type collector** A collector which uses a liquid as the heat transfer fluid.

**Longitude** Referring to a point on the earth as determined by an angle formed by the intersection of a line from the center of the earth to a particular point on the earth's surface and the plane cutting vertically through the center of the earth.

**Manifold** See Header.

**Mercator projection** A graphical depiction of altitude and azimuth onto a flat map for each variation of latitude. Used for plotting obstacles which might block energy collection in the "solar window."

**Miscible** Capable of being mixed.

**Nonselective surface** An absorber coating that absorbs most of the incident sunlight but which emits a high level of thermal radiation in return. Typically, a flat-black paint.

**Nomograph** A graph that enables one (with the aid of a straightedge) to read off the value of a dependent variable when the values of two or more independent variables are given.

**Normalized curve** Conformed to a standardized reference base.

**Opaque** Impervious to forms of radiant energy other than visible light.

**Open loop** Any loop in a system that is vented to the atmosphere.

**Ordinate** The vertical coordinate of a point in a two-dimensional plane, parallel to the $y$-axis.

**Orientation** Number of degrees to the east or west of south that a solar collection surface faces.

**Outgassing** The emission of gases by materials, usually during exposure to elevated temperatures.

**Overall coefficient of transmittance (U-value)** The reciprocal value of the sum of thermal resistances (BTU/hr − $ft^2$ − °F). This value is the combined thermal conduction value of all the materials in a cross section including air spaces and air films. The lower the U-value, the higher the insulating value.

**Passive solar system** A system that uses gravity, heat flows, evaporation, or other naturally-occurring phenomena without the use of external mechanical devices to transfer the collected enėrgy (i.e., south facing windows).

**Payback** The time at which the initial cost and annual operating and maintenance expenses of a solar DHW system equal the total savings generated by the system when compared with conventional energy sources. Both systems costs are computed at compounded interest rates of inflation for the same amount of energy generated.

**Percentage of possible sunlight** The percentage of daytime hours during which there is enough direct solar radiation to cast a shadow.

**pH** Measure of solution acidity.

**Phase change** Ability of a substance to change from solid to liquid and back to solid.

**Photochemical** The effect of radiant energy in producing chemical changes.

**Photovoltaic** Concerning the generation of an electromotive force when radiant energy falls on the boundary between dissimilar substances.

**Pitch** The ratio of vertical rise to horizontal span where rise is the distance from the attic floor to roof peak and span is the width of the house.

**Pitting corrosion** Caused by metal ions leaving localized areas causing a pitted surface or uneven corrosion.

**Potable water** Water free from impurities present in amounts sufficient to cause disease or harmful physiological effects and conforming in its bacteriological and chemical quality to the requirements of the Public Health Service Drinking Water Standards or the regulations of the public health authority having jurisdiction.

**Pound** The basic unit of force in the English system of measurement. Defined as the force that gives a standard pound (0.4535924277 kg) an acceleration equal to the standard acceleration of earth's gravity which is 32.174 ft/sec$^2$.

**Pressure** The ratio of force per unit area.

**Pressure drop** The loss in static pressure through a component such as a heat exchanger, length of pipe, or duct which may include fittings such as elbows or the combined losses throughout the entire length of fluid flow travel. Normally stated in terms of pounds per square inch (psi).

**Proton** An elementary partical consisting of a charge of positive electricity.

**Pump staging** A method of placing two or more pumps together to increase flow or overcome head losses.

**Pyranometer** An instrument used for measuring the amount of solar radiation.

**Radiant energy** The flow of energy across open space via electromagnetic waves (i.e., visible light).

**Reflectivity** The ratio of the radiant energy reflected from a surface, to the radiant energy incident upon that surface.

**Refractive index** A measure of how much a surface will bend energy beams of radiation as they pass through the material.

**Refractometer** An instrument that uses the index of refraction of a fluid to determine its freeze potential.

**Refrigerant** A liquid such as Freon, used to absorb heat from surrounding air or liquids as it evaporates.

**Re-radiation** Radiation resulting from the emission of previously absorbed radiation.

**Risers** The flow channels or pipes that distribute the heat transfer liquid from the headers across the surface of an absorber plate.

**R-value** See Thermal resistance.

**Selective surface** A surface that absorbs radiation of one wavelength (i.e., visible light) but emits little radiation of another wavelength (i.e., infrared), thereby reducing heat loss.

**Sky vault** The entire projection of the sun's path at any particular latitude. (See Sun path diagram)

**Slope** The ratio of vertical rise to horizontal run where rise is the distance from the attic floor to roof peak and the run is the horizontal distance from the roof peak to the end of the roof section.

**Solar constant** The average amount of solar radiation reaching the earth's outer atmosphere (436.5 BTU/ft$^2$-hr; $\pm$3.5 percent)

**Solar Noon** The instant of time the sun's position is true south (azimuth is 0°) and the altitude is a maximum for the day.

**Solar radiation** (Solar energy)—electromagnetic radiation emitted by the sun. The visible part of this spectrum ranges from long red to short violet wavelengths.

**Solar window** An outline of an area in the sky for a particular latitude through which a maximum amount of direct solar radiation reaches the collectors during any particular time of year and day.

**Solder** An alloy melted to join metallic surfaces.

**Solenoid** A coil of wire usually in the form of cylinder which draws a movable core when activated by a current.

**Solstice** The time at which the sun reaches its greatest declination, north or south.

**Specific gravity** Ratio of density of a material to density of water; used to determine the freezing points of fluids.

**Specific heat** The quantity of heat, in Btu's, needed to raise the temperature of one pound of material 1° Fahrenheit (BTU/lb-°F).

**Spectral distribution** An energy curve or graph that shows the variation of radiant energy in relation to wavelength.

**Stagnation** A no-flow condition.

**Standby heat loss** Heat lost through storage tank and piping walls under no flow conditions.

**Stratification** The tendency of storage tank water to remain in layers of different temperatures, with the coldest on the bottom and the warmest on the top.

**Sun path diagram** (Solar window)—A circular projection of the sky vault, similar to a map, that can be used to determine solar position and to calculate shading.

**Tedlar®** Polyvinyl floride film.

**Temperature** An indicator of the intensity or degree of heat stored in a body; a scalar quantity.

**Temperature gradient** A change in temperature in a specific direction.

**Thermal conductance (*C*)** A property of a material equal to the quantity of heat per unit time that will pass through a unit area of the material when a unit average temperature is established between the surfaces (BTU/hr-ft$^2$-°F).

**Thermal conductivity (*K*)** A measure of the ability of a material to permit the flow of heat (BTU-in./hr-ft$^2$-°F).

**Thermal inertia** The tendency of a large mass to remain at the same temperature or to fluctuate only very slowly when acted upon by external sources.

**Thermal resistance (*R*-value)** A measure of the ability of a material to resist the flow of heat; the higher the *R*-value, the greater the insulating value of the material (hr-ft$^2$-°F/BTU).

**Thermistor** A sensing device that changes its electrical resistance with changes in temperature. Used with differential controllers and control monitors to supply collector and storage tank temperature information.

**Thermodynamics** The study of heat transfer and conversion.

**Thermoelectric** Involving relations between temperature and the electrical condition in a metal or in contacting metals.

**Thermionic** Dealing with electrically charged particles emitted by an incandescent substance.

**Thermosiphon** The natural convection of heat through a fluid that occurs when a warm fluid rises and cool fluid sinks under the influence of gravity.

**Tilt angle** See Collector tilt.

**Toxic fluids** Gases or liquids that are poisonous, irritating, and/or suffocating, as classified in the Hazardous Substances Act, Code of Federal Regulations, Title 16, Part 1500.

**Translucent** Admitting and diffusing light so that objects beyond cannot be distinguished clearly.

**Transmissivity** The ratio of solar energy passed through a surface to the radiation striking it. Energy not transmitted is either absorbed and/or reflected.

**Transparent** Having the ability to transmit light without appreciable scattering so that objects beyond are entirely visible.

**Transversal** A line that intersects two coplanar lines in two different points.

**Tropic of Cancer** The latitude denoting the most northerly position of the sun in which the declination angle is +23.5 deg.

**Tropic of Capricorn** The latitude denoting the most southerly position of the sun in which the declination angle is −23.5 deg.

**Ultraviolet radiation** Electromagnetic radiation with wavelengths shorter than visible light.

**U-value** See overall coefficient of transmittance.

**Vacuum** The depression of pressure below the atmospheric level (usually expressed in inches of mercury).

**Viscosity** The readiness with which a fluid flows when acted upon by an external force (grams/cm-sec).

**Wavelength** The distance between the start and finish of an energy pulse.

# INDEX